LANCHESTER LIBRARY

3 8001 00199 3850

er Library

£40. 00

D1429029

Tenth Edition

Gem Testing

To C. J. Payne
with recollections of happy partnership
in routine and research

Tenth Edition

Gem Testing

B W Anderson
Formerly Director of the London Gem Testing Laboratory

Revised by
E A Jobbins
Formerly Curator of Minerals and Gemstones,
The Geological Museum (British Geological Survey), London

Butterworths
London Boston Singapore Sydney Toronto Wellington

 PART OF REED INTERNATIONAL P.L.C.

All rights reserved. No part of this publication may be reproduced in any material form (including photocopying or storing it in any medium by electronic means and whether or not transiently or incidentally to some other use of this publication) without the written permission of the copyright owner except in accordance with the provisions of the Copyright, Designs and Patents Act 1988 or under the terms of a licence issued by the Copyright Licensing Agency Ltd, 33–34 Alfred Place, London, England WC1E 7DP. Applications for the copyright owner's written permission to reproduce any part of this publication should be addressed to the Publishers.

Warning: The doing of an unauthorized act in relation to a copyright work may result in both a civil claim for damages and criminal prosecution.

This book is sold subject to the Standard Conditions of Sale of Net Books and may not be re-sold in the UK below the net price given by the Publishers in their current price list.

First published 1942
Fifth edition 1951
Sixth edition 1958
Seventh edition 1964
Eighth edition 1971
Ninth edition 1980
    Reprinted 1980, 1982, 1985, 1988
Tenth edition 1990

© **Butterworth & Co (Publishers) Ltd, 1990**

British Library Cataloguing in Publication Data

Anderson, B. W. (Basil William), *1901–1984*
    Gem testing.—10th ed.
    1. Gemstones. Identification. Tests—Manuals
    I. Title   II. Jobbins, E. Alan
    549.133

ISBN 0-408-02320-1   /0 00142

Library of Congress Cataloging-in-Publication Data

Anderson, B. W. (Basil William, *1901–1984*
    Gem testing / B. W. Anderson.—10th ed. / rev. by
E. A. Jobbins.
        p.   cm.
    Includes bibliographical references.
    ISBN 0-408-02320-1
    1. Precious stones—Testing.   I. Jobbins, E. Alan.
    II. Title.
TS756.A58   1990
736'.2—dc20   90-31378

Coventry University

Composition by Genesis Typesetting, Laser Quay, Rochester, Kent
Printed and bound in Great Britain by Courier International Ltd., Tiptree, Essex

Prefaces

Extract from Preface to Fifth Edition (1951)

In the period which has elapsed since the fourth edition of *Gem Testing* appeared there have been startling developments in the production of synthetic gemstones. Verneuil's 'flame-fusion' method, now half-a-century old, has been ingeniously modified to yield a spectacular new gemstone – synthetic rutile – and synthetic star rubies and sapphires. The appearance of the latter is somewhat disturbing, as it represents man's entry into a field in which Nature was thought to be inimitable.

Extract from Preface to Sixth Edition (1958)

There have been several important developments in the world of gemstones since the last edition of this book appeared in 1951. A new gem species – sinhalite – has been added to the list; and since this mineral had already been cut into gemstones fairly extensively (being accepted as stones belonging to other species) it obviously has an importance beyond the purely scientific interest that usually attaches to a 'new' gemstone.

Among the synthetic gemstones, the arrival of the long-heralded strontium titanate is also a matter of more than passing interest, since this stone is less easy to distinguish at sight from diamond than any other natural or artificial product. The synthetic emerald manufactured by Carroll Chatham in San Francisco is intruding more and more upon the world's gem markets, and is causing some perturbation amongst those who deal in emeralds. And a new, sintered type of synthetic spinel has been marketed as a substitute for lapis lazuli, which, in appearance, it closely resembles. Red synthetic spinels have also appeared.

Developments have also taken place on the determinative side. Simple yet powerful methods such as 'immersion contact' photography and its offshoots, and the 'crossed filter' fluorescence technique, both of which were first introduced by the author, provide tools which can assist the gemmologist in some of his most difficult problems at little cost in the way of apparatus.

Extract from Preface to Seventh Edition (1964)

Since the sixth edition of *Gem Testing* was published in 1958 there have been important developments in the synthetic emerald field, which has made the critical study of all emeralds more than ever necessary. Cultured pearls without the usual mother-of-pearl nucleus have emanated from Lake Biwa in Japan and have been marketed in considerable numbers, and cultured pearls without nuclei in larger sizes but smaller numbers have also been produced in Australian and Burmese waters.

In the field of natural gems, a new and radioactive gem mineral 'ekanite' has been discovered in the Ceylon gravels, and though chiefly of scientific interest, is notable in forming the sixth gem mineral hitherto unknown to science which has been discovered in the present century. Amongst more commercial gems, new sources of ruby from Tanganyika and emeralds from Rhodesia have become quite important, and the massive grossular garnet from the Transvaal which can so closely resemble jade, and which has already become familiar in its green form, has recently been found and marketed in a wide variety of attractive colours.

Extract from Preface to Eighth Edition (1971)

Today, more than ever before, the jewellery trade has need of gemmologists. This need is reflected by the increasing flow of students who enter for examinations such as those held by the Gemmological Association of Great Britain, which attract entries from almost every country of the world. This is because a trade, which in older and more stable times dealt with a very limited range of well-known gemstones from well-known localities, and with a few familiar substitutes in the form of glass and the Verneuil synthetics, is now faced not only with a much wider variety of natural gem materials but also with a bewildering number of man-made stones – not to mention new methods of treating and 'faking' natural stones to improve their appearance.

We live in an age where the needs of the electronics industry, space projects, and so on, have so stimulated the search for new substances with special optical and physical attributes that the art and science of growing crystals has reached a pitch where most of the important gem minerals can be produced in the laboratory, and in addition there are a number of crystals not found in nature, which are hard enough and attractive enough to be exploited for use in jewellery as a side-line from their intended function. As examples of such are the hard and optically pure substances conveniently called the 'rare-earth garnets' which have the same crystal structure as natural garnets but contain no silica. The author feels that a book such as this, which is intended to help the gemmologist, should contain sufficient details of such new materials to enable them to be identified, even if at present they are merely on the fringes of the jewellery trade. There have been at least two more types of synthetic emerald in production since the last edition was published, as well as new sources of natural emerald which have entered the market. The number of jade-like minerals in circulation, already considerable, has been increased by the discovery of green chrome chalcedony in Rhodesia and a range of massive green garnets (often mixed with idocrase) in Pakistan which can closely resemble jadeite. In addition, transparent green grossular has been found in Pakistan and in East Africa.

Far more important commercially and aesthetically has been the surprise discovery of a new and beautiful form of zoisite, a mineral hitherto known only in its massive ornamental forms. Near the Umba River in Tanzania magnificent violet-blue crystals (and some in shades of brown) were found by a lone worker in 1967, and the fame and exploitation of these followed so quickly that fine sapphire-like cut stones of the new gem variety are now being offered for sale in all the gem markets of the world.

Extract from Preface to Ninth Edition (1980)

Since the appearance of the eighth edition in 1971 there have probably been more changes in the science of gemmology than in any previous period of similar duration. Much of the fresh material now incorporated concerns new synthetic stones or other substitutes for natural gems. The techniques of crystal growing are now so well understood that almost any gemstone can be produced in the laboratory, provided there is sufficient incentive to warrant the time and cost of the operation. In addition many hard crystalline substances not found in nature have been manufactured and used to represent real stones – in particular diamond.

Additions to the repertoire that have appeared on a commercial basis during the last decade include synthetic turquoise, synthetic alexandrite and, most surprisingly, synthetic opal; in the field of diamond substitutes the most successful yet – cubic zirconia – manufactured and sold under a number of fancy names, has achieved a formidable reputation. In addition to the synthetic opals made by the firm of Gilson there have appeared some surprisingly effective imitation opals, originally sold under the name 'Slocum Stones' after their inventor, which the jeweller has to guard against; these artefacts, coupled with new and ingenious forms of opal doublet, render this gem (which in the past could safely be identified by inspection alone) one of those that now need to be treated with great care if expensive mistakes are to be avoided.

Other tiresome matters of concern to the honest gem trader include the legitimacy of dealing in stones that have been improved in colour by oiling, impregnating, staining or radiation. The emergence of blue beryls, too deep in colour to be properly termed aquamarine, was a recent case in point. Many of these stones were found to fade with disconcerting rapidity when exposed to sunlight, which made their high cost seem unreasonable. Deep blue topaz has also appeared on the market, and again the colour has been artificially induced. In topaz the induced colour can apparently be permanent. Testing for permanence of colour is an unhappy occupation, as clearly the client in search of reassurance must be warned that his stone may deteriorate in the course of experiment. The best information available on all such cases will be found in this edition.

Preface to Tenth Edition

To follow in the footsteps of Basil Anderson, probably the world's most distinguished gemmologist and a master of words, is not easy, However, I have been greatly helped by two friends and colleagues, Kenneth Scarratt, the present Director of the Laboratory that B.W.A. inspired for so many years, and Roger Harding, my successor as Curator of Minerals and Gemstones in the South Kensington geological museums. They have painstakingly read the manuscript and made many valuable suggestions. I thank them both.

The overall format of previous editions has been maintained and new information has been incorporated in a similar style. Two new chapters have been introduced on the manufacture and enhancement of gemstones, and the old chapter on the detection of synthetic, imitation and composite stones has been rearranged. Dr Kurt Nassau has kindly made useful suggestions here and many of the illustrations from his fine book *Gems made by Man* have been used with his permission and that of his publishers. Seventy colour plates, mostly new, are provided in this edition and many of the monochrome illustrations have been modernized.

In the past decade the development of new synthetics has proceeded apace. Nowadays, clean rubies and emeralds need very careful investigation – they could be natural or synthetic – and any doubtful stones are best sent to well-equipped laboratories where sophisticated instrumentation is available. This is not to say that the jeweller is now powerless. Careful use of the microscope can still elucidate many problems and new instruments are shown in the text. Faceted yellow synthetic diamonds have appeared on the market and irradiated diamonds may pose many problems in detection. Diamond simulants are ever with us, but new compact reflectivity/thermal conductivity meters (separate and combined types) are now available to assist in this field.

Since the last edition new gemstone deposits have been discovered in several countries. Alexandrite rivalling the best Russian material has been found in Brazil, as have several significant emerald deposits. Sapphire and aquamarine from Nigeria, garnets of many colours and compositions from Kenya/Tanzania, natural pink topaz and fine coloured emeralds from Pakistan, and large clean kunzites from Afghanistan are among the more notable newcomers. Ruby, sapphire, peridot, and aquamarine are just a few of the gemstones that China promises to produce in the next decade.

Finally, I must thank Butterworths for their patience while awaiting the completion of the manuscript.

E. A. Jobbins

Contents

Colour plates 1 to 70 between pages 278 and 279

How to use this book

This book is intended to help the jeweller, dealer, or gemmologist: someone in fact who is seriously concerned with the problem of identifying gems. It is assumed therefore that the reader is already familiar with the names and general appearance of the better-known stones.

The experienced gemmologist can identify a number of stones either at sight or after scrutiny with a pocket lens. Knowledge and skills of this kind are of great value, and should be practised and enhanced at every opportunity. In the text, simple signs and tests involving no apparatus are always considered first. But today, when so many synthetic, imitation, and faked stones abound, there are inevitably many cases where the use of instruments is essential to ensure that a correct determination is made.

The early chapters are therefore devoted to describing the more important properties of gemstones and the necessary apparatus or instruments for observing or measuring these properties. Full practical instructions, based always on the author's own experience, are given for carrying out each test, and just enough theoretical background is provided to allow intelligent use to be made of the observations. *Intelligent* use, even of simple apparatus, is very necessary: anyone who is experienced in handling or dealing with precious stones, but who has no knowledge of gemmology, is likely to make more wrong decisions through *unintelligent* use of scientific apparatus than he or she would have made by relying entirely on unaided judgement.

The order in which the more commercially important gemstones have been treated amounts to a classification on a colour basis. This is admittedly quite unscientific, since it tends to bring together stones which have nothing except their appearance in common, and to obscure the virtual identity of such stones as ruby and sapphire or emerald and aquamarine which are merely colour varieties of the minerals corundum and beryl. Also, by considering all the stones that are likely to be confused with ruby or with emerald, etc., one encourages the natural but deplorable tendency to think of all red stones merely as inferior substitutes for ruby, all green stones as inferior substitutes for emerald, and so on, whereas fine specimens of red spinel or garnet, or fine green tourmalines

1

and garnets should be admired on their own merits as beautiful gemstones. The less valuable gems are often available in larger and more perfect examples than can be commanded by even a millionaire in such stones as ruby or emerald, and can thus provide pieces suitable for pendants, necklaces, or large brooches at a reasonable price.

However, this is a book on gem *testing*, and the order based chiefly on colour is clearly the most practical one for our purpose. The possibilities to be considered when faced with a ruby-like stone, for instance, are quite different from those arising when faced with a stone resembling sapphire. The colour classification is not rigidly adhered to, however; the quartz minerals, zircon, the garnets, aquamarine, alexandrite, and several others are treated separately. A good deal of repetition and overlapping inevitably occurs, which may prove irritating to anyone attempting to read the book straight through as though it were a novel – a feat only likely to be attempted by some diligent reviewer. But in a book of this kind such repetitions are inevitable unless cross references are to be used to an extent which would be very tiresome to the reader.

The methods and apparatus dealt with in the first part of the book are chiefly concerned with the measurement of certain physical and optical properties which characterize the various gemstones. Practically all the gem materials are minerals, and the properties possessed by a given mineral species which distinguish it from other minerals depend upon two factors: (1) its chemical composition, i.e. the *kinds* of atoms with which it is built, and (2) its crystal structure, i.e. the spatial *arrangement* of these atoms within the mineral. Of these two factors, the latter is by far the more important; a fact strikingly illustrated by the two crystallized forms of carbon: graphite and diamond. These two minerals, so utterly contrasted in appearance and physical properties, are yet constructed of precisely the same kind of atoms; all the great differences between the two materials in hardness, transparency, density, and so on must be laid to the account of the different internal structure of their crystals. This fundamental importance of crystal structure, and the assistance given in recognizing gem minerals in the rough by a knowledge of the characteristic shapes of their crystals ('crystal habit') is recognized by all gemmologists, and in every textbook one of the opening chapters is devoted to elementary crystallography.

However, experience with students of all ages has led the author to realize what a difficult barrier even this simplified form of crystallography is for the beginner to surmount, and in this book it was thought better to dispense with it entirely except for a summary of crystal systems, which will be found available for reference in Appendix 2 at the end of the book. To enable the reader to make sense of certain references in the text, however, it may be said here that all crystals can be divided, according to their symmetry, into seven main groups called the 'crystal systems'. These are named the cubic, tetragonal, hexagonal, trigonal, orthorhombic, monoclinic, and triclinic systems. To the cubic system belong the most symmetrical crystals, and these are the only crystals within which light travels with the same velocity and character in all directions: for this reason they are known as optically isotropic, or simply 'isotropic'. In crystals belonging to any of the other systems the effect known as double refraction (see Chapter 3) takes place, in which a ray of light entering the stone is split into two polarized rays which travel with slightly different velocities through the crystal, and are therefore bent or refracted from their original course by different amounts. This is of high practical importance in testing stones, so that it is useful to bear in mind which are the gemstones that belong to the cubic system, and thus do *not* show this effect. These are diamond, the garnets, spinel, fluorite, and others of lesser importance. Non-crystalline substances such as opals, glass, amber, and the plastics are also isotropic.

Tables and other useful features will also be found in the Appendices, including a short descriptive list of books on precious stones for those who wish to extend their knowledge further, or form a small reference library on the subject, and names and addresses of firms who can supply the various instruments etc. mentioned in the text.

The number of different minerals which have occasionally been cut as gemstones is very large, and it would spoil the simplicity and usefulness of this book for the majority of its readers to include all such possibilities in the text. For the benefit of fellow gemmologists who are interested in these out-of-the-ordinary stones, the author has included most of them in the Appendices. With the aid of the data provided and the necessary simple instruments, it should be possible to identify these rarities when they are encountered.

Unfortunately, in these days of advanced technology, the very high prices asked for almost all natural gemstones on account of their scarcity has made manmade substitutes far more prevalent than in past years. Thus one of the gemmologist's most urgent and difficult tasks is to distinguish these from the natural stones they represent. A special chapter (Chapter 9) is devoted to a description of synthetic and imitation stones and the best means of detecting them.

In a number of cases the colour of gemstones has been altered by heat treatment, staining, or radiation to make them more attractive before putting them on the market. It is important for the jeweller and gemmologist to know in which of such treated stones the colour is stable and the process accepted as legitimate by the trade, as opposed to those in which the improvement is of only short duration or is intended to deceive. This topic is reviewed in Chapter 8, and is also mentioned when discussing the testing of the individual gem species.

1

Collecting, handling and housing gemstones

Though the majority of readers may be already familiar with the elementary equipment which the dealer or the trained gemmologist uses in handling and housing gemstones, a brief description and some advice on the fundamental tools and techniques may be valuable to those who are as yet unaccustomed to coping with loose stones.

To watch an expert sort through a collection of stones on a cloth-topped desk – picking up selected specimens with tongs to scrutinize closely with a powerful lens, scooping up those required with a small shovel, and shooting them neatly into a stone paper – one might consider the whole process extremely simple. But anyone attempting for the first time to follow this routine would find each stage full of unexpected snags, and would feel clumsy and inept. There can, of course, be no real substitute for experience in such matters, for when it comes to manual dexterity it is the hand as well as the brain that has to learn. But at least one can make things easier by some advice on the selection of tools and suggestions on the best way to use them.

Before dealing with tools, however, one must stress the importance, in any kind of assessment of gemstones, of absolute *cleanliness* of the stone to be examined: a matter which is too often disregarded through carelessness or impatience. With unmounted stones cleaning is usually a simple matter: a quick rub over in the folds of a clean handkerchief will often suffice. Mounted stones are far more likely to have gathered a considerable quantity of dirt, particularly on the rear facets, and are, of course, also less easy of access. A small pot of water to which some drops of liquid detergent have been added and a small soft toothbrush will usually enable a good cleaning job to be done. As a drying-off agent warm boxwood dust is traditional: but a few puffs with a watchmaker's 'blower', or simply leaving the jewel on clean blotting paper under the warmth from a desk lamp for a few minutes will prove quite effective. The difference in appearance and the ease with which features in the stone can be examined after cleaning are often quite astonishing.

Ultrasonic cleaning is now extensively used in the jewellery trade and is extremely effective when applied to metalwork. There are a number of gemstones, however, in which the high-frequency vibration is extremely damaging, especially when the mineral concerned has a ready cleavage. The attractive new 'tanzanite' variety of zoisite may be mentioned as a particular example, but to be on the safe side any gemstones should be removed from their setting to avoid subjecting them to an ultrasonic ordeal.

It is largely to prevent stones under examination from being soiled by handling that dealers use tongs (tweezers) of one sort or another. The choice of tongs ('corn tongs' as they are sometimes called) is very important, as many one sees are too finely pointed or too strongly sprung. For general usefulness, tongs about 12.5 or 15 cm (5 or 6 in) in length, made of stainless steel, with rather blunt, rounded tips, scored inside to prevent the stone slipping, and with a mild spring, are ideal. Tongs with a 'slide' which enable the grip on a stone to be maintained without continued external pressure can also be obtained, and are useful when a stone is to be handed from one person to another or tilted in different orientations under the microscope – but they are heavier, clumsier, and more expensive than the simple type. The author must confess that, after 45 years of stone handling, he still finds tongs difficult to manage. It only needs a little too firm a grip and the stone to be ever so slightly on the tilt for it to spring suddenly away to some awkward and often unknown destination. The 'pick-up' position for a stone is when it is resting flat on its table facet. If the tongs, also resting flat on the table, are slid into position until their tips flank the stone and project a little beyond it, and the specimen is then lightly gripped, it should be in a correct position to be handled and inspected with safety. The late Sir James Walton, in whose safe surgeon's hands tongs were instruments of precision, found it useful to mark with a file a longitudinal groove on the inner faces of the

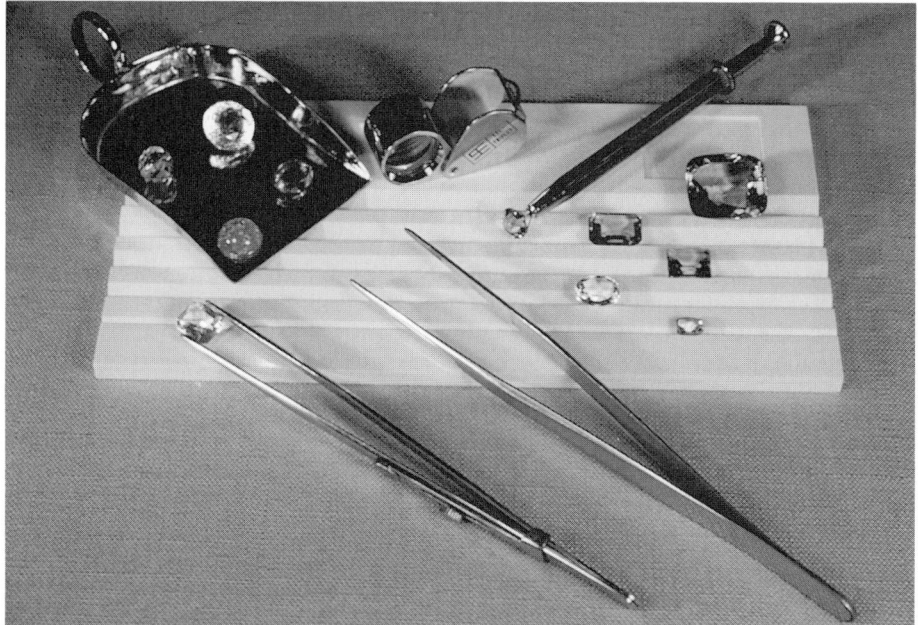

Figure 1.1 Lens, tongs, scoop, and tray for gemstone sorting

tongs, into which the girdle of the stone to be handled fitted snugly. Special spring tongs with strongly grooved lips are, in fact, much used in the USA and in Switzerland for holding stones securely while under observation, particularly under the microscope, where they can be mounted on a holder fixed on the stage of the instrument. There is another useful form of spring tongs in which three wire prongs project and retract in a pencil-like tubular holder. Round, brilliant-cut stones of no great size are very securely held in tongs of this type, but the position in which the stone can be held is limited to that in which the table facet is at right angles to the length of the holder, and large or rectangular stones cannot be accommodated. The spring in these tongs may be quite strong, and if they are used care should be taken to avoid scratches being made on the gemstone.

Apart from their function in picking up stones for inspection, tongs are really essential for sorting or counting stones on a desk, and also for counting or handling pearls. The labour of counting large quantities of pearls or small stones can be shortened by separating them from the spread-out mass in groups of three with the tips of the tongs held tightly closed. The tip of a propelling pencil can, for this purpose, function efficiently in much the same way.

For transferring a number of stones or pearls from desk to packet speedily and safely, a small shovel or scoop is almost a necessity. Suitable scoops can be obtained from jeweller's suppliers. Tongs, lens, and scoop of the types recommended can be seen in Figure 1.1.

The pocket lens or loupe

Without question, the jeweller's or gemmologist's most important aid to gem identification is a pocket lens. Gemstones, and still more the detailed features in these which one needs to observe, are small objects, and some form of magnification is necessary if they are to be critically inspected. The normal low-power 'reading glass' familiar to the public and beloved of Sherlock Holmes is far too clumsy and too low in magnifying power for the purpose. The usual 'watchmaker's loupe' is also too low in power, though it is highly convenient for its proper purpose, since it can be held in the eye like a monocle, leaving both hands free.

An ideal lens for the gemmologist is one magnifying 10 times (usually written '×10') which enables most of the desired features to be clearly resolved. A ×20 lens is occasionally useful, but the field of view is limited and the focus very critical. The question of how the 'magnifying power' of a lens is assessed is rather a confusing one. When the eye is very close to the lens (as it should be in practice) the magnification can be defined as $M=d/F$, where d is the least distance of distinct vision and F is the focal length of the lens. Since the distance d for normal eyesight is usually taken as 25 cm (10 in), this means that the focal length of a ×10 lens, for instance, would be 2.5 cm (1 in). In more understandable terms, the magnification of a lens is the number of times greater that the diameter of an object appears when viewed through the lens held close to the eye compared with its diameter as it appears at the nearest distance of distinct vision. This means, in effect, that the magnification given by a lens to a short-sighted person is less than that for a long-sighted subject. The high degree of curvature needed to produce lenses of such high power as 10 diameters brings with it considerable marginal distortion and colour-fringed images unless the lens is a compound one carefully made to overcome these aberrations. So important is a good pocket lens to the student of gemstones that it is well worth spending more in buying a Zeiss anastigmatic lens, or one by some other high-class optical firm. In its neat plastic housing a Zeiss lens weighs very little and slips easily into a small pocket.

Until one is accustomed to using a high-power lens it may seem awkward to handle, as the focal length is so short. It is essential to hold the lens close against the eye with one hand (normally the right hand and the right eye) while the specimen, held in tongs or fingers, is brought to within an inch or so of the glass, when careful, steady-handed adjustment should ensure an exact focus for any observed part of the stone. During the operation the hands are in contact, which enables sharp focus, once found, to be maintained without involuntary wavering. The lighting is very important. If one is seated at a desk or table, an adjustable shaded lamp should be available to provide light slightly above, to the front and to the right of eye level. To beginners it will seem that, whatever attitude they may adopt, they are always in their own light. Figure 1.2 may do more than the above description to enable beginners to adopt the comfortable posture which experts use without thinking.

When one has gathered sufficient knowledge and experience, it is astonishing how large a number of stones can be identified with certainty merely by careful inspection with a lens. Diamond, zircon, peridot, demantoid, amethyst, hessonite, tourmaline, kunzite, and sphene are among the transparent stones which one should be able to identify under the lens, as well as doublets, pastes, and many of the synthetic stones. Among the non-transparent gems, lapis

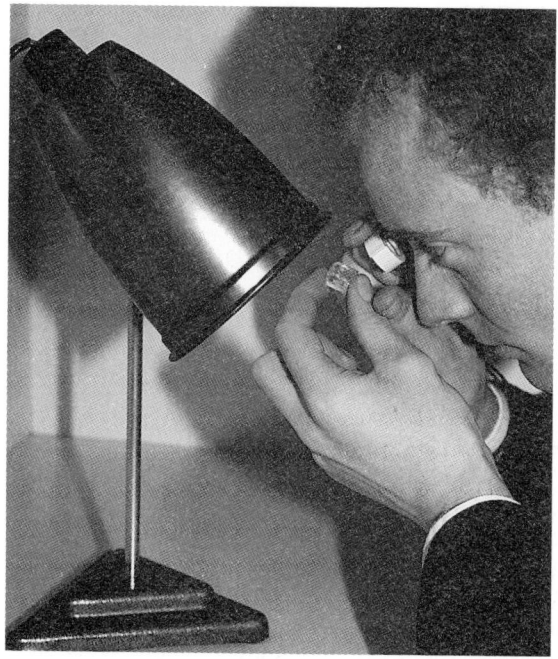

Figure 1.2 Relaxed pose of an expert using a ×10 lens

lazuli, 'Swiss lapis', real and false aventurine, ivory, pearl, cultured pearl, imitation pearls, and pink (conch) pearls will almost always show distinctive signs. The beginner should be able to gain much more knowledge with a lens after carefully studying this book and practising with specimens along the suggested lines than formerly – and this is true not only of the beginner but of the average supplier, who has had no time for gemmology. For prolonged work, and where a lower degree of magnification is sufficient, a head loupe or visor will prove a valuable accessory. This allows binocular vision, and leaves the hands free for any manipulation required.

Making a collection

It is obvious that without specimens to work with, one cannot hope to learn practical gemmology, and any reader who has not already done so is urged to start putting together a collection of gems, however modest. Fine gems are, of course, very expensive, but fortunately even inferior or broken stones will show the characteristic properties of their kind just as well as (and sometimes better than) show specimens, and anyone connected with the jewellery trade can usually acquire a variety of damaged or commercially unacceptable stones which will serve admirably as 'instrument fodder', and form the basis of a *working* collection, invaluable for experiment and comparison. Synthetic and imitation stones should not, of course, be neglected, as it is vital to become familiar with the appearance, range, and properties of all these. Here even quite large specimens can be purchased very cheaply. A real lover of gemstones will probably not rest content with inferior bits and pieces of this kind, and will seek opportunities of purchasing gemstones of fine quality. If good sense and judgement are used, money so spent is by no means wasted: good specimens are always in demand, and continuously rising in value, and meanwhile collectors have the pleasure of showing their collections to fellow-enthusiasts, with whom they may be able to exchange some of their surplus specimens for stones absent from their own collections.

Housing specimens

For the working collection, for carrying stones around in the pocket and universally in the trade for buying and selling loose gems and pearls, goods are most conveniently kept in folded stone papers. These can be purchased from jewellers' suppliers in various sizes, machine folded and lined with tough transparent tissue, or they can be home made from good-quality A5 (148 × 210 mm) typing paper by proceeding as follows:

1. Fold over the bottom part of the sheet until the edge is 40 mm (1½ in) from the top edge and parallel to it. Hold it in this position while making a firm horizontal crease along the fold (Figure 1.3).
2. Make a faint vertical crease in the paper by folding the right-hand edge over to meet the left and pressing only very lightly – just enough to indicate the centre of the sheet. Restore to previous position.
3. Fold the right and left edges over to coincide with this central line and crease each firmly in that position (Figures 1.4 and 1.5).
4. Fold the bottom edge to within 40 mm (1½ in) from the top, and crease firmly.

Figure 1.3 Making a stone paper from an A5 sheet – the first fold

Figure 1.4 Making a stone paper – the second fold

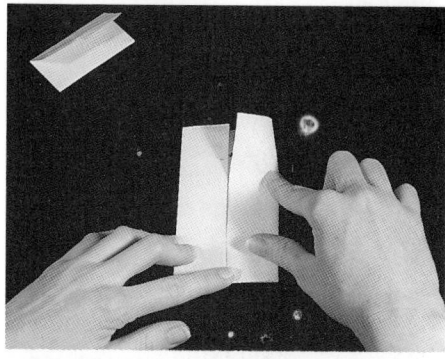

Figure 1.5 Making a stone paper – the third fold

Figure 1.6 Making a stone paper – the final fold

5. Fold the top edge over this as far as it will reach, and crease firmly into position. The packet is now complete (Figure 1.6) and measures about 75 × 45 mm. By using A4 paper (210 × 297 mm and twice the size of A5) and changing the 40 mm measurements above to 55 mm a stone paper of about 105 × 60 mm may be made.

To prevent the sharp culets of stones from piercing the paper, the packet can be lined with tough tissue folded in the same manner or a square of lint can be folded to act as a 'bed' for the stones. This, or a thin piece of cotton wool, will also help to prevent stones from jostling together and becoming 'paper worn' in consequence. Brittle stones, such as zircon, should be housed in separate papers, or each stone wrapped in a small screw of tissue, if chipping of the facet edges is to be avoided.

Handling stone papers correctly needs a little practice: indeed, when lay people have been using such a packet they seem invariably to hand it back wrongly folded – a point which might be used with telling effect in some detective story. If only a single stone is being placed in a paper it is safe enough to open the packet completely on a desk or table, place the stone therein, and refold the packet. However, where a large number of stones is being handled there is a danger that some may spread beyond the line of the side folds and leak through the side flanges of the packet. It is therefore safer in such cases to open the right-hand side only of the packet, keeping a good grip with the thumb and forefinger on the closed end, tilting the open end upwards. Into the bag so formed a large number of stones can be shot from a shovel in perfect safety. The open end of the packet should then be firmly folded into the closed position while the stones are still kept by gravity at the far end of the packet. Only when the paper is closed is it safe to redistribute the stones within the packet more evenly by giving it a gentle shake.

Until thoroughly practised, the reader is advised not to open a stone paper unless it be resting on a table or desk, and then to do it slowly and watchfully. If this is not possible, open one end only, bag fashion, to assess the size and number of the enclosed stones or pearls before proceeding further. Stones can only escape from a properly folded packet by bursting through a hole in the paper or by having been allowed to spread beyond the side creases.

An obvious advantage of stone papers in addition to their portability is the ease with which the number, nature, and weight of the enclosed goods – to which the dealer can add the price per carat in code and the gemmologist any technical details he or she wishes to record – can be determined. It is very convenient for one's 'working collection' to be housed in packets of this kind and arranged in alphabetical order in one or more long boxes or a card index cabinet.

With finer specimens or for display purposes a collection of gemstones can be attractively and conveniently housed in shallow plush-lined boxes provided with a number of round or oval depressions in which the stones can rest. When obtained commercially such boxes are usually made to take stones of fair size – say, three or four carats upwards. For smaller stones, boxes with grooves to receive the stones can be used.

Sir James Walton used to display his stones in shallow trays filled with plaster of Paris into which the gems had been pressed while the plaster was still moist to make a mould for their exact reception. When set, the plaster was varnished to make a more pleasing finish and colour.

Small glass-topped aluminium or plastic boxes can be obtained from jewellers' suppliers in different sizes, and these also provide convenient and quite attractive housing for a gem collection. Stones can be pressed, individually or in groups, into pads of cotton wool of sufficient thickness for the stones to be held firmly against the glass when the lid is closed. Details of the stones can then be written on labels stuck to the backs of the boxes.

Perhaps the most attractive manner of displaying a collection of small gemstones is that adopted by one keen connoisseur who presses the culet of each stone gently onto a bed of pure beeswax such as that used by manufacturing jewellers for their 'arranges' of stones for a jewellery design. Quite a small bed of wax – say, 150 × 100 mm (6 × 4 in) can accommodate a considerable number of small stones arranged in rows, and the texture of the wax makes an ideal background against which to display their beauty.

2

Refractive index and its measurement

The first task in identifying a gemstone is to make quite sure to which mineral species it belongs: to determine, for instance, whether it is a corundum (ruby or sapphire), a beryl (emerald or aquamarine), quartz (amethyst or citrine), topaz, tourmaline, spinel, peridot, or zircon.

Though one may often make a shrewd guess as to the nature of a stone by reason of its colour, lustre, and general appearance, it is often only by measuring one or other of its optical or physical 'constants' that one can be really sure of one's ground. In this chapter methods are described for measuring the most important of these constants.

Man is a profoundly lazy creature and it has been said that most of his inventions have been born of the urge to save himself trouble. Presumably, therefore, jewellers who are not scientifically minded would welcome an apparatus which would enable stones to be identified in some really simple manner, quickly, clearly, and without calculation.

If, merely by placing the table facet of an unknown stone on an instrument, its name could be read off by means of a pointer on a calibrated scale, one would expect such an apparatus to be widely popular. Instruments closely corresponding to this 'ideal' have in fact been designed and marketed during recent years. These *reflectivity meters* aim to identify gemstones, where these have a clean and well-polished table facet, by measuring their reflecting power when a narrow beam of infra-red light impinges on this facet. Such instruments have certain distinct advantages over the traditional types of jeweller's *refractometer* that have been the mainstay of gem identification for more than three-quarters of a century. But they also have grave limitations, which are explained later in this chapter.

The refractometer is still an essential instrument for the gemmologist, and the principles of its construction and how to make the best use of it are fully described later. First, however, we must explain the meaning of refraction and refractive index and the basic principles on which the refractometer operates.

What happens when a ray of light falls on the surface of a transparent solid such as a gemstone or a sheet of glass? Some of the light is *reflected* at the surface of the stone, the reflected light leaving the surface at an angle equal to that at which it falls upon the surface ('incident angle' or 'angle of incidence'). It is this reflected light which provides the surface lustre of the stone. The greater proportion of the light, however, passes into the stone, but in this denser medium it travels much more slowly than in air.

The effect of entering the denser medium, in which their velocity is diminished, upon trains of light waves striking the surface obliquely is to alter their direction and make them follow a new path nearer to the perpendicular (or 'normal' as it is called) to the interface between the two media.

This deviation of rays of light on entering a new medium is called *refraction*. In Figure 2.1 the ray of light, IO, is refracted along OR on entering the denser medium below the surface PQ. The broken line shows the path followed by the *reflected* light.

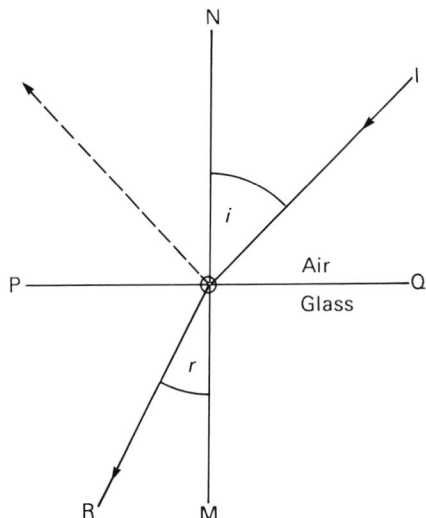

Figure 2.1 The ray of light IO is refracted on passing from air into glass along OR. NOM is the 'normal' (perpendicular) to the surface. The angle *i* is called the angle of incidence and the angle *r* is called the angle of refraction. The extent of refraction shown is given by ordinary window glass (RI = 1.52)

The extent of the bending or refraction of light on entering the stone depends upon its refracting power or 'refractive index', and this is inversely proportional to the velocity of light within the substance. Put differently, the refractive index of a medium may be defined as the velocity of light in air* divided by the velocity of light in the medium.

The velocity of light in air is approximately 300 000 km (186 000 miles) per second, and light from the sun and stars travels to us at this immense speed. In quartz (rock crystal, amethyst, etc.) the velocity is reduced to approximately 193 000 km (120 000 miles) per second and in diamond to only 124 000 km (78 860 miles) per second.

Thus diamond, in which light travels, as mentioned above, at 124 000 km/ second compared with the 300 000 km/second in empty space, has a refractive index of 300 000/124 000=2.42; higher than the refractive index of any other

* Or, more strictly, a vacuum.

gemstone used in jewellery – accounting for the brilliant, adamantine lustre of the stone.

It has already been stated that each mineral has a definite refractive index by which it can be identified on the refractometer; lists of these indices will be found in Table 2.1 and in the Appendices.

For the benefit of those who like to have at least a rough idea of the working of an instrument which they are using, a short description of the basic principle upon which all gem-testing refractometers depend follows. However, those who would like to know how to use a refractometer, but who prefer to shirk such explanations, can skip the next section and await the strictly practical directions which will be given later in this chapter.

What follows can best be understood by considering Figure 2.2. Here rays of light are considered passing from a dense medium into a rarer one, say from glass into air. The rays will then be refracted *away* from the normal, NOM (the reverse process of the former case considered in Figure 2.1). Thus the ray AO is refracted along OA′, apart from the small fraction of light which is reflected at O, as indicated by the dotted line.

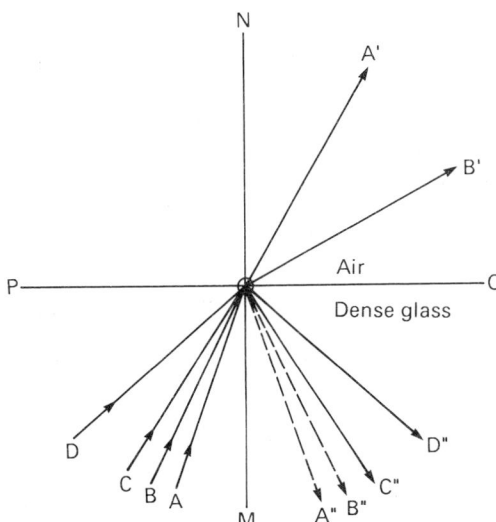

Figure 2.2 Passage of light from dense glass (RI = 1.8) below into air above. The rays AO and BO are refracted along OA′ and OB′ and also partially reflected. Beyond the critical angle COM = 34° rays are totally reflected, being unable to pass into the rarer medium

Similarly, the ray BO is refracted along OB′. As we consider rays which form increasingly greater angles with the normal, NOM, we reach an angle where the refracted ray only just grazes along the surface, OQ, between the two media.

This is known as the *critical angle*, and all rays reaching O from the denser medium at an angle greater than the critical one are totally reflected back into the denser medium, it being physically impossible for any light from such rays to penetrate into the rarer medium.

In the figure the angle COM can be considered as the critical angle of incidence; all the light is thus totally reflected back along OC″, and any other ray, such as DO, with a still larger angle of incidence, is, of course, also totally reflected. It is important to realize that the *size of the critical angle will depend upon the relation between the refractive indices of the denser and the rarer medium*. If, therefore, in any two media which are in optical contact we can measure the angle where total reflection begins (i.e. the critical angle), and we know the refractive index of the denser medium, it is possible to calculate the refractive index of the rarer medium.*

This is the underlying principle of all total reflection refractometers (sometimes called 'total reflectometers'), but in the instruments designed for gem testing all calculation is ingeniously avoided. In such refractometers the optically dense medium of known refractive index is in the form of a polished hemisphere of heavy lead glass, or a segment of such a hemisphere, or a truncated 60° prism of the same material, the flat upper surface in each case forming the 'table' of the instrument. If the flat, polished surface of any gemstone of lower refractive index than the hemisphere is placed in optical contact with this table, rays passing through the glass to the stone will be mostly refracted into the stone, and thence escape into the air, when they strike the surface at less than the critical angle, but *totally reflected* back from the surface of the stone when they strike at an angle exceeding the critical angle.

There is a point here which may usefully be cleared up, since it causes confusion to students. What is known as *the* critical angle of a stone is the angle at which rays of light passing *from the stone into air* are refracted at 90° to a line perpendicular to the surface. In other words, the refracted ray just grazes the surface between the two media. Rays striking the inside surface of a stone at angles greater than this critical angle cannot emerge, and are totally reflected back into the stone. Here, the stone is the denser medium, and the *higher the refractive index of the stone, the smaller is its critical angle.*

In the case of a refractometer the principle is exactly the same, but here the glass of the refractometer is the denser medium, and the stone is the rarer medium. Thus, in this case *the higher the refractive index of the stone, the greater is the critical angle between it and the glass of the instrument.*

The totally and the only partially reflected rays are projected by a lens system onto a transparent scale which is viewed through an eyepiece. The part of the scale illuminated by the totally reflected rays will be brightly lit, while the rest of the scale will be relatively in shadow. The scale is calibrated by the makers of the instrument to read directly in refractive indices, so that by simply observing the position of the edge of the shadow on the scale the refractive index of the stone tested can be ascertained.

Refractometers of the total reflection type were in use during the latter half of the nineteenth century, but the first satisfactory low-priced instrument for testing precious stones was devised by Dr G. F. Herbert Smith in 1907. The more recently designed and very compact Rayner refractometer (Figure 2.3 (*left*)) has a truncated prism of dense glass in place of the usual hemisphere, but the principle of the instrument is exactly the same. The prism form was originally evolved to allow isotropic minerals to be used in place of glass. Special Anderson–Payne models incorporating synthetic spinel, blende, and diamond have been made which have certain advantages as well as certain limitations compared with standard refractometers. In Rayner refractometers the scale extends from 1.30 to 1.86. In the spinel model the shorter range of 1.30 to 1.68 allows a more open scale to be used, while the long upward range of the

---

* The formula is $n = n' \sin Ic$, where n is the unknown refractive index, n' that of the denser medium, and Ic the critical angle.

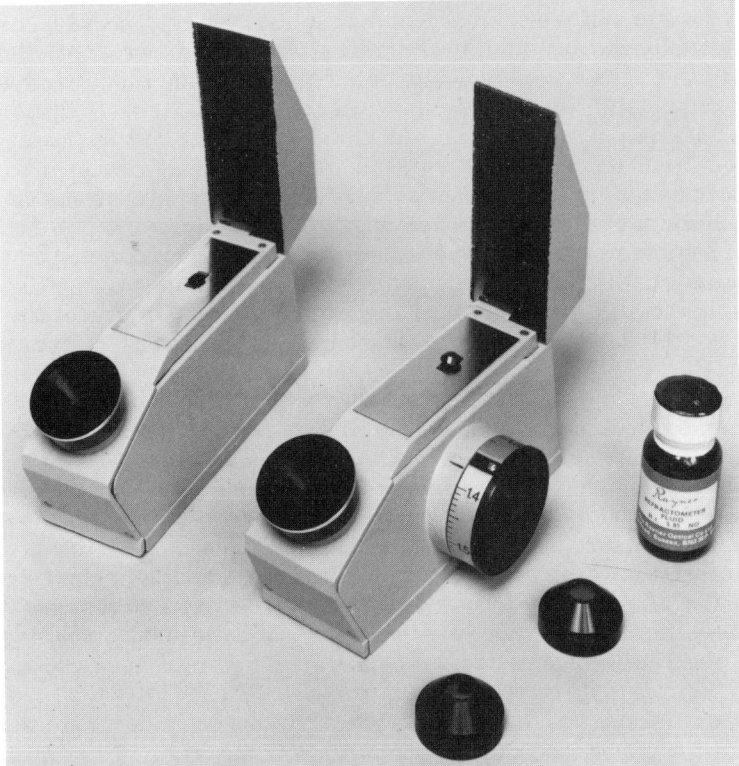

Figure 2.3 The Rayner standard (S) (*left*) and Dialdex (*right*) refractometers

diamond refractometer, 1.55–2.05, necessitates a sliding eyepiece for scanning the scale. Refractometers using cubic zirconia prisms are now available; their range is from 1.40 to 2.10.

Refractometers manufactured in several other countries are widely available. These operate on precisely the same principle as the British types described, and the instructions given will apply to them equally well.

The Dialdex refractometer

In 1972 Rayner succeeded in producing a refractometer of quite a new type (Figure 2.3 (*right*)) although it acts upon the same principles as all others. In place of the customary scale of the conventional refractometer, the user (having placed the usual small drop of contact liquid and the stone to be tested on the table of the instrument) sees through the eyepiece a blank screen crossed with the faint shadow edge or shadow edges due to the stone, the position of which will depend, as usual, on the refractive index or indices of the stone. By turning the calibrated dial at the side of the instrument a black ribbon-like marker can be adjusted to coincide with each shadow edge. The corresponding refractive index can then be read very easily from the dial – hence the name 'Dialdex'. Readings accurate to one or two units in the third decimal place can be assured on this instrument rather more readily than on the previous standard Rayner refractometer. A further advantage has been gained by the employment of a new type of glass for the prism of the instrument, which has a lower dispersion than is normal for such high-index glasses. This means that the shadow edges are sharper in white light, and by using the dark yellow filter provided in combination with a strong light the edges are so sharp that the use of a sodium vapour lamp (see later) becomes an unnecessary luxury.

Instrument makers have now utilized the unique properties of the high RI isotropic diamond simulants for use as prisms in refractometers instead of soft glass. A refractometer known as the 'ER 602 Riplus' is being commercially produced by A. Krüss of Hamburg. The prism is strontium titanate ($n_0 = 2.418$), while the contact difficulty is overcome by using a melt of high RI liquefied by heating the prism electrically. The scale extends from 1.75 to 2.21 (see Figure 2.16).

A further development in the early 1980s was in the use of cubic zirconia ($n_0 = 2.17$) for the prism. In this instrument, with an optimal white light/LED yellow light source built in, high RI liquids are used as contact fluids. These have RIs of 1.80, 1.90, 2.00, and 2.11. The range is from 1.40 to 2.10. This instrument was developed by S&T Electro-optical Systems of California, but other manufacturers are now producing refractometers with CZ prisms.

The refractometers using strontium titanate prisms have the higher RI range, but suffer from having a softer prism ($H = 5\frac{1}{2}$) which may need careful cleaning when using a melt. The harder prism ($H = 8\frac{1}{2}$) of the CZ instruments probably more than offsets its slightly lower maximum reading.

Most gemmologists, however, will find that their standard 1.81 refractometer will serve them well in the great majority of determinations.

How to use the refractometer

Though refractometers are essentially very simple to use, the beginner will do well to read carefully the instructions and warnings that follow, most of which will apply to any present-day type of jeweller's refractometer. Remember, above all, that the surface of the prism on which the stone is placed consists of a very soft form of glass and is therefore easily scratched by a corner or edge of the stones that are being tested. Good readings can only be expected if the surface is preserved in an unblemished state.

The standard type of refractometer will be dealt with first. The instrument should be placed on a steady bench or table, and should be mounted on some stand or block to allow for easy reading. A bench lamp or adjustable reading lamp fitted with a 60-watt pearl bulb provides adequate illumination and should be so arranged that the light can pass direct from the lamp to the 'window' of the refractometer, the lamp itself being positioned a little lower than the window.

In every case, if the illumination is correct, the observer should see on looking through the eyepiece a brightly and evenly illuminated scale. If this is not sharply in focus, the eyepiece should be adjusted suitably.

We have now reached a point where all is in readiness to take a refractive index reading. A supply of highly refractive liquid in a dropping bottle is provided with each instrument. Thoroughly clean the stone to be tested, then place a *small* drop of liquid on the glass table of the refractometer and position the stone carefully so that its table facet rests on the prism of the refractometer, the liquid being flattened out into a thin layer which serves to exclude the film of air which would otherwise prevent the stone from making 'optical contact' with the glass of the instrument. If the stone is of fair size – say, 3 carats upwards – it can safely be handled with the fingers, special care being taken to avoid scratching the soft glass of the refractometer. With smaller stones it is wise to use 'corn tongs' in order to be quite sure that the stone is lowered into position with its table facet parallel to the glass. In removing the stone from the instrument, slide the stone gently from the glass to the metal, whence it can safely be picked up without any risk of damage to the glass. After each reading, any surplus contact fluid should be carefully wiped away to avoid tarnishing the glass surface. The best reading should be obtained when the stone is placed at the exact centre of the glass, but in some instruments better results may be obtained when the stone is slightly off-centre.

Now let us suppose that the stone has been correctly placed on the instrument (with liquid to make optical contact) and that the stone in question is a spinel.

Looking through the eyepiece, the observer should see part of the scale brightly illuminated, the remainder relatively in shadow. It will be seen that the point where the shadow cuts across the scale will be near 1.72, which is the refractive index for spinel (see Figures 2.4 and 2.5). The *exact* position of the shadow edge on the scale is not easy to estimate when using white light, as the

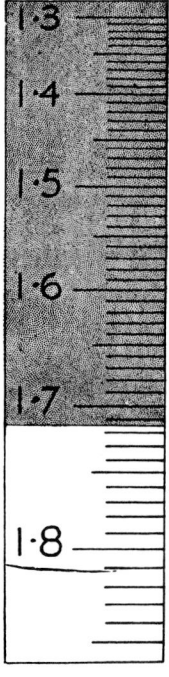

Figure 2.4 A diagrammatic representation of the observer's view of the scale of a standard Rayner refractometer on which a spinel is in position, using sodium light. The reading is 1.715. The faint edge at 1.81 is due to the liquid

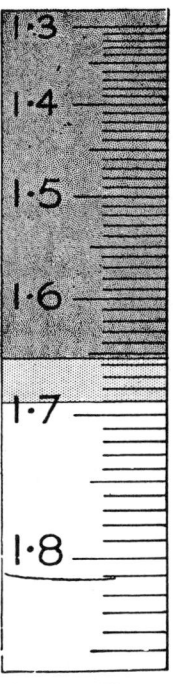

Figure 2.5 The two edges which are seen with a peridot at their maximum possible separation (1.653–1.690)

edge is not a sharp one, but consists of a narrow spectrum due to the fact that the relation between the refractive indices of the stone and the glass of the hemisphere is not the same for all colours (wavelengths) of the spectrum, the glass having a higher 'dispersion' than any of the gemstones which can be tested on it.

The reading should be taken at that part of the shadow where the green passes into the yellow and at the point of the slightly curved edge where the refractive index recording is at its highest. It should be mentioned here that in addition to the shadow edge due to the stone, a fainter edge should also be visible due to the liquid employed. This should read a full 1.81 if the liquid used is the standard fluid devised by Anderson and Payne in the London Gem Testing Laboratory. It may be noted that this liquid shadow edge is much more sharply defined than that due to the stone, due to the closer approximations of the dispersons of the liquid and the glass of the hemisphere. The scale is an inverted one; that is, the lower parts of the scale represent the higher refractive indices. It may be mentioned that within its limited range the Anderson–Payne spinel refractometer gives sharper and more accurate readings in white light than the other instruments. It was indeed designed for this purpose. The synthetic spinel used for the prism has a very similar dispersion to those of the stones tested, so that the critical angle between prism and stone (upon which the position of the shadow edge depends) does not vary appreciably, whatever the wavelength of the light. Photographs of shadow edges on a spinel refractometer for obsidian and for tourmaline are reproduced in Figure 2.6.

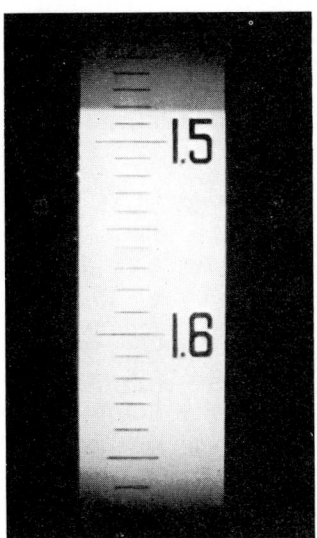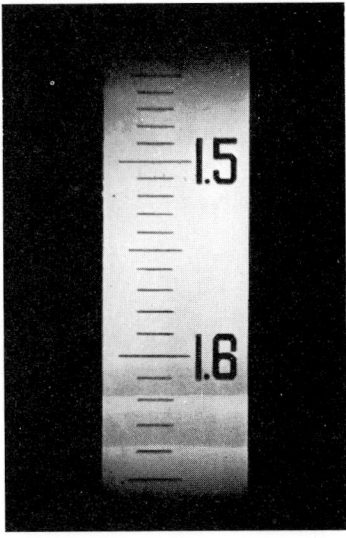

Figure 2.6 (*Left*) Photograph of single edge shown by obsidian; (*right*) photograph of shadow edges shown by tourmaline (both on a Rayner spinel refractometer)

To obtain an absolutely sharp edge to the shadow with a standard refractometer one must use 'monochromatic' light, i.e. light of one colour or wavelength only; the standard and most easily produced light of this nature is the pure yellow light produced by glowing sodium vapour (wavelength

589.3 nm). This yellow sodium light is easily produced if a Bunsen gas burner is available. Any substance containing sodium, such as soda (sodium carbonate), salt (sodium chloride), or window glass (a sodium calcium silicate), which is placed on the edge of the hot blue flame at its base, will produce the desired yellow sodium flame. A sodium flame can be produced in a similar way with a spirit lamp by smearing salt on the wick, or adding some salt to the methylated spirit used as fuel. Such sodium flames are not very intense, but if used in a darkened room give quite clear results.

Once having seen the beautiful sharpness of the shadow produced with sodium light, the observer may well be reluctant to return to the use of white light again, though with the latter refractive index readings sufficient to identify a given gem are usually obtainable.

For those who can afford it, the most brilliant source of sodium light is obtained by use of a sodium discharge lamp. Rayner have provided for many years a compact sodium source designed for use with their refractometers, but the replacement lamp is very expensive. Units using larger but more readily available commercial sodium discharge tubes are now marketed by British (including Rayner), American, and West German manufacturers at much more reasonable prices.

The use of light-emitting diodes (LEDs) as substitutes for sodium discharge tubes was first suggested by D. Minster of South Africa. Some sources use six LEDs housed in a compact unit and these have a much longer life than filament lamps. The yellow emission spectrum of the LEDs is wider than that of a sodium lamp, but it produces a much sharper shadow edge on the refractometer than a white light source. Where sodium light is not available shadow edges can be improved by using a red glass or gelatine filter either between the light source and the instrument or over the eyepiece. The readings will be slightly higher than normal, but sharper than with any known *yellow* filter because more nearly monochromatic. A dark yellow filter is made by Rayner to fit over the eyepiece of their refractometer: if used in conjunction with a strong white light this is very effective.

As an example of refractive index measurement we have so far mentioned only spinel, which gives a single shadow edge near 1.72. If now the observer places a peridot on a refractometer not one shadow edge but two (in addition to the constant faint edge produced by the liquid) should be seen.

The precise positions of the edges will vary according to the orientation of the specimen, but it will be found that by carefully turning the stone on the table (keeping contact all the time) a minimum reading is obtained with the lower edge and a maximum with the other – *not necessarily in the same position*. With peridot these two important critical readings will be rather above 1.65 and about 1.69 respectively (see Figure 2.5). This effect of two edges showing on the refractometer is due to *double refraction* and is shown by a large number of gemstones. It is of immense importance.

A single ray of light passing into any gemstone except those which crystallize in the cubic system is, in general, split into two rays which are refracted to a different extent. For all such stones there is a maximum and a minimum refractive index, and the difference between these two measures the extent of the double refraction. Thus for peridot the double refraction is normally $(1.690 - 1.654) = 0.036$, for tourmaline $(1.638 - 1.620) = 0.018$, and for yellow topaz only $(1.637 - 1.629) = 0.008$.

In the last two examples we at once see the importance of the extent of double refraction from the testing point of view, since both topaz and tourmaline have a very similar mean refractive index; and, since a deep pink topaz is difficult to distinguish by eye from a tourmaline of similar colour, it is valuable to be able to distinguish between them on the refractometer. This is easily done by observing the nature of the shadow edges as the stone is rotated. With tourmaline the separation of the two edges can distinctly be seen even in white light, whereas with topaz the double refraction is too low to be distinguished clearly except on the spinel refractometer or in sodium light. A 'Polaroid' disc assists considerably in detecting and measuring double refraction on the refractometer. Suitable polarizing caps to fit over the eyepiece are provided by the various makers. A polarizer only allows light to pass which is vibrating parallel to a certain direction. The rays corresponding to the two refractive indices in a doubly refractive stone are vibrating at right angles to one another, so that by turning the polarizer to the correct position one can see only the shadow edge due to one ray, then by rotating the polarizer through 90° the shadow edge due to the other ray is seen. Thus, even where the double refraction is insufficient to allow two edges to be seen in ordinary light, there will be noticed a slight shift of the edge when the polarizer is rotated, whereas with a singly refracting stone, such as garnet or spinel, no such variation should be observed.

The only important singly refractive stones which come within the range of the refractometer are fluorite, opal, spinel, and the garnets – hessonite, pyrope, almandine, and spessartine. The green demantoid garnet, zinc blende (sphalerite), diamond, and several diamond imitations are also singly refracting but are beyond the range of the ordinary refractometer. Glass imitations also give a single edge on the refractometer, but this may vary in position from about 1.50 up to 1.70 according to the composition of the glass, and it is worth noting that *no singly refractive natural gemstone comes within the normal glass range, 1.50 to 1.70, just mentioned.*

When a singly refractive reading is seen in a region of the scale where no gemstone has its refractive index, one may at once suspect a 'paste'. The nature of the shadow edge in white light is also revealing. Pastes with refractive index between 1.60 and 1.70 are lead glasses having a higher dispersion than gemstones of comparable index. This results in a *sharper* shadow edge with the standard refractometer and a coloured edge when using a spinel refractometer. Doublets which have an almandine garnet top (as they often do) will given an almandine reading of about 1.79, which aids in identifying the fraud.

A list of the refractive indices of the principal gemstones is given in Table 2.1 for convenience. A more comprehensive list will be found in Appendix 5. The figures are given to two places of decimals only, but the double refraction is quoted to three places to enable small but often important distinctions to be made clearer.

Stones having refractive indices above the range of the refractometer are also included in the table. Although these do not give a reading (i.e. a shadow edge at the limit imposed by the contact liquid) this in itself gives valuable evidence as to the nature of the specimen. In practice (apart from some synthetic materials used to imitate diamond) it limits the possibilities to certain almandines, to demantoid, zircon, diamond, and, less probably, sphene. These can usually be separated by their appearance, and by other characters described later.

In certain species there is considerable variation in the value for the indices

Table 2.1　Refractive indices of important gemstones

Gemstone	RI		DR
Fluorite		1.43	–
Opal		1.45	–
Quartz	1.54	1.55	0.009
Beryl	1.57	1.58	0.006
Topaz, yellow, pink	1.63	1.64	0.008
Tourmaline	1.62	1.64	0.018
Andalusite	1.63	1.64	0.010
Spodumene	1.66	1.68	0.015
Peridot	1.65	1.69	0.036
Zoisite	1.69	1.70	0.009
Spinel		1.72	–
Chrysoberyl	1.74	1.75	0.009
Hessonite		1.74	–
Pyrope		1.74	–
Corundum	1.76	1.77	0.008
Almandine-pyrope		1.77	–
Spessartine		1.80	–
Almandine		1.81	–
Demantoid		1.89	–
Sphene	1.90	2.02	0.120
Zircon	1.93	1.99	0.059
Diamond		2.42	–

due to variations in chemical composition. This is notably the case in the series of red garnets which vary more or less continuously from an 80 per cent pure pyrope, RI = 1.732, to an 85 per cent pure almandine, RI = 1.810. Typical representative figures are given in Table 2.1.

By using sodium light, the skilled observer can obtain much more detailed information concerning the optical nature of the stones tested – whether they are 'uniaxial' or 'biaxial', 'positive' or 'negative', etc. Such factors can often be of value in practical testing. A summary is therefore given of the effects seen.

1. With singly refractive stones (amorphous or cubic), only one shadow edge is seen, which remains immovable when the stone is rotated.
2. With uniaxial stones (tetragonal, trigonal, and hexagonal), in general two shadow edges are seen, one of which, corresponding to the extraordinary ray, moves, as the stone is rotated, towards and away from the immovable edge due to the ordinary ray. The index for the extraordinary ray is read at its greatest divergence from the ordinary; the difference between the two readings is then the birefringence of the stone. When the extraordinary index is higher than the ordinary the stone is optically positive; when less, optically negative.

   Only if the table facet happens to be cut exactly at right angles to the optic axis does the shadow edge due to the extraordinary ray remain fixed (in its position of full birefringence) when the stone is rotated.

   When the optic axis lies in the plane of the table facet a single edge will be seen when the stone is turned into a position where the axis lies parallel to the 'axis' of the instrument (i.e. the line between the eyepiece and the window through which light enters the instrument). The full birefringence will then be seen when the stone is turned until the optic axis is at 90° to the axis of the instrument.

3. With biaxial stones (rhombic, monoclinic, and triclinic), in general two shadow edges will be seen, both of which may vary in position as the stone is turned on the refractometer. The maximum index obtainable with the higher shadow edge should be read, and this corresponds with the value known as γ. The lowest index obtainable with the other edge should be noted separately, and corresponds to the value known as α. Then γ − α is the full birefringence.

Measuring the full birefringence of a biaxial stone on the refractometer is not always an easy process, but one that every serious student should practise and master. Some instructors encourage the noting down of pairs of shadow edge readings taken with the stone in a series of orientations, hoping that by subtracting the lowest reading of the series from the highest the full birefringence figure will be obtained. Experience in marking thousands of examination papers has shown the author that this is an unsafe assumption. The only sound method (which is also the quickest and simplest) is to concentrate on the movement of one shadow edge at a time as the stone is slowly rotated on the table of the refractometer by gentle nudges with the forefingers of each hand – chasing the upper shadow edge to its highest limit and then the lower to its lowest limit, recording each of these critical readings, and subtracting the lower from the upper.

A critical intermediate value, β, is not so easily obtained. It corresponds either with the lowest value for the higher edge or the highest value for the lower edge. The facet *may* happen to be cut in such an orientation that an optic axis (or even both axes) lies in the plane of the facet. In this case a single edge (corresponding to the β value) will be observed in certain positions.

The convention is that when β is nearer in value to α than to γ the stone is optically positive; when nearer to γ than to α it is negative. In practice it is seldom necessary to ascertain the exact value for β. If the higher shadow edge is seen to move beyond the half-way position between maximum and minimum readings it must be positive. If the lower edge moves past the half-way position, the stone is negative.

Advanced gemmologists may find it interesting to deduce the crystal orientation of a cut stone by the careful observation of shadow edge movements, but for the purposes of identification maximum and minimum readings are all that are normally necessary. *It can be stressed that for all stones, uniaxial or biaxial, the difference between the maximum and minimum readings obtainable on any facet must always represent the full birefringence for that stone.* This is a fact of the highest diagnostic importance. Certain gemstones are 'cryptocrystalline', that is, they consist of masses of tiny crystals in more or less random orientation. Jadeite, nephrite, and members of the chalcedony (agate) family are well-known examples of this. Such stones may be expected to give only a single generalized shadow edge on the refractometer, and not a very sharp one. However, with some of the agate group (onyx, cornelian, for instance) the edge can be seen as distinctly double, with an apparent birefringence of about 0.006, which does not vary on rotation of the specimen. This curious but usefully diagnostic feature has been called 'form birefringence', and ascribed to the effect of particles of one kind (here small fibres of quartz) being embedded in a substance of different refractive index (here, opaline silica).

Figures 2.7 and 2.8 demonstrate typical values of refractive indices when uniaxial and biaxial stones are turned through 180°.

Useful and simple methods for detecting double refraction without the aid of a refractometer will be discussed in the next chapter.

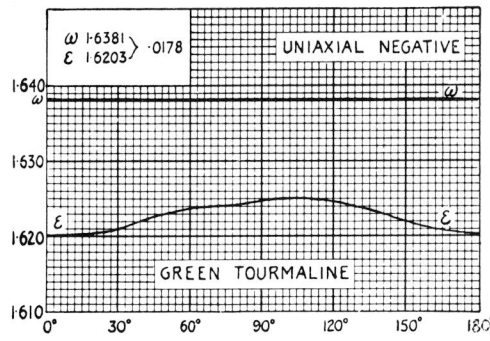

Figure 2.7 Movement of shadow edges observed in a specimen of green tourmaline

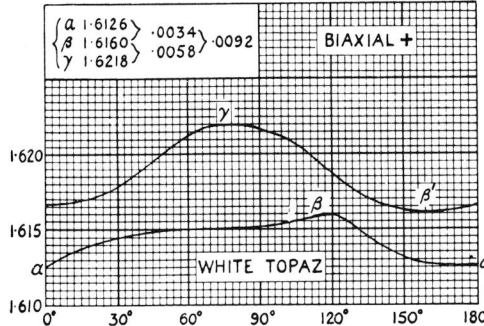

Figure 2.8 Movement of shadow edges observed in a specimen of white topaz

'Distant vision' method for cabochon stones

To obtain accurate readings on the refractometer it is essential that the stone should have a flat, polished facet to make good contact with the glass of the instrument. However, the late Lester Benson of the Gemological Institute of America developed and described an ingenious and important new technique whereby modified refractometer readings can be obtained even with cabochon stones or with faceted stones which are too tiny to give a shadow edge in the normal way. Benson's method (known in America as the 'spot' method and in Britain as the 'distant vision' method, for reasons which will be obvious in what follows) is neither so easy to apply nor so critically accurate as the standard method, but it so enlarges the scope of the refractometer that the keen gemmologist is strongly recommended to take the necessary trouble to become proficient in its use, as it will enable identifications where the normal procedure fails.

Chrysoberyl and quartz cat's eyes, for instance, sometimes so difficult to distinguish with certainty when mounted in jewellery, show their great difference in refractive index in clear-cut fashion, as do cabochons of nephrite or jadeite and the several massive green minerals with which they may be confused. Turquoise, which in the normal way is reluctant to show a shadow edge even from a flat surface, also yields quite a distinct effect by the distant vision method, and quite tiny faceted stones are also made to a give a reading.

The procedure, when using a standard Rayner refractometer, is as follows. The first essential is to limit the amount of the contact fluid to the smallest possible droplet. This can be done by applying it with a fine glass capillary tube or a fine wire or needle, or by touching the stone on an ordinary small drop, withdrawing it, and wiping the main drop away, using only the minute quantity adhering to the stone. The stone (we will assume it to be a cabochon) is now allowed to rest in the centre of the refractometer table: the point of contact will be slightly enlarged by the trace of contact liquid, and if the eye be withdrawn about twelve inches away from the eyepiece and the head moved slightly until the correct position is found, it should be possible for the observer to see the point of contact of the stone and liquid droplet as a little disc in the middle of the limited portion of the scale still visible at such a distance.

If the eye (and the attention) be focused on this contact spot the scale readings will be, unfortunately, slightly out of focus, but by a slight lateral movement of the head and an effort of will the position of the spot on the scale can be gauged fairly closely. It will be found that if the contact spot be viewed against index readings which are below the refractive index of the specimen, it will appear completely dark.

On gradual alteration of the vertical line of vision towards the higher index readings, beyond a certain point the contact spot will be seen to change from dark to light (as total reflection sets in). With care and practice the critical position can be found where the spot is exactly bisected by the dark shadow. *A reading of the scale at this point will give the refractive index of the stone.*

Practice in the method should be gained with such objects as beads or cabochons of quartz, moonstone, or other known minerals, or with faceted stones which, though small, are just large enough to show a shadow edge when using the refractometer in the normal manner. Thus confidence in the new technique will be gained before an attempt is made to tackle an unknown specimen.

An ingenious modification using a pinhole for the distant vision method has been devised by G. S. Walker of Australia. He uses a pinhole of rather less than one millimetre in diameter (made by pushing a pin or needle through exposed black (photographic) film or other suitable material) held as close as possible to the eye and no more than 6 cm (2½ in) from the refractometer eyepiece. It is then possible to see both the scale and the contact spot in focus at the same time. More of the scale is visible (with the closer viewing plane) and the spot is much less elusive. Holes of various sizes may suit different eyes and a little practice may be needed, but the method is both elegant and inexpensive.

Care of the refractometer

When new, the glass surface of the refractometer hemisphere or prism is beautifully flat and perfect, and will give very dense shadow edges. It is well worth trying to maintain this surface in good condition by protecting it from mechanical damage due to careless placing of the stone thereon in such a way that an edge or corner of the stone scratches the soft glass, and from what may be called chemical damage from allowing the highly refracting liquid (which is usually di-iodomethane with one or more solids dissolved in it) to remain on the glass after the test is made. Strips of clean blotting paper should be kept at hand to wipe away the surplus liquid and the glass then rubbed with a clean wash leather when the test has been completed. A smear of Vaseline on the glass when the instrument is not in use is an additional protective measure. If the glass should have become affected and is giving poor readings, a useful tip is to clean it with a little wad of cotton wool dipped in a paste of jeweller's rouge and water. This will produce sparkling results. If the surface has become seriously damaged, however, it will be best to return it to the makers for repolishing.

Heavy pressure on a specimen while taking a reading should be quite unnecessary, and often gives rise to damage in the soft glass of the refractometer. Normally the weight of the stone itself is quite sufficient to make good optical contact with the table of the instrument.

It is worth noting that some pastes fail to give a shadow edge on the refractometer, due to their having a thin surface film of refractive index slightly different from that of the main body of the glass. Such a 'coating' is probably accidentally induced during manufacture. Deliberately 'coated' stones have been recently introduced (see end of Chapter 9) which also fail to give refractometer readings. In either case, a brisk rub with a rouged cloth will usually clean the surface of the specimen sufficiently to enable clear readings to be obtained.

'Bright line' technique

In cases where readings are difficult to obtain under the ordinary lighting conditions, C. J. Payne has often found the following technique to give good results. The hood of a Rayner refractometer is pulled out of its socket, and the window through which the light normally enters the instrument is blocked. A microscope lamp (or sodium lamp) is then so adjusted as to provide light at grazing incidence on the stone when it is placed on the refractometer table. If these conditions are correctly carried out it is usually possible to see a *bright line* crossing the scale in the position where the shadow edge would normally appear. A thin paper tissue placed between the refractometer and the light source often helps when using this technique. It may be necessary to rotate the stone for the effect to be clearly seen. With birefringent stones, of course, two bright lines should be visible.

Identifying stones of high RI

It is generally recognized that a refractive index of 1.81 represents the upper limit for readings on any of the standard refractometers, however high their scale may extend. This limit is imposed by the contact liquid commonly used, consisting of di-iodomethane saturated with sulphur and tetra-iodoethylene. Other fluids such as phenyldiiodoarsine or selenium bromide have been used in the Gem Testing Laboratory of Great Britain to enable higher readings to be obtained with a refractometer fitted with a diamond prism, but such liquids are chemically highly reactive and would quickly ruin the surface of the lead glass used for the prism or hemisphere of normal refractometers. Even were this not so, the small increase in range made possible would not be of much practical value, since there are few gemstones with indices within the 1.8–1.9 range.

On the other hand, the ability to make refractive index readings on stones having indices between, say, 1.9 and 2.6 has become more and more desirable for gemmologists in recent years, since a whole range of artificially made products has been successively promoted as possible substitutes for diamond, all of which have high refractive indices. A list of these, together with their refractive indices and other properties, will be found in Chapter 12, Table 12.1.

Fortunately, there is a fairly simple method available capable of making such high readings with a fair degree of accuracy, and this will now be described. An account will then be given of several new 'reflectivity meters', which also enable the gemmologist to discriminate between stones that have refractive indices above the range covered by conventional refractometers.

Direct-measurement method

For this method a good standard monocular microscope (see Chapter 6) is the essential requirement. The microscope to be used for the necessary measurements should preferably have a scale attached to the fine adjustment control, in order to register the amount it has been turned to achieve an exact focus (the actual units represented do not matter); alternatively, a millimetre scale should

Figure 2.9 Vernier attached to simple microscope reading to 0.05 mm for 'real and apparent depth' method

be fitted to the body tube and read by a vernier attached to the fixed limb of the microscope (see Figure 2.9). Such attachments are admittedly rare, but can be fitted by a good instrument maker without undue trouble. A suggestion made by F. S. H. Tisdall provides a simple and effective alternative. This makes use of a sliding millimetre gauge with vernier scale reading to a tenth of a millimetre and incorporating a depth gauge. Gauges of this sort are not expensive and are useful to the jeweller as well as the gemmologist for establishing the exact dimensions of notable stones. The depth gauge attachment enables the microscopist to measure differences in focal position by taking readings of the distance between the top of the focusing rack and the fixed block through which it moves.

In carrying out the method, the following procedure is recommended. The stone to be tested is cleaned and fixed on a glass slide by means of a small ball of wax or modelling clay against which the girdle of the stone can be gently pressed so that the table of the stone is uppermost and horizontal while the culet rests on the slide. The stone is then placed directly under the objective, which should be of moderately high power to enable the exact positions of sharp focus to be established, and the exact focus obtained for the surface of the table facet, making use of fine dust particles, polishing marks, etc. for the purpose. A reading should be taken and noted on the focusing wheel, vernier, or millimetre depth gauge at this point. The focus should then be lowered until an exact focus is obtained of the culet as seen looking through the stone, and the measurement of the focal point again noted. The difference between the first and second readings should give the 'apparent' depth of the stone. The slide can then be moved a little along the stage, to enable a third reading to be obtained when sharp focus is achieved on the surface of the slide without the stone intervening. The difference between the third reading and the first will provide a measure of the real depth of the stone concerned. The following simple formula can be applied:

$$\text{Refractive index} = \frac{\text{Real depth}}{\text{Apparent depth}}$$

The measurements should be repeated several times and an average taken to ensure maximum accuracy. Under good conditions, if the stone is of reasonable size the result should be accurate to within 0.02, which is usually quite adequate for identification when one is dealing with the single index of a cubic substance. The method can only be expected to give good results with singly refractive stones or with the ordinary index of uniaxial stones.

While the method is simple and fairly rapidly carried out with unmounted stones, it becomes more difficult or even impossible when the stone to be measured is mounted. When one is working in millimetres a Leveridge diamond gauge can sometimes be used to provide the real depth of the stone, while the 'apparent' depth, of course, is as easily obtained with mounted as with unmounted gems.

It is worth noting that the depth of an inclusion below the surface of the table facet can be estimated with some accuracy by measuring the difference between the focal positions for the surface of the table and the inclusion, and then multiplying by the refractive index of the stone – e.g. by 2.42 if the stone be diamond.

Reflectivity (reflectance) meters

In gemmology, as in other walks of life, a new threat often stimulates people's ingenuity and enables them to counter that threat. In recent years jewellers, and even trained gemmologists, have been increasingly bothered by a succession of manmade crystals which, when cut, resemble diamond more or less closely, and all of which have refractive indices far beyond the range of the standard refractometer. However, a number of instruments have appeared on the market that have broken through the '1.81 barrier' imposed by the customary contact liquid and enabled some estimate to be made of the refractivity of any transparent stone having a well-polished flat surface. No contact liquid is required in operating these instruments, hence the ability to break through the barrier mentioned.

As in all kinds of gem-testing instruments, there are snags and limitations in the use of these new reflectivity meters, of which the 'Gemeter' was the forerunner. Reflectivity meters will lead to many a misidentification in the hands of those who do not properly understand what it is they are measuring, and who expect answers as exact and reliable as those obtainable on lower-index stones with a standard form of refractometer. But at least they should provide a rapid and safe means of separating diamond from all its would-be competitors.

The connection between the reflective power of a transparent isotropic solid and its refractive index for sodium light was given long ago by the French physicist Fresnel, who found that in the unique case of perpendicular incidence the formula

$$\frac{(n-1)^2}{(n+1)^2} \times 100$$

equals the percentage of light reflected (n being the refractive index of the specimen). It is interesting to note that, according to this law, diamond reflects 17.2 per cent of light falling at right angles to its surface, whereas with quartz the amount is little more than 4 per cent.

The late L. C. Trumper deserves credit for having devised and constructed an instrument that made use of this principle and enabled him to measure refractive indices, no matter how high, to an accuracy of about 0.02. However, his apparatus was not made commercially available, and probably required more skill devoted to its use than an impatient age was willing to give it. The same might be said of similar measurements made by A. F. Hallimond, using a special microscope objective and a photoelectric cell to record the intensity of the reflected light.

The several reflectivity meters now available do not depend to any degree on the Fresnel equation. The 'light' involved consists of a broad band in the infra-red region approximating to radiation of some 950 nm, and is provided by a light-emitting diode powered by a small battery. The light reflected from the surface of the specimen being tested, which is placed on a small opening a few millimetres above the diode, is received by a photo-transistor positioned at an angle of up to 30° to the beam of infra-red light. The differences between the versions employing the same principle seem to rest on the convenience of the shape and size of the orifice on which the stone must rest; on the type of scale against which the moving needle records results; and on the stability of the activating current. Some instruments have two ranges, one for low-index and another for high-index minerals, according to which button is pressed by the operator. The numerical accuracy of the scale, where this is calibrated in terms of refractive index, is less important than the constancy of results for stones of any one kind: in some of the early instruments this constancy was woefully lacking, but most instruments are now reasonably reliable.

Reflectivity meters *per se* appear to have lost ground in the diamond-testing market to instruments measuring thermal properties, and these instruments are now more generally available. They have commonly been labelled thermal conductivity meters, but certain anomalies in their operation suggest that in fact they measure thermal inertia. To the practising gemmologist the important point is that they can help in differentiating between diamond and its simulants. The early instruments were not able to operate on small stones on account of the size of the probe which is pressed onto a facet of the stone under test. However, the probes are now so fine that stones down to 0.02 carat (or smaller) can be tested.

A wide variety of instruments is now on the market, and they can be mains or battery powered. Benchtop models have now been joined by instruments which fit into the pocket. All have a probe which is pressed into contact with the stone to be tested and the result may be visual (on a dial or by lamps) or audible (buzzer or pips). Some require a longish time interval between testing successive stones, others need seconds only. Many give audible warnings if the metal of the setting is accidentally touched. Any instrument purchased should be able to operate with mounted stones.

In all cases the surface of the stone to be tested must be thoroughly clean and both the stone and the instrument should be allowed to reach room temperature. It is preferable to check the instrument against a known diamond before starting operations and a double check on the stone under test is desirable.

Dual-purpose instruments measuring both reflectivity and thermal properties are now on the market and their use is to be commended since they provide a

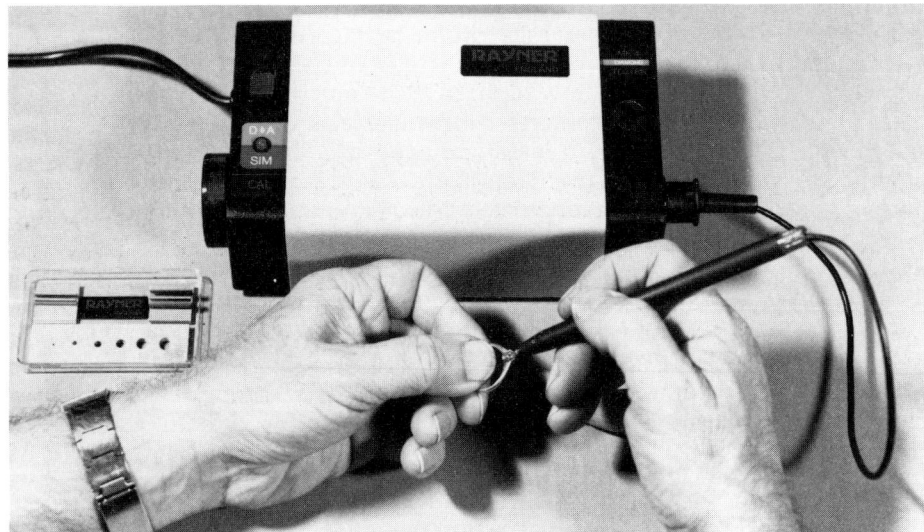

Figure 2.10 The Rayner diamond tester (thermal probe)

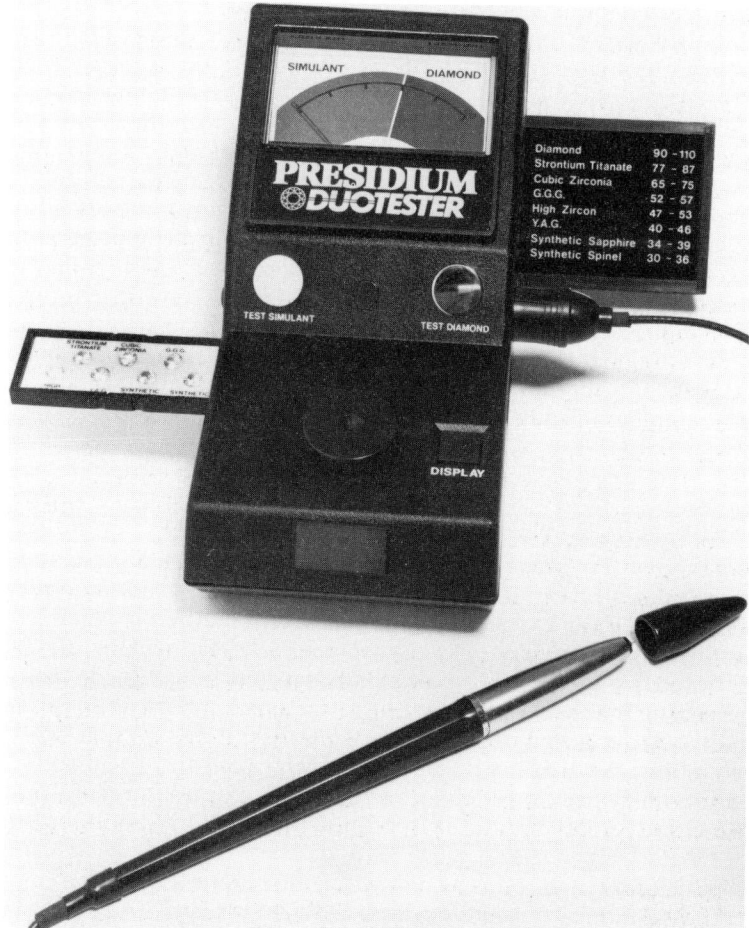

Figure 2.11 The Presidium 'Duotester', a dual-purpose instrument

rapid double check. They are expensive, but probably cost little more than buying two separate instruments. Before purchasing any of these it would be wise to consult with colleagues who already possess them and to take the advice of the various gemmological associations and reputable dealers. There is a continuously developing range of instruments in this field, and any description of a model here could rapidly become out of date. However, the Rayner Diamond Tester (thermal) (Figure 2.10) and the Hanneman 'Diamond Eye' (reflectivity) have long been used by the writer and are still giving effective service. A modern dual-purpose instrument is the Presidium 'Duotester' (see Figure 2.11).

Brewster Angle

In 1941 B. W. Anderson showed that the RI of a gemstone could be derived by utilizing the Brewster Angle of polarization, but did not develop his ideas into a usable instrument because of practical difficulties. Brewster's Law states that complete polarization of a ray reflected from the surface of a denser medium occurs when it is at right-angles (normal) to its associated refracted ray within the medium.

In 1979 P. G. Read built an experimental Brewster Angle refractometer but did not bring it to function on account of the low intensity of the imaged light in the instrument. By coincidence, Dr R. M. Yu of Hong Kong was independently developing another Brewster Angle refractometer at the same time. Neither instrument went into production, but with the advent of small helium/neon lasers new light sources became available, and a redesigned version of the refractometer is now being developed by P. G. Read. The new instrument should have an RI range from 1.40 to 3.3, be easy to use, and require no contact fluid. A Brewster Angle refractometer with a digital readout has been announced by a French firm as this edition was being revised. With these sophisticated instruments cost may be an inhibiting factor.

Immersion methods

Although, as we have seen, by means of the 'distant vision' method it has now been found possible to use the refractometer with stones having curved or very small flat surfaces, there are still cases where some different method of refractive index determination may be necessary. For instance, where uncut or carved stones are being tested, or numerous small stones set in a brooch or eternity ring; or where the setting makes contact with the refractometer glass impossible. The immersion methods described below are particularly suitable in such cases. All that is required in the way of apparatus is a glass dish or cell and small bottles containing various liquids, specified later. A microscope increases the scope and accuracy of the method greatly, but the required information can often be obtained without its aid.

The theoretical basis of the method is as follows. Transparent objects can only be 'seen' by the refraction and reflection of light at their surfaces, and where a transparent solid is immersed in a liquid of nearly the same refractive index, refraction and reflection are reduced to a minimum and the solid becomes virtually invisible. Lumps of ice in a tumbler of water are a familiar example of this effect, and it is easy to prove by experiment that the degree of visibility (or 'relief', as it is called) of a transparent object immersed in a transparent fluid depends entirely upon how nearly the refractivity of solid and liquid approximate to one another.

Thus if we can find a liquid in which the outline of the stone, when immersed, becomes so indistinct that it almost vanishes, we can safely assume that the refractivity of the stone is very near to that of the (known) index of the liquid. Good examples of this are fire opals in carbon tetrachloride, moonstone or orthoclase in chlorobenzene, and quartz in 1,2-dibromoethane.

Of the many possible liquids, Table 2.2 provides a useful range, and includes as far as possible those which are readily available and least obnoxious. Some liquids may affect the cements of doublets and benzyl benzoate is very unkind to many plastics. Care and discretion should be exercised at all times.

Mixtures containing di-iodomethane, which are necessary for higher indices, are expensive and not very pleasant to handle. Phenyldi-iodoarsine (RI = 1.85),

Table 2.2   Liquids suitable for RI determination by immersion methods

With all these liquids great care is necessary. Always use in a well-ventilated area and for prolonged use a spot-ventilated system or a fume cupboard is recommended.

Older name	Newer name	Chemical formula	RI
Carbon tetrachloride	Carbon tetrachloride	CCl_4	1.46
Toluene	Toluene	$C_5H_5.CH_3$	1.50
Monochlorobenzene	1-Benzene chloride	$C_6H_5.Cl$	1.526
Ethylene dibromide	1,2-Dibromoethane	$Ch_2Br.CH_2.Br$	1.54
Monobromobenzene	1-Phenyl bromide	C_6H_5Br	1.56
Benzyl benzoate	Benzyl benzoate	$C_6H_5.COO.CH_2.C_6H_5$	1.57
Monochloronaphthalene	1-Chloronaphthalene	$C_{10}H_7Cl$	1.63
Monobromonaphthalene	1-Bromonaphthalene	$C_{10}H_7Br$	1.66
Methylene iodide	Di-iodomethane	$CH_2.I_2$	1.74

though useful in the laboratory, is very poisonous, and has a violent blistering action on the skin: it is thus emphatically *not* recommended for any but the skilled worker who is prepared to take the necessary precautions.

Where stones are sizeable and have a polished facet available the refractometer, of course, provides the most convenient, rapid, and accurate test. But there are many cases where the refractometer cannot be applied, and immersion methods then come into their own, particularly when used in conjunction with the microscope. One may be faced with a brooch, for instance, set with tiny diamonds,, some of which, it is suspected, have been replaced by synthetic white sapphires. If the whole brooch be immersed in di-iodomethane and the stones viewed under a low-powered microscope the diamonds can at once be distinguished by their high relief (every facet showing clearly) in contrast to the white synthetics, which have hardly distinguishable edges. Any intruders, indeed, into a group of small stones mounted or unmounted, purporting to be all of the same kind, can very easily be seen by immersion in an appropriate liquid.

When stones are free from their setting it is quite easy to see whether they have an index higher or lower than that of a liquid in which they are immersed by placing a sheet of white paper under the glass dish in which the stones are resting (table facet down) and viewing them by light from a single overhead lamp. If a stone has an index *higher* than that of the fluid, it will cast a dark-rimmed shadow on the paper below, and the projection of the facet edges will appear white. If the stone has a *lower* index than that of the fluid, it will show a bright margin and the facet edges will appear as dark lines. The breadth of the border margin will give a good idea of the degree of the difference which exists between the indices of stone and fluid in each case. Where there is nearly a perfect match, a coloured margin will appear, showing that the fluid index coincides for some wavelengths with that of the stone, but (as it has higher dispersion than the stone) not for others.

The effects could more easily be seen if viewed from below; and it is quite easy to devise a means for doing this. Two wooden blocks, or hollow cubes made from cardboard, are placed so as to flank a third, wedge-shaped block of wood or card, the sloping (45°) face of which confronts the observer. On this sloping surface is fixed a piece of mirror (of the type that women carry in their handbags). The two outer blocks are bridged by a sheet of finely ground glass, and on this the glass dish containing the immersed stones is placed. When this is illuminated by an overhead light the observer sees a beautiful display of 'immersion contrast' in the mirror, which, of course, reflects the image of the undersides of the immersed stones as projected on the ground-glass screen. Figure 2.12 gives a clear idea of this simple and effective arrangement.

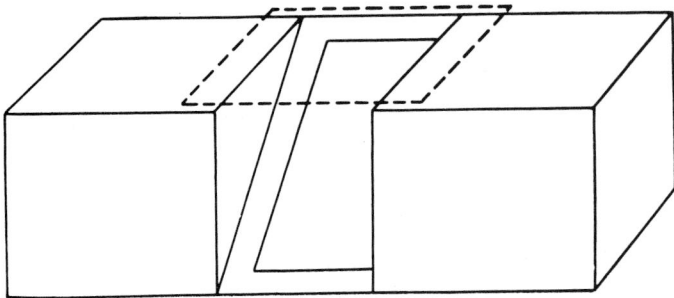

Figure 2.12 Diagrammatic representation of two cardboard cubes and wedge-shaped support for mirror, as recommended for a study of immersion contrast effects. The dotted line indicates the position of the ground-glass 'bridge' on which the immersion cell containing the stones is to be placed. A single overhead light reveals striking differences in contrast when the observer views the immersed stones in the mirror

Immersion contact photographs

Another development of the same general technique is to place the dish of immersed stones over a piece of slow 'line' or 'process' film in a dark-room, and expose for a second or two to light from an overhead lamp. The developed negative gives a beautifully clear pattern of the stones showing their index in relation to that of the immersion fluid in no uncertain manner. Figure 2.13 shows an example of such an immersion contact photograph, in which stones with a wide range of refractive index are immersed in 1-bromonaphthalene.

Figure 2.14 is another immersion contact photograph of a stone necklace, reputedly of variously coloured tourmalines, immersed in bromobenzene. The presence of five aquamarine beads, nearly matching the fluid, is clearly revealed.

Such photographs will also serve to record the exact size and shape and facet distribution of a cut gemstone. For this purpose, however, a liquid should be chosen which has a refractive index near that of the stone. A further practical use of such photographs to the gemmologist is that synthetic stones are often found to reveal the curved 'structure lines' by which their synthetic origin can be assured in immersion contact photographs, even where these lines are not visible to the eye or under the microscope. This will be further discussed in Chapter 9.

More advanced workers should take note that when making immersion contact photographs the operative light for the slow film suggested is in the violet region of the spectrum. For light of these wavelengths the refractive index of the immersion liquids is not, of course, the same as that given in Table 2.1,

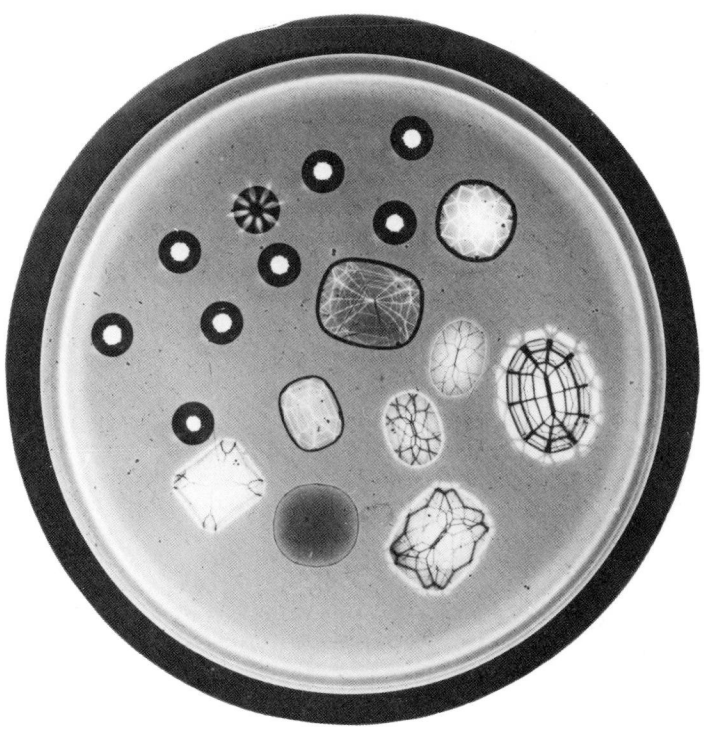

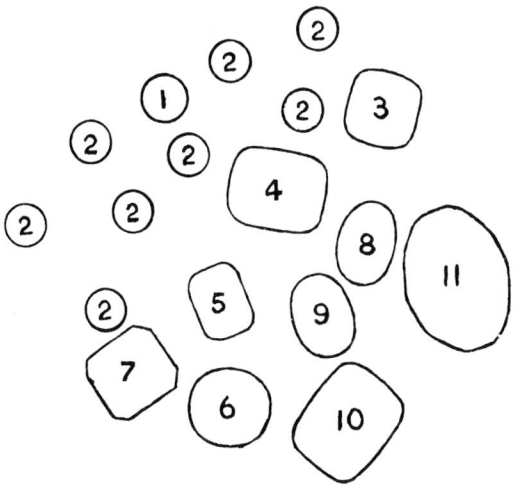

Figure 2.13 Immersion contact photograph of stones immersed in 1-bromonaphthalene (courtesy: Gemmological Association). 1 Synthetic rutile; 2 white zircon; 3 white sapphire; 4 golden chrysoberyl; 5 pink spinel; 6 brown sinhalite; 7 green spodumene; 8 phenakite; 9 hambergite; 10 tourmaline; 11 citrine

Figure 2.14 Immersion contact photograph of a tourmaline necklace in bromobenzene (index 1.56) revealing the presence of five aquamarine beads (in low contrast) (courtesy: Gemmological Association)

which is for the standard yellow sodium light. The index will in each case be much higher, and since the liquids have a higher dispersion than that of the stones being tested, the relation between the two may shift appreciably for different wavelengths of light.

Since actual figures for the liquids concerned are not easily obtained from the literature, some determinations made by the author are given in Table 2.3 for sodium light and for the wavelengths of the strong lines emitted by any mercury vapour lamp. These were chosen because they offer a useful range, and because any one of them can be used in isolation if so desired simply by passing light from a mercury lamp through the appropriate 'mercury green', 'mercury violet',

Table 2.3   Refractive indices of liquids at different wavelengths

Wavelength (nm)	CH_2I_2	$C_{10}H_7Br$	$CHBr_3$	C_6H_5Br
Hg 623.4	1.7364	1.6564	1.5940	1.5615
Na 589.3	1.7414	1.6590	1.5962	1.5620
Hg 579.1	1.7434	1.6614	1.5977	1.5652
Hg 546.1	1.7504	1.6676	1.6017	1.5689
Hg 491.6	1.7662	1.6820	1.6091	1.5769
Hg 435.8	1.8006	1.6998	1.6207	1.5906

etc. filter. Gelatine filters specially made for this purpose are inexpensive and can be obtained from any of the large photographic firms. The figures for di-iodomethane, 1-bromonaphthalene, bromoform, and bromobenzene are given in the table.

When dealing with stones in the rough – amber or imitation amber beads, jade ear-rings, and the like – it is often feasible to detach a tiny fragment without harming the specimen, and the chip can be laid on a plain glass slide, a drop of liquid being placed on it before it is examined under the microscope. In addition to noting the relief or lack of relief shown by the fragment in the particular liquid used, information can readily be obtained as to whether the chip or the fluid has the higher index, in the following manner. The sub-stage condenser (if used) is lowered and the microscope sharply focused on the edge of the chip. The focus is then raised and lowered alternately above and below the true position and the effect observed. If the lighting has been correctly arranged, a bright line, following the contour of the specimen, will be seen alternately to move into and spread outwards from the fragment.

The rule is simple to remember, that when the focus of the microscope is *raised* the bright line passes into whichever medium has the *higher* index. Conversely, when the focus is *lowered* from the position of sharp focus the bright edge passes into the medium of *lower* index. It is wise to practise the effect with fragments and liquids of known index before attempting to work on an unknown specimen.

Various fluids can be tried in succession, draining away one drop with a small piece of blotting paper before applying the next. Observation of the bright line can also be carried out at the edges of quite large faceted specimens immersed in liquid.

Faceted gemstones, however, are not really suitable subjects for the 'Becke line' method in its original form. R. K. Mitchell has given a good account of a modification of the method which can be very easily used when examining an immersed stone. Mr Mitchell's observations were made independently of the earlier work by G. O. Wild and by Professor Schlossmacher in West Germany.

The effect is best seen with the sub-stage condenser lowered and the diaphragm at least partially closed. Then, as the focus is lowered towards the stone, it can be noticed that either the facet edges first appear white and then dark as the focus is lowered into the stone, or the reverse, according to whether the stone or the liquid has the higher index.

The stones should always be examined table facet down, and the facet edges observed should be the pavilion junctions. When the stone has a lower index than that of the liquid, the facet edges appear first black and then turn to white as the focus of the microscope is lowered. When the stone has a higher index than that of the liquid the edges first appear as white lines and then turn to black as the focus is lowered.

Such observations are of considerable practical value as, for instance, when testing a number of small rubies under the microscope; any stone in which the inclusions are not clearly those of ruby can be checked, if the immersion fluid be di-iodomethane, simply by lowering the condenser, closing the diaphragm, and observing the appearance of the facet edges when racking the focus up and down. If the stone is a red spinel which has strayed into the parcel the reverse effect to that for ruby will be clearly evident. Photographs taken by Mr Mitchell (Figure 2.15) demonstrate this very clearly.

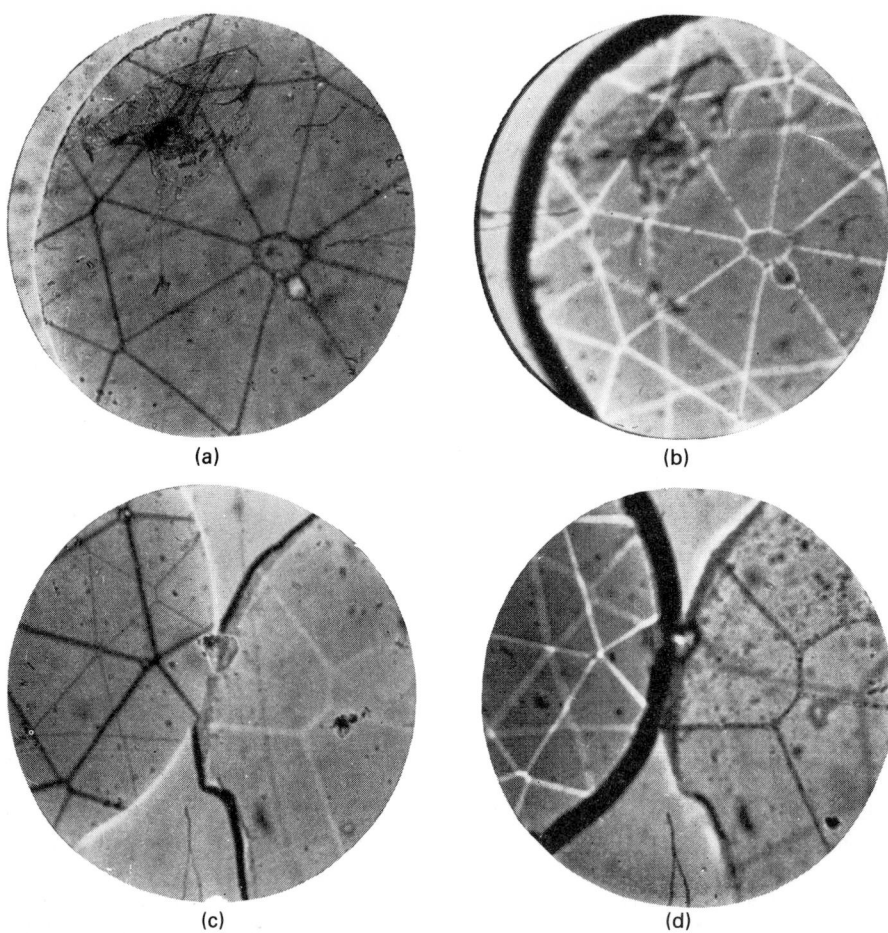

Figure 2.15 (a) Ruby in di-iodomethane (focus inside stone); (b) the same ruby with focus raised into the liquid above the stone; (c) ruby and spinel in di-iodomethane (focus inside the stones); (d) the same stones with focus raised into the di-iodomethane (photos: R. K. Mitchell)

The Becke test also provides a means of determining whether the microscopic crystals to be found as inclusions in most gemstones have a higher or lower refractive index than the stone which encloses them. Here again the degree of relief gives information as to how different or how similar are the indices of the included and including minerals. Unfortunately, since the smaller crystal is permanently 'immersed' within a solid, only the one experiment can be carried out on these lines.

As with the heavy liquid method for determining SG, which is described in Chapter 5, immersion methods can either be used as a rapid means of gaining an approximate value for the refractive index of a stone (which is often all that is necessary to discriminate between two possible alternatives) or it can be used with care and skill to obtain quite accurate measurements. Mineralogists have elaborated the technique to a high degree of perfection.

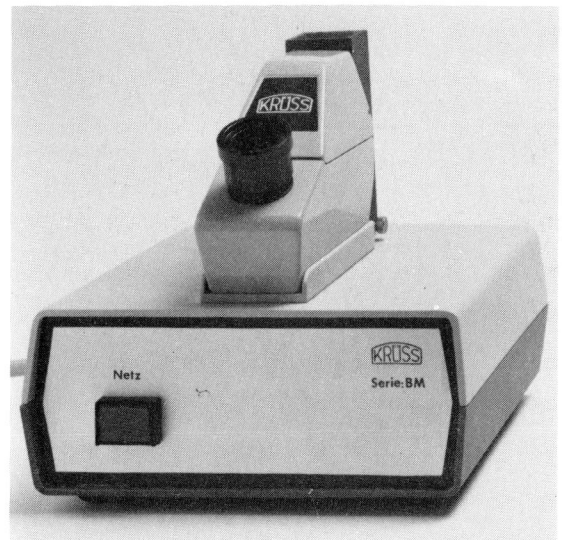

Figure 2.16 The 'ER 602 Riplus' refractometer

Immersion in liquids will be found exceedingly valuable in revealing the internal features, distribution of colour, etc. in gemstones, whether cut or uncut. A lapidary would be well advised to inspect rough potential gems under these conditions before deciding whether they are worth cutting and how best to deal with them.

Although not strictly a method of determining refractive index, the method of shadowing developed by John Koivula can be of considerable assistance in the viewing of inclusions in gemstones. These inclusions are seen as a result of differences between their refractive indices and those of the host mineral; shadowing can exploit and accentuate these differences and greatly increase their three-dimensional appearance (their 'relief'). The technique requires the use of a standard gemmological microscope, preferably with a stage, but this is not absolutely necessary. The gemstone is viewed by transmitted light and the inclusions are focused and illuminated in the best possible way. A light shield is then gradually introduced from one side and a dark shadow will begin to appear from one side of the microscope field; this is the out-of-focus image of the shield. As the shield approaches, the inclusion shadowing will take place until, at a particular point, the inclusion will take on a virtually three-dimensional appearance and appears to be lifted out of its surroundings.

The shield may be rectangular, say 6 × 2 cm (2½ × 1 in) in size, or it can be scythe shaped or a piece of wire gauze or indeed any other shape at the desire (and experience) of the operator. It needs to be rigid and it is best if it is painted black to avoid distracting reflections. The shield may be inserted at any level between the light source and the gemstone. A shield glued to a microscope slide makes for easy sliding across a microscope stage when the stone is supported in tongs.

The principle behind the method is fairly simple. As the shield is inserted into the light path it interferes with the upward passage of light, causing diffraction

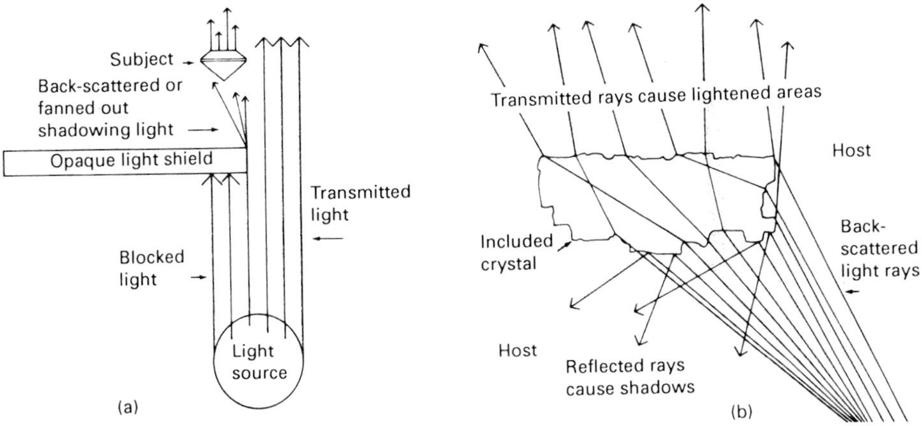

Figure 2.17 (a) An exaggerated diagram showing the back scattering of light as the opaque light shield is gradually inserted into the transmitted light path during shielding; (b) a simplified theoretical situation showing the effect of both reflection and refraction transmission of the back-scattered light as it contacts an included crystal with a refractive index different from that of its host (after J. I. Koivula)

and scattering at the edge. This fanning out of the light (see Figure 2.17) results in transmission of the light in some parts of the inclusion and darkening or shadowing in others, thereby increasing the relief in the inclusion. The method was developed to allow a more intensive study of growth zoning in flux-grown synthetic rubies which might reveal subtle undulations in zoning which appeared straight under the normal illumination provided by standard techniques. Shadowing is particularly helpful when examining multi-phase inclusions such as those in Colombian emeralds.

3

Double refraction and dispersion

When discussing the use of the refractometer in the preceding chapter, reference was made to the curious phenomenon of *double refraction*, which is characteristic of all minerals which crystallize in systems other than the cubic system. This means that the splitting of a single ray into two polarized rays which have different velocities and therefore different refrangibility is observable in all the common gem minerals except the garnets, spinel, and diamond, which crystallize in the cubic system. The magnitude of the effect in the various stones differs considerably, and this fact is highly important as an aid to identification.

Double refraction is measured numerically as the difference between the least and greatest refractive indices for the stone; some figures will be found in Table 2.1, but for convenience a selection of values is given in Table 3.1 in diminishing order of birefringence (birefringence is another term for double refraction).

Table 3.1 **Birefringence of some gem minerals**

Rutile	0.287	Spodumene	0.015
Calcite	0.172	Quartz	0.009
Sphene	0.120	Chrysoberyl	0.009
Zircon	0.059	Topaz	0.008
Peridot	0.036	Corundum	0.008
Tourmaline	0.018	Beryl	0.006

In some minerals the figures vary somewhat, thus in zircon, while the blue, white, and golden 'fired' types most popular in modern jewellery are almost constant in their birefringence of 0.059, many of the Sri Lankan zircons, particularly the green varieties, have a much lower double refraction, sinking in some cases practically to zero. This peculiarity is very rare among minerals, and will be discussed at greater length in Chapter 17. Synthetic rutile has been manufactured in the USA (since 1949), and its enormous double refraction will be one means of identification. Calcite, though not a gemstone, plays an important part in gemmological instruments in its optically pure form of 'Iceland spar'.

How, apart from the refractometer, can one test whether a given stone is doubly refractive or not? The simplest method of all, which with a little practice is quite easily carried out, is to examine the stone carefully with a powerful pocket lens – say, one giving ×8 to ×12 magnification. For example, look through the table facet of a zircon with a lens, and focus sharply on to the edges of the back facets where they adjoin the culet, and it will be noticed that instead of a *single* sharp line where the facets join, as would appear in a singly refractive stone such as diamond, each edge appears as a *double* line (see Figure 3.1).

Figure 3.1 Photomicrograph of zircon taken through the table facet showing doubling of the back facets

A word of warning, however, should be uttered here; according to their crystal structure, all doubly refracting stones have either one or two directions in which only single refraction obtains. Those gemstones belonging to the tetragonal, trigonal, or hexagonal system of crystals have only one such direction; one 'optic axis', as it is called. These minerals, which are grouped together optically as 'uniaxial', include zircon, corundum (ruby and sapphire), beryl (emerald and aquamarine), tourmaline, and quartz (amethyst and citrine). The minerals crystallizing in either the orthorhombic, monoclinic, or triclinic systems have two optic axes, and are called 'biaxial'. These include peridot, chrysoberyl (alexandrite), topaz, and spodumene (kunzite) among the commoner gemstones.

Thus it must be realized that if, say, a zircon is cut with its table facet at right angles to its optic axis it may be necessary to tilt the stone and make observations through the bezel facets to see the double refraction well developed. Also, as R. T. Liddicoat has pointed out, in directions at right angles to the optic axis of a uniaxial mineral, although the double refraction is at its greatest it cannot be directly observed, as the image due to the extraordinary ray is either directly in front or directly behind that due to the ordinary ray.

With zircon, sphene, and peridot the birefringence is so strong that even the novice should have no difficulty in seeing the effect mentioned even in small stones, but it needs a practised eye to detect the doubling effect in quartz and corundum unless the stones are large. (J.-P. Poirot, Director of the Paris gem-testing laboratory, has made the important observation that for a given birefringence the doubling effect is more marked in stones of low RI than in those of high index.) As with any other test, practice and perseverance are needed, and when beginners have become accustomed to getting results in easy specimens they should strive to develop their skills to the limit and learn to gauge approximately the strength of the double refraction shown. The author has found this to be a most valuable accomplishment. The more things that can be learned with a pocket lens, the better – even if one possesses other apparatus this will not usually be available in an auction room or on other people's premises. This same test can, of course, be extended and rendered easier by using a microscope, which functions as a much more powerful lens, magnifications of 25 to 60 diameters being quite comfortably usable with the stone held between the thumb and forefinger of the left hand, supported on the microscope stage.

When a doubly refractive stone is too small or has too low a birefringence to show 'doubling' at all easily by the straightforward lens inspection described above, another simple but more sensitive technique may yield valuable information. This has been practised independently in the past by Max Bauer, G. O. Wild, G. R. Crowningshield, and R. K. Mitchell, and in recent years has been systematized into a routine process of **visual optics** by Alan Hodgkinson, who uses it with exceptional skill and success.

The method is best applied in a darkened room with a single light source, which may be either a clear electric bulb with 'C' filament or (more conveniently) a small 'slit' torch of the type produced by Hanneman Lapidary Specialties (Figure 3.2). The stone to be examined, if large or mounted with an open backing, can be held in the hand, or if small, in tongs. The light source is viewed through the table facet of the specimen, held as close to the eyeball as possible, the viewing eye being partially closed and the other eye shut. If the stone has

Figure 3.2 Hanneman–Hodgkinson slit assembly on a 'Penlite'

been correctly positioned a number of images of the light source can be seen, showing spectrum colours, and where the stone is birefringent the images are also doubled. The higher the refractive index of the stone, the further will be the images from the centre of vision, until with stones of really high index (such as zircon, diamond, and its modern substitutes) the images are refracted so far from the centre that they can only be seen if the stone is slightly tilted. It is as well to practise with a large quartz, in which the images are numerous and easily seen.

By this simple technique the practised observer can gain information not only on the strength of the double refraction (by the distance apart of the twin images of the light source) but also on the strength of the dispersion (by the width of the coloured band) and on the refractive index of the stone.

Such techniques as the above can, in practised hands, be of great value in giving strong indications of the nature of any faceted gemstone in circumstances where none of the standard instruments are available.

A more sensitive test for double refraction than those mentioned hitherto is possible by using polarized light. A ray of ordinary light, vibrating in all possible directions at right angles to that of the ray, becomes transformed when passed through a Nicol prism or a sheet of 'Polaroid' and emerges in a polarized condition. Polarized light is indistinguishable from ordinary light to the unaided eye, but the fact that it is vibrating in one plane only gives it peculiar properties which are of great value in the study of minerals. A detailed description of the uses of polarized light would be beyond the scope of this book, but since the arrival of the artificial compound 'Polaroid' has made polarized light available to all at a low cost, a brief account of its use in detecting double refraction will be attempted here. Those who would like further information on the subject are advised to consult any good textbook on mineralogy.

A ray of light entering a doubly refracting mineral is in general split at once into two polarized rays vibrating in planes at right angles to each other. These are each slowed down to a different extent in passing through the orderly atomic field of the crystal, and are thus refracted differently. What is important to realize in the present connection is that along any one direction in a doubly refracting stone only two sorts of ray can travel, and these are vibrating at right angles to each other and in directions which are rigidly fixed by the orientation of the ray to the structure of the stone.

Now a beam of light which has passed through a Polaroid disc (or Nicol prism) and is vibrating, shall we say, north and south, will be quite unable to pass through another disc or prism which is set with the vibration direction at right angles to this, i.e. east and west. Such a system through which virtually no light can pass is referred to as 'crossed Nicols' or 'crossed Polaroids', while the term 'crossed polars' can be used for either. If one of the Polaroid discs or Nicols is turned ever so slightly from the 90° position some light will be able to pass through, and more and more light will come through on further turning until the maximum is reached with the two discs in a 'parallel' position (see Figure 3.3).

Now if between two crossed Polaroids a piece of glass is inserted the darkness of the field will remain unaltered (unless the glass has been badly annealed) since glass, being amorphous, imposes no change in the vibration direction of light from the first Polaroid disc. But this is not the case when a doubly refractive gem is placed between the discs.

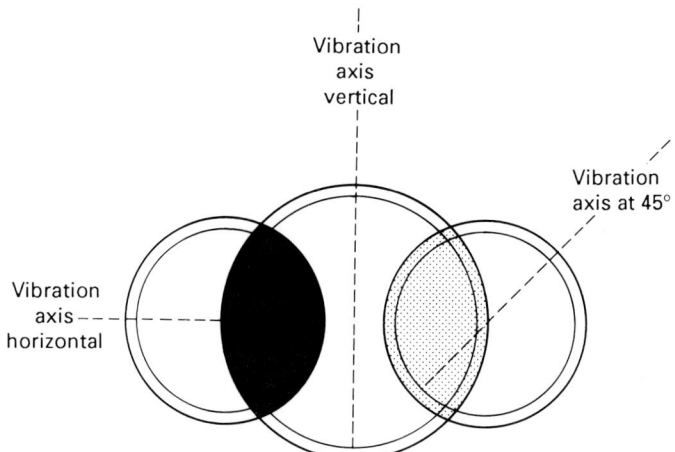

Figure 3.3 Three Polaroid screens showing the effect upon the transmitted light by crossing the vibration axes (after H. S. B. Meakin Ltd)

Unless the vibrations of the rays passing through the crystal are parallel to those of the polarizers, a certain amount of light can now pass through the second Polaroid disc.

Thus the effect of rotating a doubly refractive gem between crossed Polaroids is in general to give four positions of darkness 'extinction' (at intervals of 90°) in a complete revolution and four positions of maximum brightness.

This provides the most usual means by which the mineralogist distinguishes between singly refractive (amorphous or cubic) substances, on the one hand, and doubly refractive stones, on the other. Glass or cubic minerals under strain show 'anomalous' double refraction between crossed polars, but the effect is not likely to be confused with true double refraction since the extinction is not clear-cut all over the field, but mottled or grid-like. Almandine garnet and diamond often show anomalous double refraction. Natural spinel is usually quite isotropic, but synthetic spinel always displays anomalous effects resulting in a 'tabby' extinction which is a helpful feature in identification. Due to the stresses involved in moulding 'pastes' to the form of faceted gemstones, these often show a crude extinction cross in convergent light between crossed polars which is very distinctive.

Undoubtedly the most comfortable and efficient way of studying these effects is by means of a polarizing microscope with rotating stage. Such instruments are expensive, and hardly worth while for the jeweller, but with the aid of pieces of Polaroid an ordinary microscope can be made to function quite well, one Polaroid disc being mounted below the stage and the other used as an eyepiece cap which can be turned into the extinction position before making the test. Even without a microscope, the test can be made either by mounting Polaroids after the fashion of the old 'tourmaline tongs' or with a short length of tubing with Polaroid at either end and an aperture for inserting the stone between the Polaroids.

The Rutland polariscope, made by Messrs Rayner, is one of the least expensive and most convenient of the many models of hand polariscope using

Polaroid commercially available. A convenient polariscope with built-in lighting is also made by Rayner and can be strongly recommended. This has sufficient space between the crossed polars to allow for the examination of stones mounted in jewellery, and 'interference figures' of characteristic pattern may be observed by the use of a pocket lens or strain-free glass bulb made for the purpose (see Figure 21.1). Polariscopes, with built-in lighting, are now available with a special 'conoscope' lens which serves the same purpose as a strain-free glass sphere (Figure 3.4). The stone is placed between the lower polar and the 'conoscope' and, by suitable adjustment, interference figures are easily obtained.

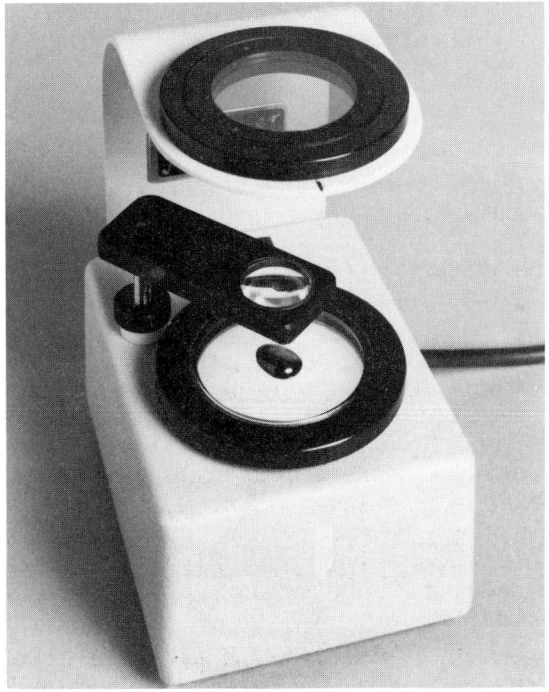

Figure 3.4 Rayner polariscope with built-in lighting and 'conoscope' lens (courtesy: Gemmological Association)

The presence or absence of double refraction often provides us with an easy and certain means of discriminating between the real and the false. Glass or 'paste' may have local strains within its structure which will cause it to show 'anomalous' double refraction to some extent, as noted above, but in no glass or singly refractive gem will such strain effects be sufficient to cause any of the 'doubling' phenomena described above. To give a practical example: the reader armed with this knowledge will be able to distinguish quickly between a true crystal (quartz) ball, of the kind used for crystal gazing or for ornament, and one of the spheres of glass which are commonly sold as substitutes. If the corner of a card be placed immediately behind a quartz crystal ball and viewed through it, a ghostly double image of the edge of the card can be seen in most positions. On rotating the ball it will be noticed that the doubling effect is more pronounced first on one edge and then on the other, while in one direction through the ball (along the optic axis of the original quartz crystal) and round the great circle or 'equator' at right angles to this no doubling will occur. In the case of a glass ball the edges of the card will appear single in all directions. A little practice will do more than any amount of reading to engender skill and confidence in using the test. Once these 'doubling' effects have been clearly seen and understood, they will be found quite easy to observe, and of great practical value in the discrimination of precious stones.

It is worth noting that, when scrutinizing with a lens a brown or deep green tourmaline or a brown sphene, the expected doubling of the back facet edges may not be apparent. The reason for this is that in such stones, one of the two rays may be totally absorbed. This is an extreme example of the effect known as *dichroism*, which will be considered in the next chapter.

Dispersion

Whereas double refraction is due to the variation in the refractive index of a crystal according to the direction of vibration of the light passing through, dispersion is the name given to effects due to variations in index according to the colour (wavelength) of the light employed. Since the figures representing the magnitude of these two properties happen to be rather similar, beginners are sometimes apt to confuse them. While double refraction, unless exceptionally large, has little effect on the appearance of the stone, it has a very great significance as a diagnostic feature. Dispersion, on the other hand, being the basis of 'fire' in a faceted gemstone, has a considerable effect on the appearance of a stone, but is of small importance as a testing factor.

The extent of double refraction in a mineral bears little relation to its refractive index, depending very largely on its atomic structure: carbonates and nitrates, for example, may have rather low mean indices of refraction but display enormous birefringence. Dispersion, however, with a few exceptions such as diamond, is found to increase fairly steadily with the refractive index of the stones concerned.

As far as visible light is concerned, the refractive index of all minerals becomes progressively higher in passing from the red to the violet end of the spectrum. If a graph is drawn, plotting the refractive index as ordinate and the wavelength as abscissa, the relationship is seen to be represented not by a straight line but by a curve which becomes increasingly steep as it approaches the ultra-violet region. Dispersion curves, in fact, may be expected to become infinitely steep as they approach the wavelength at which a fundamental absorption edge for material in question begins. Thus most highly dispersive solids and liquids will be found to absorb light completely beyond the end of the violet, or in the near ultra-violet, while materials with a low dispersion such as fluorite, quartz, or pure corundum transmit light down to perhaps 200 nm, which makes them useful for spectrographs or other optical instruments.

Table spectrometer and measurement of dispersion

All figures for refractive index given in texts or tables are those for yellow (sodium) light of wavelength 589.3 nm, which is universally accepted as a standard; and refractometer scales are calibrated to give correct readings for this wavelength.

Effects due to dispersion have been mentioned several times in the previous pages, but nothing has so far been said about its measurement. This can be carried out on apparatus known alternatively as a table spectrometer or reflecting goniometer, using the 'minimum deviation' method.

Essentially, the table spectrometer consists of a horizontal circular table graduated in degrees around the perimeter of which a telescope, focused on infinity and with an eyepiece provided with cross hairs, can swing from a pivot which forms the axis of the instrument (see Figure 3.5). A vernier fitted to the telescope should enable its angular position to be assessed to within one minute of arc.

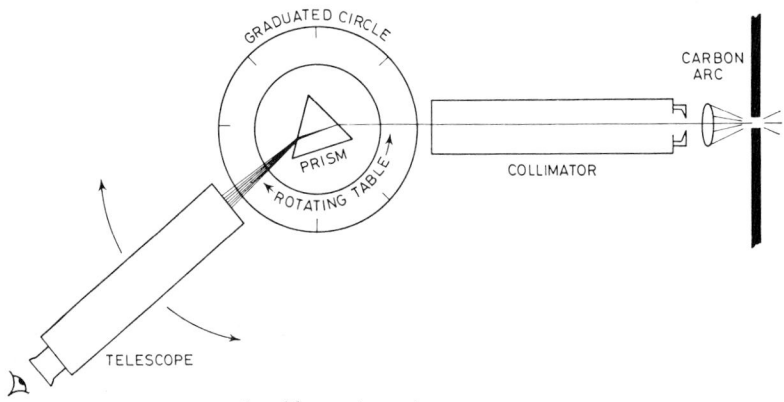

Figure 3.5 Arrangement of a table spectrometer

Opposing the telescope, and fixed to the instrument, is a similar-looking tube, known as the collimator, at the outer end of which is a narrow adjustable slit which is at the principal focus of a lens at the inner end, which ensures that the narrow ray of light passed by the slit emerges as a parallel beam. The usual light source for dispersion measurements is a carbon arc or a mercury vapour lamp. Mercury vapour emits a series of bright lines which are well distributed throughout the spectrum, while light from a carbon arc always contains the yellow sodium doublet which is the standard light for refractive index measurement, and can be 'fed' with lithium or other salts which will provide other bright lines of any required wavelength.

In the centre of the graduated table is a raised circular platform which can be rotated at will, and on which the prism (or stone acting as a prism) can be mounted. If the refractive index of a gemstone is to be measured at all accurately (say, to three places of decimals) on this instrument, the specimen must be fully transparent and have flat, well-polished facets, so that two of these (usually the table and one of the pavilion facets), which are inclined to one another at a suitable angle (about 40°) can be used as a prism for the experiment. It is not, of course, necessary that these facets should actually meet in a prism edge, but step-cut stones, in which a prism edge can in fact be seen, are undoubtedly easier to mount than brilliant or mixed-cut stones where there is no such easy guide to the eye. The most tricky part of the whole operation is mounting the stone on wax in the centre of the raised platform in such a way that the facets forming the refracting prism are truly vertical to the axis of the instrument. Finishing touches can be carried out with adjusting screws fitted to the platform, but the main adjustments are usually made by hand. When the stone is correctly adjusted, it is a fairly simple matter to measure the angle between the chosen facets by picking up with the measuring telescope the narrow beam of light transmitted by the collimator as it is reflected from one of the chosen facets and then from the other. Half the angle between these two readings will give the angle of the prism.

The central platform on which the stone is mounted is then turned into a position where light from the collimator is refracted by the prism formed by the two chosen faces and can be seen as a spectrum through the measuring telescope. The platform with the prism is then slowly turned and the resulting movement of the spectrum followed through the telescope until the critical position of 'minimum deviation' is found and measured: first for the sodium line, which is always to be seen in the carbon arc spectrum, and then for any lines of known wavelength introduced into it.

Having obtained these figures the following formula can then be applied:

$$RI = \frac{\sin \frac{1}{2}A + D}{\sin \frac{1}{2}A}$$

where A is the angle of the prism and D the angle of minimum deviation for the particular wavelength concerned.

The spectrum seen through a gemstone used as a prism is seldom very perfect, as it is only in diamond that the facets are likely to be optically flat. It is thus important to introduce substances into the arc which give only a few powerful lines, and these in significant regions of the spectrum. Two useful substances to have on hand for the purpose are small pieces of lepidolite mica and ordinary blackboard chalk. These can be introduced into the arc while it is burning by an assistant using crucible tongs and wearing dark goggles to obviate the risk of conjunctivitis. Lepidolite fuses readily in the arc and yields a strong and persistent lithium and sodium spectrum. The line to be measured is the prominent lithium red line at 670.8 nm, while the chalk will provide the useful calcium line in the violet at 422.7 nm. The difference in the refractive indices of a stone for these two wavelengths gives a figure for its dispersion which differs very little from that given by the range from 686.7 nm to 430.8 nm, which are the B and G Fraunhofer lines of the solar spectrum usually taken as standard. When sunlight is used the B and G lines are quite easily seen; but with a carbon arc source their use is impracticable. Incidentally, it is essential that a screen should cut off any direct light from the arc lamp: it would otherwise be quite impossible for the operator to pick up the required faint signal through a telescope. Only the slit at the end of the collimator tube should be exposed to the light, through a made-to-measure hole in the screen, with which its surface should be flush. Light from the arc can be concentrated on to the slit through a lens held in an adjustable clamp, so arranged that the image of the arc flame is projected onto the slit. As in all such experiments, good results can only be expected if the optical train is strictly in line and on the level.

It can be seen from the above short description that the minimum deviation method is extremely time consuming and very limited in its practical application for the average gemmologist. With specially prepared prisms it can be accurate to four places of decimals, and there is no upper limit to the refractive indices that can be measured, provided that the prism angle is not more than twice the critical angle for the stone. For randomly cut, uniaxial minerals a true value can only be expected for the ordinary ray, while with biaxial minerals the values obtained can be anywhere between those for the α and the γ indices.

It should perhaps be added that the method is very well suited for accurate measurements on the refractive indices and dispersions of immersion fluids. All that is needed is an optically worked, hollow glass prism into which the liquid to be measured can be introduced.

It is exact measurements by such methods which enable figures for dispersion to be given in the textbooks. In practical gemmology dispersion measurements are seldom used as a means of distinguishing one gemstone from another, as there are many simpler methods which are far more discriminative.

It is important to realize that one cannot directly obtain a true refractive index reading for a stone on the refractometer, except for sodium light, for which the instrument was calibrated. The position of the shadow edge depends upon the critical angle between the material forming the refractometer prism and the stone under test; and unless the dispersion of these two media is the same, the angle will vary for each wavelength of light. In the standard instruments, the dense glass prism has a dispersion much higher than that of any stone tested, with the curious result that a reading for *red* light will appear to be higher than that for violet light – the reverse of what is really true. Also it will be noticed that stones having a low dispersion will show a wide band of spectrum colours instead of a sharp shadow edge when white light is used, while readings for lead-glass 'pastes' will be relatively sharp because their dispersion more nearly approaches that of the refractometer prism, and the edge due to the contact liquid will be almost free from colour. With a spinel refractometer the reverse effect is true. Stones within the instrument's range usually have a dispersion only a little below that of the synthetic spinel used for the prism (0.021) and thus give edges in white light which are almost sharp and free from colour, while pastes give an edge with a pronounced colour fringe, and contact liquids such as bromonaphthalene will show a very wide spectrum of colours.

In everyday practice, these marked differences in the appearance of the shadow-edges of pastes in the range 1.60–1.70 and those for stones of similar index are the only instances where differences in dispersion can be considered to provide a useful accessory test, since the stones within the refractometer range have such very similar degrees of dispersion. An acute observer, however, can notice that tourmaline gives distinctly sharper images on a spinel refractometer than does topaz, due to the abnormally low dispersion of the latter.

When the dispersion of an uncommon gemstone is not available in the literature (e.g. brazilianite), it is found that a close approximation to the true value can be obtained by taking refractometer readings with light passed through good colour filters (e.g. red and blue). Differences between refractometer readings employing light passed through first one filter and then the other will not give direct dispersion readings, as explained above, but can be carefully compared with readings for stones of known dispersion and similar refractive index, and an estimate for the unknown dispersion thus obtained.

While it is true that dispersion seldom provides a test that can be quoted in actual figures, the effect on the appearance of cut stones due to this property can

Table 3.2   Dispersion of gemstones for the B–G range

Synthetic rutile	0.30	Pyrope	0.022
Strontium titanate	0.19	Spinel	0.020
Blende	0.156	Peridot	0.020
Lithium niobate	0.130	Corundum	0.018
Cassiterite	0.071	Spodumene	0.017
Cubic zirconia	0.066	Tourmaline	0.017
Demantoid	0.057	Scapolite	0.017
Sphene	0.051	Danburite	0.016
GGG	0.045	Chrysoberyl	0.015
Diamond	0.044	Phenakite	0.015
Benitoite	0.044	Topaz	0.014
Zircon	0.039	Beryl	0.013
Paste (1.635)	0.031	Quartz	0.013
'YAG'	0.028	Feldspar	0.012
Spessartine	0.027	Beryllonite	0.010
Scheelite	0.026	Silica glass	0.010
Almandine	0.024	Fluorite	0.007

be highly important as an aid to identification. This is especially true for colourless stones where, for example, the high dispersions of strontium titanate and synthetic rutile enable them to be distinguished from diamond. In rutile the colour effects are so marked as to make the stone look almost like an opal. With strontium titanate skilled judgement is needed if one is to be quite sure that the stone is not a diamond. Sample specimens of strontium titanates are valuable in such cases to enable direct comparisons to be made.

Even in coloured stones, the liveliness conferred on a stone by high dispersion adds greatly to its attraction – as, for example, in demantoid, sphene, and the rare and beautiful benitoite, and it enables the trained eye to distinguish them from other stones of similar colour but less notable dispersion. With doubly refractive stones the different rays may show different degrees of dispersion. In Table 3.2 the higher figure for the species is quoted in such cases.

4

Colour, colour filters and the dichroscope

Colour plays an extremely important part in contributing to the beauty and popularity of precious stones. It was their colour that made such stones as turquoise and lapis lazuli among the first to attract the cupidity of early civilized man, and it is the magnificent crimson red of ruby, the deep cornflower blue of the finest sapphires, and the verdant green of emerald which (added to their transparency, hardness, and rarity) have caused them always to rank among the supremely precious gems. Only in the case of diamond, unique in this as in so many other ways, has a complete *lack* of colour been regarded as the standard of perfection.

Those unversed in gemmology find it hard to realize that in most of the mineral species used as gems a wide colour range is possible, and that in their 'pure' state they would be colourless and thereby of relatively little value. While ruby and sapphire are the red and blue varieties of the mineral corundum, there are also white, yellow, pink, and green sapphires. Emerald and aquamarine are the rich green and pale blue varieties of beryl, but there are also white, red, pink, and yellow beryls, and green beryls that do not rank as emeralds. Tourmaline, quartz, zircon, and topaz are further well-known examples of such polychrome minerals, and tourmaline in particular is well known for frequently appearing as parti-coloured crystals, the boundaries between the colours (usually green and pink) being quite sharp either across the length of the crystal or concentrically in zones.

Granted that there are a considerable number of red, blue, and green gemstones, there are very few that can approach ruby, sapphire, and emerald in their rich shades of these colours, and a colour-sensitive and trained eye can go a long way towards distinguishing them from other natural gemstones, glass imitations, or even synthetic counterparts. It is indeed worthwhile for a beginning gemmologist to practise at every opportunity the sight identification of the various gems and to pay special attention to the particular grades of colour typical of each species.

When even the most practised eye may find it difficult to distinguish between stones of different species but of very similar colour, simple instruments can sometimes be used to 'analyse' the colours in one way or another and make the separation simple.

What we call 'white' light, i.e. light from the sun or other incandescent bodies, is composed of a mixture of all the colours of the rainbow. Newton, in the year of the Great Fire of London (1666), was the first to show that sunlight has a composite nature. He analysed a narrow beam of light that passed through a chink in the shutter into his darkened room by the simple expedient of placing a glass prism in the path of the ray.

Light is variously refracted by a transparent solid according to its wavelength, the red rays, of longer wavelength, being less deviated than the shorter, violet, waves. Thus these rays, having been deviated by different amounts on passing through the prism, were separated, and could be seen to fall upon a screen as overlapping patches of pure colour – in fact what we call a spectrum band of colours – red, orange, yellow, green, blue and violet merging imperceptibly into one another.

Now some stones appear coloured to the eye simply because they absorb some of the white light when it passes through them, and some wavelengths (colours) are more strongly absorbed than others. This is known as 'preferential absorption'. Those colours which are least absorbed pass onto the eye and mingle to form what we call the 'colour' of the stone. Actually, the colour was in the original white light and has not been created by the stone, which has merely robbed the white light of the colours complementary to those we see. It is a fact that two stones may absorb quite different sets of wavelengths from white light and yet present precisely the same colour effect to the naked eye so long as what we may call the 'mean effective wavelength' of unabsorbed colours is the same in each case.

The spectroscope (which analyses light in a manner similar in principle to Newton's original experiment) will often reveal this difference in the 'make-up' of two similar colours, and thus provide us with a beautiful and rapid means of identifying coloured stones. Details of the use of this valuable instrument are given in Chapter 10.

Chelsea colour filter

A similar and often very effective means of revealing underlying differences in colour is by use of suitable 'colour filters'. These are especially successful in discriminating between emerald and its imitations.

Emerald is almost alone among green stones and glasses in transmitting an appreciable amount of deep red light and absorbing to some extent the yellow-green portions of the spectrum. Many emeralds also emit a deep red fluorescent glow when strongly illuminated. When viewed through a coloured filter which transmits only deep red and yellow-green light, most emeralds will appear distinctly red, whereas most of the imitation emeralds or real stones resembling emerald retain their green appearance through the filter. Many such filters have been marketed in Britain and on the Continent, but by far the most successful is the 'Chelsea' filter, developed in the Gem Testing Laboratory of the London Chamber of Commerce (now the Gem Testing Laboratory of Great Britain) and at the Chelsea College of Science and Technology. The Chelsea filter has other uses besides the separation of emerald from its counterfeits, but users are apt to draw false conclusions unless they bear in mind carefully the various possibilities, as sometimes 'red through the filter' is a danger sign and in other cases it is a sign that all is well. A careful perusal of what follows should prove helpful.

Green stones

Most emeralds, as already stated, appear red through the filter, varying from a fine ruby-red with stones of good colour to pale pink with paler emeralds. Reddest of all are certain synthetic emeralds. Some emeralds, notably those from South Africa and India, show practically no colour change. On the other hand, green glass imitations, most green doublets, green tourmaline, and green jadeite retain a greenish colour through the filter. Demantoid garnet and green zircon give a pinkish effect, and so do some specimens of stained green chalcedony, but these are sufficiently unlike emerald in ordinary light to cause little confusion. Unfortunately, some types of 'soudé emerald' triplets show a distinct red, and a reddish tinge may be seen in emerald-green fluorite.

Blue stones

Materials coloured with cobalt blue transmit a considerable amount of deep red light and appear red through the filter. Thus blue synthetic spinels and blue cobalt glasses show a telltale red colour which assists in distinguishing these counterfeits quickly from sapphire, aquamarine, or blue zircon, which appear a dirty green. It must be noted, however, that many Sri Lankan sapphires contain a trace of chromic oxide; these appear purple in artificial light and show distinctly red through the Chelsea filter. Natural blue spinels also show a reddish tint but in neither of these natural minerals does the colour approach the full-blooded red seen with deep blue synthetic spinel. Aquamarine is notable for showing a markedly green appearance under the Chelsea filter.

Red stones

The filter is not of much help here, but the bright fluorescent red effect seen when Burma and synthetic rubies are viewed through the filter is worth noting. No other red stone shows quite this appearance.

To obtain the best results with the Chelsea filter, the stones should be held close to a strong artificial light and viewed through the filter held close to the eye so as to cut out any extraneous glare.

Dichroism and the dichroscope

The colour coming from a gemstone may be composite in another sense from that already described. In almost all doubly refractive coloured stones there are *two differently tinted rays* reaching the eye together, inextricably mingled. The fact that a stone is doubly refractive means in general that rays travelling through the stone in a given direction are of two kinds, one vibrating in one plane and the other in a plane at right angles to the first. Not only are these rays as a consequence of their different vibrations travelling with different velocities (hence the effect of double refraction) but they usually suffer a *different colour absorption*. This effect is known as 'dichroism' ('two-colour effect') or, more generally, 'pleochroism' ('many-colour effect'), since some stones show three colours, though there can be only two in any one direction.

The effect cannot be seen with the unaided eye except by turning the stone in different directions and noticing the change in tint that strongly dichroic stones may show according to the direction in which light traverses the crystal.

To see *both colours at once* as they come from a stone, we use a simple instrument called a dichroscope. This, in its essentials, is merely a tube with a window at one end and a lens at the other; in between there is mounted a piece of calcite so chosen that it causes (by its strong double refraction) two images of the window to appear side by side as the observer looks through the eyepiece. Now light from these adjacent images of the window is vibrating in two planes at right angles to each other, one plane for each image. By holding a coloured stone in front of the window of the dichroscope, so that light passes through the stone into the instrument, we are able to see side by side the colours appropriate to each of the two polarized rays from the stone (see Figure 4.1).

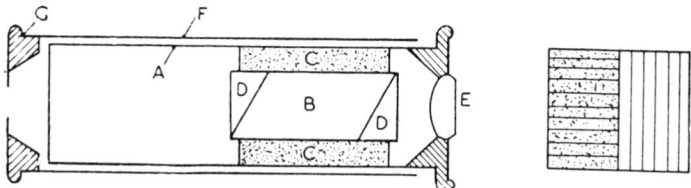

Figure 4.1 The construction of a dichroscope. A is a metal tube fitted with a rhomb of calcite B in a cork setting C and an eyepiece E. Prisms of glass D are cemented to the calcite to enable the rays to be directly transmitted. The tube A slides into another tube F fitted with a rectangular 'window' G. On the right is a representation of the two images of the dichroscope window as seen through the eyepiece

Usually it is merely a matter of a different depth of tint of the same colour – sapphire, for instance, commonly shows dark blue and light blue (Oxford and Cambridge) respectively in the two windows, but in some stones amazing differences in colour are seen; e.g. in Siberian alexandrite the colours purple, green, and orange may be seen (two at a time) if the stone is turned in front of the dichroscope.

Turning the stone is indeed an important factor in testing for dichroism. In all doubly refracting stones there are either one or two directions of single refraction, known as optic axes, and in these directions there can be no dichroism. This is well known, but it is not so commonly realized that in other directions also no dichroism will be visible if the vibration directions of the two images of the aperture happen to be at 45° to those of the two rays from the stone.

An important point to note is that the mere fact that dichroism is seen assures the observer that the stone is doubly refracting and therefore not a paste or a cubic mineral. Thus we have another test for double refraction (see Chapter 3). Ruby can be distinguished from red spinel and garnet, blue spinel from blue tourmaline of similar shade, sapphire from blue synthetic spinel and so on. Descriptions of the dichroic colours for the various stones are given in books, but even stones of the same kind vary a good deal and the effects are best learned by experience. When the effect is faint it is difficult to be sure that the imagination is not providing the supposed difference in tint.

For the observation of dichroism, daylight is the best illumination. Hold the stone up to the light in corn tongs, using the left hand, then observe it through the dichroscope held in the right hand, using it as you would a short-focus lens. The 'window' of the instrument is small, and beginners find it difficult to place the stone exactly in front and at the same time to get sufficient light through the stone to enable them to judge the colour effects. Most instruments are fitted with a weak magnifying lens as eyepiece, and two slightly enlarged images of the stone, or part of the stone, should be seen, one in each image of the 'window'. As a rule, only one or two facets will be transmitting enough light to see the colour, but with even only a small patch of colour in each image, one can readily compare the two tints seen and observe any differences of colour.

As already stated, the stone should be viewed from a number of different positions to ensure that the maximum effects are seen. In a well-cut ruby, in which the optic axis should be perpendicular to the table facet in order to get the best colour effect, little dichroism should be seen when viewed through the table. In a synthetic ruby, the peculiar colour of which is often partly due to the fact that it is wrongly cut from the point of view of colour, the dichroism as seen through the table is usually strong.

The more expensive types of dichroscope are fitted with a holder on which the stone can be fixed and easily rotated in front of the window of the instrument. This makes observations more easy for the beginner to carry out, but is not, of course, essential for good results.

If a transparent piece of Iceland spar 2.5 cm (1 in) or more in length is available, a 'dichroscope' of a kind can be constructed merely by gumming a piece of black paper onto one end of the spar and scraping a small hole in the paper. The only skill required is in making the hole of such a size that the two images of it produced by the double refraction of the spar are contiguous.

Effective dichroscopes can also be made with Polaroid in various ways. One suggestion is to cut two circular discs of the material (using a safety razor blade) into quadrants as shown in Figure 4.2, later reassembling them in such a manner that light passing through the north and south quadrants is vibrating at right-angles to that passing through the east and west sectors (see Figure 4.2).

The assembly can then be mounted on glass and fitted to the end of a short tube made of cardboard or metal, at the other end of which is fixed a lens of appropriate focal length, to act as an eyepiece. Light from a dichroic gemstone viewed through the eyepiece of this simple gadget will in general show a contrast in colour between the north and south sectors, on the one hand, and the east and west, on the other, due to their different vibration directions.

Beautiful dichroic effects can be seen in such rather rare stones as violet-blue iolite, green andalusite, blue apatite, sphene, and benitoite. Better-known gems in which dichroism is strong or distinct include ruby, sapphire, emerald, alexandrite, most tourmalines, kunzite, aquamarine, and blue zircon. The most spectacular colour changes of all can be seen in the violet-blue form of zoisite (marketed by some as 'tanzanite') which is a relatively recent addition (1967) to the ranks of coloured stones. In a good crystal of this variety, deep blue, purple-red, and pale green rays can be seen in pairs through a dichroscope. The original effect is often diminished by heat treatment after the stone is cut as a

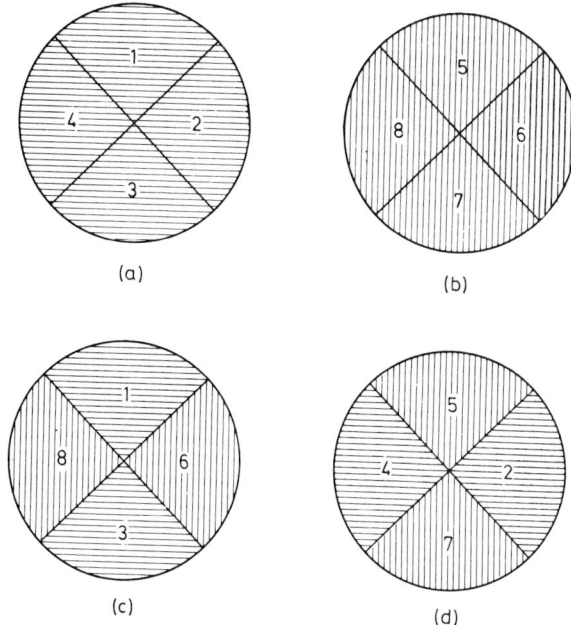

(a)   (b)

(c)   (d)

Figure 4.2 Assembly of a Polaroid film to function as a dichroscope. (a) and (b) Polaroid discs sectioned with a razor blade and reassembled as in (c) and (d). Such an assembly is mounted at one end of a tube with a suitable lens for viewing mounted at the other end

gemstone. In some stones the change of colour due to dichroism when they are turned and viewed from different directions is quite apparent to the unaided eye, and skill in discerning dichroism by this kind of direct observation is well worth cultivating.

The keen gemmologist may find that the dichroscope provides more interest and yields more information if the vibration directions of the light coming from each image of the 'window' is known. This can easily be ascertained by experiment, and marked once and for all on the rim of the eyepiece. One way of determining the vibration directions is to look through the dichroscope at light reflected from a polished desk or table. Such light is largely polarized, and its vibration direction is parallel to the reflecting plane. The dichroscope image which is the brighter of the two will be that transmitting rays which are vibrating parallel to the table, while the vibration direction of rays from the darker image of the window will be at right angles to this.

5

The specific gravity of gemstones and how to measure it

Every jeweller knows by experience that some gemstones 'weigh heavier' than others: that, for instance, a white zircon weighs more than a diamond of equal size and a sapphire more than an emerald. Scientists long ago provided a precise definition for this quality of 'heaviness', which provides a valuable means of distinguishing one substance from another when worked out on a numerical basis. This was done by taking water as a standard and comparing the weight of each substance with the weight of an equal volume of pure water. The figure thus arrived at is called the 'specific gravity' or 'relative density' of the substance in question. The specific gravity of a body, then, is simply the ratio of its weight to that of an equal volume of pure water. For accurate work, water at 4°C is specified as the standard.

Thus, when we say the specific gravity of aluminium is 2.7 and that of ruby 4.0, we mean that, volume for volume, aluminium weights 2.7 times, and ruby 4 times, as much as water. Materials used in jewellery show a very wide range in specific gravity, from amber, 1.08, to platinum, 21.5.

Table 5.2 at the end of the chapter gives the specific gravities of the substances of most importance to the jeweller, and it will be recognized that if this property can be easily determined (and it can), it will provide a very certain means of distinguishing one species of stone from another.

Provided the stone to be tested is free from any kind of setting (and admittedly this proviso implies a serious limitation for the jeweller), the method is universal in its application – be the stone rough, faceted, carved, cabochon cut, large, or small, its specific gravity may be determined by one method or another, and without the use of expensive apparatus.

In saying 'without the use of expensive apparatus' one is assuming that every jeweller has a good diamond balance on the premises as part of his or her essential equipment. For accurate determinations by the first method to be described below, the balance should be in perfect order and the weights really accurate.

The type of balance illustrated in Figure 5.1 (in which weights of 1 carat and over are added to the right-hand pan to counterpoise the goods in the left-hand

Figure 5.1 A balance arranged for hydrostatic SG determination

pan, and fractional weights are avoided by the use of a sliding 'rider' along the calibrated top arm of the balance to enable weights accurate to 1/100 carat or better to be obtained) will strike modern jewellers and diamond merchants as old-fashioned, since during the last twenty years or so there has been an almost complete conversion to the more convenient 'electric' or 'aperiodic' type. In this form of balance, at least for weighings up to 100 carats or so, weights in carats or multiples of a carat are added automatically by turning a calibrated knob, and any fractions above the weights added are shown on an illuminated scale without delay. There is none of the long-drawnout business of observing the slow swinging of a pointer about the zero position. If the reader is fortunate to own or have access to the use of one of these balances, so much the better. But the older types are still in use in many schools and colleges, and are quite adequate in 'home' laboratories where costs must be kept low. For those who only need a balance for specific-gravity work the 'Hanneman' balance described later can be recommended.

Hydrostatic method

Any good diamond balance may be readily used for measuring the specific gravity of a stone by employing a principle which was stated by the Greek mathematician, Archimedes, in about 250 BC, and is thus known as 'Archimedes' Principle'. This states that bodies when immersed in a liquid are buoyed up by it with a force equal to the weight of the displaced liquid. This means that when (say) a gemstone is suspended in water it will apparently suffer a loss in weight equal to the weight of the water it displaces. Obviously, the gem displaces its own volume of water, and thus, simply by weighing a stone first in air and then immersed in water we have all the data necessary to calculate its specific gravity (abbreviated to SG). For, by definition,

$$SG = \frac{\text{Weight of stone}}{\text{Weight of equal volume of water}}$$

and this, as we have seen, may also be expressed:

$$SG = \frac{\text{Weight of stone}}{\text{Loss of weight in water}}$$

The weight of the stone is, of course, determined in the usual manner; to determine the weight of the stone in water some simple contrivances are necessary.

A small bridge or stool must be arranged to straddle over the left-hand balance pan in such a way that the free swing of the balance is unimpeded (Figure 5.1). Such a bridge can be bought cheaply from chemical suppliers, or it can be easily made by an amateur carpenter. An effective improvization is to place a matchbox upright on either side of the pan with a thin strip of wood, such as a 15 cm (6 in) protractor rule, acting as a bridge between the boxes. Upon this bridge is placed a beaker (or a small tumbler), some three-quarters full of distilled or boiled water. Using copper wire of fairly heavy gauge (about 1 to 1.5 mm), make a spiral coil in which a stone of any size from about 2 to 30 carats can comfortably rest, and suspend this from the lower hook of the balance arm by a much finer wire of copper, brass, or (best of all on account of its tensile strength) tungsten, of such a length that the coil which is to hold the stone is well immersed in the water, yet not touching the bottom of the beaker. If it is too much trouble to make a separate heavy cage supported by a finer wire, a simpler procedure is to use one piece of brass wire just thick enough to be sufficiently stiff (about 0.5 mm), coiling one end into a spiral and bending the other end into a loop to fit over the balance hook (see Figure 5.2). Since a coarse wire impedes the free swing of the balance to a serious degree for accurate work, the finer the suspending wire, the better, as long as it is strong enough to stand the strain of a few ounces. Tungsten wire of only 0.05 mm will do this, or thin nylon thread. All is now ready for the experiment.

The stone must be weighed in the ordinary way very accurately, estimating, if possible, to the third decimal place (in carats). Then, having the bridge and vessel of water in position, the stone is placed in the wire spiral and suspended in the water: it is then weighed while thus immersed. Then, with the stone removed, the spiral alone is weighed while immersed as before, and this weight subtracted from the combined weight of stone and spiral in water. We now have

Figure 5.2 Spiral coils of wire in which stones can be rested for the hydrostatic weighing test

the data we need (weight of stone in air and in water) to work out the SG by the formula already given.

It may be preferred to have the wire spiral immersed in the water throughout all the weighings, counterpoised exactly with another piece of wire (or any suitable weight) in the other pan. Then no allowance for the wire need be made, but the free swing of the balance will be somewhat impeded for the weighing in air.

It is important to *see that no air bubbles are clinging to the wire or the stone when immersed in water.* Fresh tapwater contains a lot of air; thus, to avoid trouble it is better to use distilled water or water which has been boiled and subsequently cooled. A camel-hair brush is useful in this connection. Immerse the wire cage in the beaker of water before placing it in position on the balance and, having thoroughly wetted and squeezed the brush itself to get rid of any air trapped between the hairs, gently rub over the wire coil with the brush until no air bubbles can be seen. Then, if the stone is first thoroughly wetted by dipping in water and rubbing with the fingers no bubbles should be present when it is placed in the cage. The presence of a few air bubbles may make a difference of several hundredths of a carat to the weighing and vitiate an otherwise accurate result. If the bubbles were on the cage and could be guaranteed to 'stay put' during the whole operation, and if the total weight of the cage (bubbles and all) could be allowed for, it would not matter, but bubbles are unreliable things and may become dislodged half-way through the experiment.

To illustrate the working of the method, the following is an actual example. A red stone, weighing 7.535 carats, was thought to be a garnet on account of its colour and the fact that it showed no dichroism. A refractometer was not available, so the specific gravity was measured by the above method (which is called the method of 'hydrostatic weighing').

The weight of wire cage and stone immersed in water proved to be 12.196 carats, and the wire cage alone in water weighed 6.753. The weight of the stone in water was therefore

12.196 − 6.753 = 5.443

The loss of weight in water shown by the stone is thus

$7.535 - 5.443 = 2.092$

Using the formula SG $= \dfrac{\text{Weight of stone}}{\text{Loss of weight in water}}$

we have SG $= \dfrac{7.535}{2.092} = 3.602$

This stone was therefore not a garnet but a *red spinel*, since a garnet of the almandine–pyrope series of this colour would have a SG near 3.80.

Results obtained by hydrostatic weighing are less accurate for small stones than for large, and for stones of equal weight results are less accurate for stones of high SG than for those of low SG, as their volume is smaller. It is easy to calculate what effect in the final result a given small error in the 'weighing in water' will have. Suppose a SG determination be attempted on two rubies, one weighing 8 carats and the other weighing only 1 carat, the true SG in each case being exactly 4.00. For an identical error of 0.01 carat in the 'weighing in water' the final result will be 4.02 in the case of the larger stone – not a very serious matter – but 4.16 with the smaller stone, which is so high as to make one suppose that the stone must be almandine garnet and not ruby.

Now when using water the error of 0.01 carat in the immersion weight postulated above is quite possible, since the high surface tension of water tends to vitiate the even swing of the balance due to the drag on the suspending wire where it cuts the surface of the water. Hence the desirability of a very fine wire. Further, the formation of tenacious bubbles on wire or stone is also an effect of the high surface tension. A single drop of liquid detergent or of one of the 'wetting fluids' supplied by photographic dealers added to the water will help matters considerably. But if it is desired to obtain the best possible results with small stones (say, under 3 carats) it is advisable to use not water but some other fluid having a lower surface tension. Toluene is a stable and fairly inexpensive liquid which can be recommended for the purpose.

It is, of course, essential when using such fluids to multiply the result in the simple formula by a factor representing the SG of the fluid used at the temperature of the experiment, thus:

SG $= \dfrac{\text{Weight of stone}}{\text{Loss of weight in liquid}} \times$ SG of liquid at $T°C$

Even when using water this factor should strictly be used, since the SG of water is only exactly unity at 4°C. But the value for water is so little below this, even in hot weather (e.g. 0.998 at 70°F) that this correction can be safely disregarded in ordinary determinative work. The SGs of the pure liquids can be obtained from tables, but in practice these vary slightly from sample to sample, and it is wise to check the SG of each batch (which will last for years if kept in a stoppered bottle), a task which is quite simple if one proceeds as follows.

Take a large specimen of pure quartz (say, 30 or 40 carats) and carry out a careful SG determination with this in the liquid which you wish to check. Then, knowing that pure quartz has an invariable SG of 2.651, calculate the SG of the liquid thus:

$$\text{SG of liquid} = 2.651 \times \frac{\text{Loss of weight in liquid}}{\text{Weight of stone}}$$

The temperature of the liquid should be noted, and the SGs at other temperatures can be assessed (in the case of toluene) by subtracting 0.0007 for each rise in temperature of 1°C. It should be noted that, due to recent warnings from US health authorities of the toxic effects of ethylene dibromide (dibromoethane), the use of this liquid can no longer be recommended. Almost all the fluids used by gemmologists are in fact poisonous to a greater or lesser degree and should be handled with care. Inhaling the vapour should be avoided, and hands should be washed after contact of fluids with the skin.

When hydrostatic SG measurements are going to be carried out frequently it is worth preparing a table giving the SG of the liquid used for each degree Centigrade (for each tenth of a degree for really accurate work). Previously, seven-figure logarithms were recommended for these tedious calculations, but the advent of inexpensive pocket calculators provides a much quicker and easier method of obtaining accurate results. Even for a straightforward calculation when using water these gadgets undoubtedly save a great deal of time, and are probably less liable to error if carefully used. Any mistake in the calculation is usually signalled by a wildly improbable result.

Although it is always assumed that stones must be unmounted if a SG test is to be carried out, the author has found it quite practicable to use the hydrostatic method as a test for identifying necklaces of jade, amber, or plastics, without removing beads from the string. The weight of the thread is normally so small compared with that of the beads that the result is usually quite accurate enough to help materially in identifying the beads. A necklace can be assembled into a compact 'bunch' and tied together in this more wieldy form with thread to enable the beads to be accommodated on a normal balance, which is far more accurate than any spring balance could be. Water, with a few drops of liquid detergent added, should be used in such cases.

It may also be noted that when the need arises to check the SG of quite large carvings of jade and similar materials a good spring suspension balance may usefully be pressed into service if one can be found which covers (but which does not much more than cover) the desired range of weights. A sling of strong thread can be made for the specimen, which is suspended from the hook of the balance and weighed first in air and then when the carving is completely immersed in water – which can be in a sink, a developing tank, a bucket, or any convenient large container. Even when using crude methods such as this, it is important to take every care over the readings – the 'zero' position being checked, the eye being level with the pointer and scale, etc. – if any kind of accuracy is to hoped for.

The author has found that a camera tripod makes a convenient and easily adjustable 'derrick' from which the balance can hang.

The Hanneman balance

For those who do not feel the need for any form of standard balance but who would like to be able to make accurate specific-gravity measurements, an

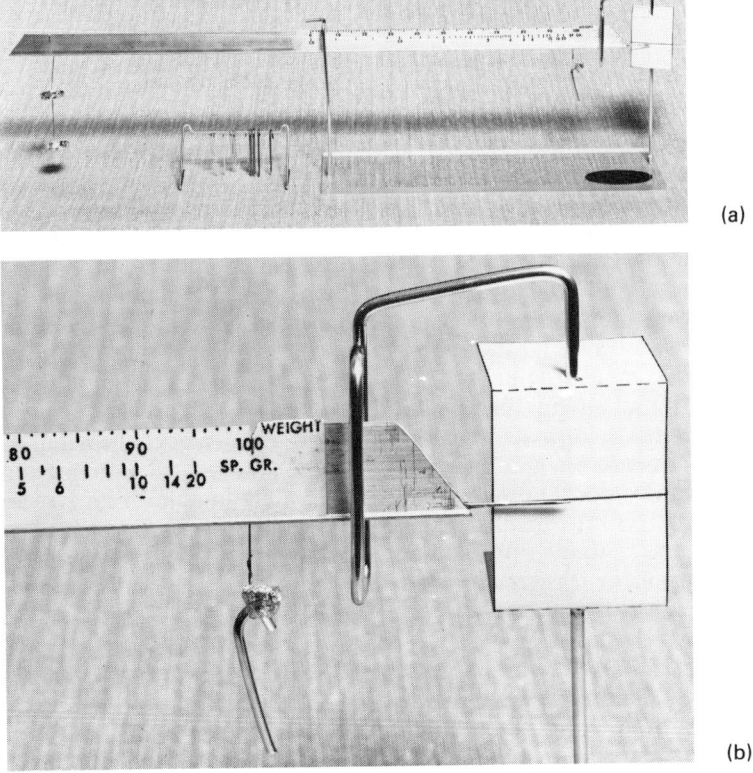

Figure 5.3 A Hanneman density balance. (a) Complete assembly; (b) detail

ingenious and simply constructed apparatus devised and manufactured by Dr W. W. Hanneman may be recommended (Figure 5.3). In addition to its considerable accuracy, the Hanneman balance has the advantage of giving the operator a SG figure without any tiresome calculation being necessary. The balance consists essentially of an aluminium beam, suspended by a fine nylon thread at its centre (to avoid the expense of a knife-edge pivot). The specimen to be tested is placed in one of two small pans at the left-hand of the beam, either dry or immersed in water, and counterpoised by hooked weights placed at positions marked by a scale on the right-hand beam. It is claimed that the balance can weigh a stone to an accuracy of 0.002 carat, which should enable reliable results to be obtained with stones of fairly small dimensions. The apparatus can be obtained in kit form from the makers; it has been tried out in the UK and found to be very satisfactory.

Heavy liquid method

It often happens that only an approximate value for the SG of a specimen is required, to differentiate between stones of similar appearance, such as topaz and yellow quartz, chrysoberyl cat's-eye and quartz cat's-eye, etc. In such cases a very rapid SG test can be made by the use of **heavy liquids**.

In principle, the 'heavy liquid method' is exceedingly simple. Obviously, a stone will sink in a liquid of lower SG than itself, float in one of higher SG, and will remain suspended in a liquid having an equal SG. Provided, therefore, that one can obtain a series of suitable heavy liquids it is clear that by placing the stone under test in liquids of different known SGs and observing whether it sinks or rises to the surface, one can quickly arrive at an estimate of its SG.

Before giving practical details for the employment of this method, let us consider what liquids are suitable. Some ten fluids have been proposed from time to time, but of these, only three need concern us, as they cover the necessary range and are the most convenient in use.

Firstly there is *bromoform*, a mobile, slightly yellowish liquid with a SG of 2.9; second, *di-iodomethane*, mobile and yellow when fresh (SG 3.33), and third, *Clerici solution*, colourless and rather viscous (SG 4.15).

The first two are organic liquids, and will mix in any proportions with 1-bromonaphthalene, which has a SG of 1.49, thus lowering their SG to any required extent. Benzene* or toluene will serve, but are inflammable and cannot be sent by post. Clerici solution is conveniently so called after its discoverer, an Italian chemist. It consists of a concentrated aqueous solution of two thallium salts, the malonate and formate, and the SG of this solution can thus be readily lowered from its maximum value of about 4.15 simply adding a little distilled water.

After mixing or diluting one liquid with another, it is essential to stir thoroughly with a clean glass rod. This is particularly necessary in the case of the viscous Clerici solution.

At this point it must be emphasized that **Clerici solution is highly dangerous**. It is easily absorbed through the skin and persists in the body. Thallium salts have been used as poisons. The use of Clerici solution is not now recommended by the Gemmological Association of Great Britain (who do not sell it) and extreme care must be exercised in its use, which is best confined to experts in laboratories.

Di-iodomethane and Clerici solution are rather expensive, but since only small quantities are required and they last a long time this does not greatly matter.

Heavy liquids can be kept, pure or diluted to a definite SG value, in small cylindrical specimen tubes about 1.9 × 7.5 cm (¾ in × 3 in), securely corked and placed in holes bored in a block of wood.

The number of liquids kept ready for use and their SG values must depend upon individual needs. The following are some suggestions derived from the author's own use of heavy liquids for all manner of gemstones:

1. Bromoform diluted to 2.65;
2. Bromoform diluted to 2.71;
3. Di-iodomethane diluted to 3.06;
4. Di-iodomethane, pure, 3.33;
5. Clerici solution diluted to 3.52;
6. Clerici solution diluted to 4.00.

---

* Benzene is a possible carcinogen and should not come into contact with the hands, nor should its vapour be inhaled.

The bromoform and di-iodomethane, as already stated, can be diluted with 1-bromonaphthalene or one of several other liquids of low SG. The Clerici solution should always be diluted with distilled water; tapwater will cause cloudiness if used, and as Clerici solution is expensive it is worth taking a little trouble to keep it in a clear condition.

If the SG of a liquid is inadvertently made too low by dilution, it can, of course, be raised by addition of the pure liquid or, in the case of Clerici, of the concentrated solution. A dilute solution of the latter can also be readily reconcentrated by careful evaporation with or without heat.

To obtain a liquid of SG 2.65, a small fragment of quartz or small cut stone of the quartz group such as amethyst, citrine, or rock crystal is placed in a tube about half filled with pure bromoform, and 1-bromonaphthalene then added drop by drop, accompanied by continuous stirring with a glass rod, until the specimen acting as an 'indicator' just begins to sink or only rises very slowly if pushed below the surface with the rod.

One then knows that the liquid has a SG of 2.65, and the indicator can be kept in the tube to enable the user to check the SG of the solution each time it is used, since this may vary due to one constituent evaporating more rapidly than the other. The second liquid (2.71) proposed is similarly prepared, using a small piece of calcite (Iceland spar) as indicator.

The third liquid (3.06) is prepared by diluting di-iodomethane until a small piece of green tourmaline remains suspended; the fourth liquid is simply pure di-iodomethane and if kept uncontaminated needs no indicator.

For the fifth suggested liquid (3.52) concentrated Clerici solution is diluted with drop after drop of distilled water (stirring thoroughly after each added drop of water) until a small diamond used as indicator remains practically suspended in the liquid; and finally the liquid of SG 4.00 is similarly prepared, using as indicator a small specimen of synthetic ruby.

If the Clerici solution as purchased contains a deposit of solid thallium salts, a little distilled water can be added and the whole be brought into solution by careful warming in a vessel containing hot water.

The process of matching the liquids exactly with the SGs of the indicator specimens is decidedly a tricky one and requires skill and patience, but as long as the indicator rises or falls in the liquid only slowly this is a sufficiently close match to enable an accurate idea to be obtained of the SG of the liquid in question, which, of course, is the purpose of the indicator.

Many workers prefer to employ two indicators in each liquid, one of which should remain floating and the other stay at the bottom of the tube. It is then known that the SG of the liquid lies between those of the two indicators. This is quite sound as long as the two indicators are so chosen that their SGs are close to one another. If their SGs lie too far apart only a very rough idea of the SG of the liquid is possible.

It must be emphasized once more that the six liquids listed above are merely suggestions which can be varied according to the type of testing the user is most likely to want to carry out. The jeweller with little time to spare can learn a great deal by the intelligent use of three liquids only – pure bromoform, pure di-iodomethane, and bromoform diluted to match quartz. The first two will need no indicators, since the pure bromoform can be reckoned as 2.9 and the di-iodomethane 3.33 at ordinary room temperature and partial evaporation will not affect these SGs; the third liquid will serve as a guide not only for all the quartz varieties but also for separating synthetic emerald from natural emerald.

A few strips of copper kept in the bottle or tube of di-iodomethane will serve to prevent it from darkening as the copper combines with the free iodine which is the cause of the trouble. The clarity of a dark sample can also usually be restored by shaking it with a liberal supply of copper scrap and leaving it in contact with the metal for some days before filtering it off into a clean bottle.

It is wiser not to test porous stones such as inferior turquoise or opal in heavy liquids as they may be harmed. Stones with flaws which reach the surface may also suffer in appearance if immersed in liquids. Stones and tongs should be cleaned after use in a heavy liquid test and before being passed from one liquid into another, since any liquid adhering to the stone would alter the SG of the next solution it enters. This rule is particularly important when using Clerici solution, as this, being an aqueous mixture, does not mix with any of the other liquids. After a Clerici immersion, stone and tongs should be rinsed with water and dried; in the case of the other liquids toluene is perhaps the best cleanser.

Minerals in which the SG is the most constant are those with simple chemical compositions, such as diamond, quartz, and corundum. Apart from lapis lazuli and turquoise, which are not properly homogeneous, there are three gemstones which show large variations in SG: these are zircon, red garnet of the almandine–pyrope series, and blue spinel. The zircons used in modern jewellery, that is, the blue, white, and golden stones mined in the area formerly called Indo-China (now Cambodia, Laos, and Vietnam) and heat treated, cut, and shipped from Thailand, have a constant SG of near 4.69. Certain green zircons from Sri Lanka, however, may have a SG of 4.00 or even lower, and a whole range of SGs is found in Sri Lankan stones between these limits.

The reason for this extraordinary variation was first established by Chudoba and Stackelburg in 1936. It is due to a breakdown in the crystal lattice: zircons of lowest SG being almost completely metamict (see Chapter 17).

The red garnets, the SG of which may range from 3.65 to 4.20, owe this variation to their being mixtures of two minerals of different composition and properties: pyrope garnet, which is a silicate of magnesium and aluminium, and almandine garnet, in which the magnesium is replaced by iron.

Pure pyrope and pure almandine are not known in nature, but stones with SGs up to about 3.85 and refractive indices up to about 1.76 are usually classed as pyropes, while the stones of higher SG are known as almandines.

Blue spinel normally has a SG near 3.60, but zinc sometimes replaces part of the magnesium, raising the SG markedly without affecting the colour or appearance. Values of 3.70 or 3.80 are not very uncommon, while 4.06 was recorded in one exceptional case.

One great advantage of using heavy liquids compared with the hydrostatic method is that results are just as reliable with very small stones as they are with large specimens. Further, a number of stones can be tested at one time; for instance, a parcel of small emeralds can be checked for any 'duds' by pouring the stones into a liquid of SG 2.71. Any stones which sink or rise at all rapidly can be taken out with the tongs and regarded with grave suspicion, their true nature being determined by other means. True Colombian or Siberian emeralds will either rise or fall quite slowly in such a liquid. South African emeralds are of rather higher SG, but in any case will not sink fast.

As a test for synthetic emeralds a liquid matching the SG of quartz (2.651) is useful, since Chatham emeralds have a SG very near to this, while all natural emeralds will sink with some rapidity in this liquid.

For distinguishing amber from its imitations among the plastics, such as Bakelite, Erinoid, etc., a liquid which can hardly be termed 'heavy' must be employed. According to R. Webster, ten level teaspoonfuls of cooking salt in a tumbler of water will make a brine solution of sufficient SG to float amber, while the various plastics so far used to imitate amber are all of decidedly higher SG and will sink in this solution. Necklaces of amber or its substitutes can be tested hydrostatically without unstringing, as suggested earlier.

Although heavy liquids are mostly used by gemmologists to obtain only an approximate value for the SG of a specimen, the method can be extended, by taking a little trouble, to the measurement of the SG of a stone with considerable accuracy. If one of the liquids be carefully diluted until an exact match is reached with the stone to be tested, that is, so that no perceptible rise or fall of the specimen in the liquid can be noted, then one may be quite sure that the SG of the stone and the liquid is the same (*at that particular temperature*) to at least the third place of decimals. If, therefore, one can accurately measure the SG of the liquid, the SG of the stone is also known. This can be done by filling a small 'specific gravity bottle' with the liquid and carefully weighing. Then if the weight of the empty dry bottle and its weight when filled with water are also ascertained the comparative weight of the bottleful of the liquid and of the water will give one the SG of the former.

A more rapid, but not quite so accurate, method of finding the SG of a heavy liquid is by using a small hydrometer. For this, a fair quantity of liquid is needed, and it is thus chiefly useful for the bromoform series. The author has found a small hydrometer, giving the range 2.5–3.0, particularly useful.

Another useful accessory is a set of glass indicators, each marked with its appropriate SG (Figure 5.4), made by Rayner and obtainable through the Gemmological Association of Great Britain. Alternatively, a set of indicators can be made with very little trouble or expense from small specimens or pieces of those minerals which are readily procurable in a pure state and can be relied on to have a constant SG. In Table 5.1 the figures given for materials in bold type can be taken as correct to the second place of decimals provided transparent flawless pieces are chosen. The others may vary by as much as ±0.02, but this does not greatly vitiate their usefulness as indicators.

Table 5.1   Gem materials for use as SG indicators

Quartz (rock crystal)	2.65
Calcite (Iceland spar)	2.71
Tourmaline (pink)	3.05
Fluorite (transparent)	3.18
Peridot	3.34
Diamond	3.52
Topaz (colourless)	3.56
Chrysoberyl (yellow)	3.72
Demantoid garnet	3.85
Corundum (synthetic white)	3.99
Sphalerite (transparent)	4.09

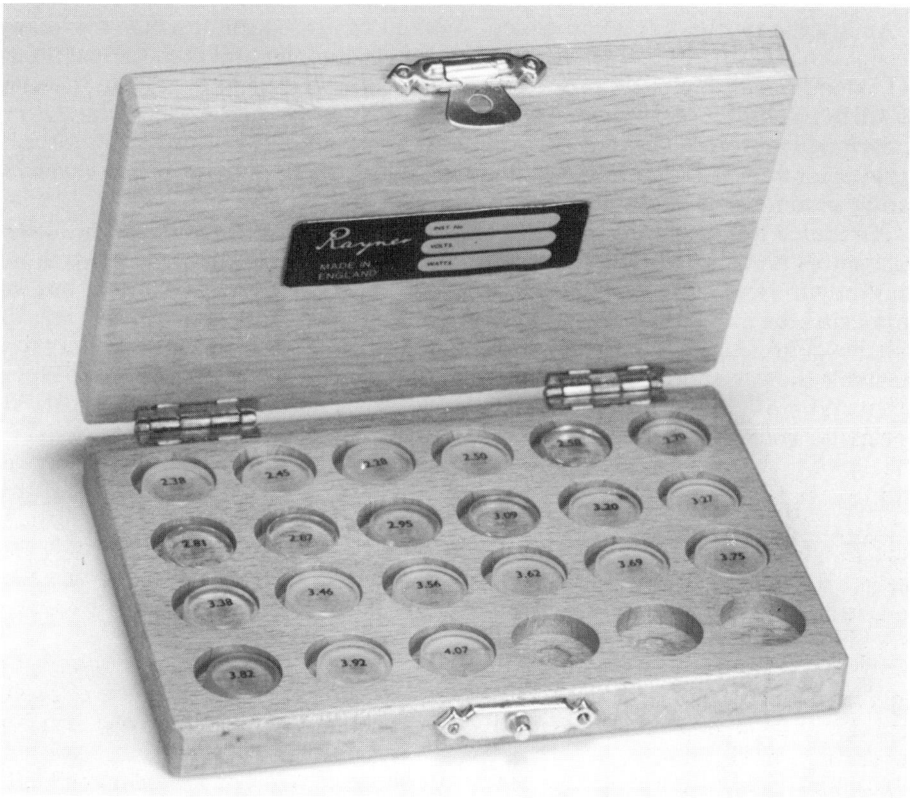

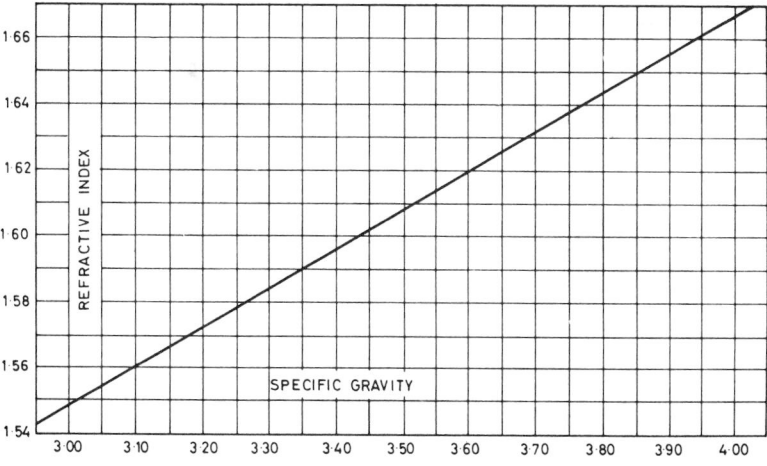

Figure 5.4 A set of Rayner density indicators

Figure 5.5 The relation between refractive index and specific gravity of Clerici solution

Another very rapid and convenient method of estimating the SG of a heavy liquid is to measure its refractive index on a refractometer and then to read off its SG value from a previously prepared graph. This is very readily done in the case of Clerici solution. Samples vary slightly in their properties, but the graph shown in Figure 5.5 is quite representative, and should certainly give results quite accurate enough for identification purposes. Being a straight-line graph, it can be extended *ad lib.* by the use of ordinary graph paper.

SG values for the more important gemstones and certain metals and other substances handled by the jeweller are given in Table 5.2. The figures are those found by the author to be most characteristic for the substances named, but for simplicity the range of variation found is, in most cases, not given. This is usually quite small, and where the SG is given in the table to two places of decimals the variation is generally confined to a few units in the second decimal place. Where there is a wider variation the SG is quoted to only one place of decimals. The more important gem materials are printed in bold.

Table 5.2   Typical specific gravities

Amber	1.08	Spodumene	3.18
Bakelite	1.26	**Jadeite**	3.33
Tortoiseshell	1.29	**Peridot**	3.34
Erinoid	1.33	**Zoisite**	3.35
Celluloid	1.38	Sinhalite	3.48
Vegetable ivory	1.40	**Diamond**	3.52
Ivory	1.80	**Topaz** (pink)	3.53
Bone	2.00	**Topaz** (yellow)	3.53
Fire opal	2.00	**Topaz** (white)	3.56
Opal	2.10	Sphene	3.53
Obsidian	2.40	**Spinel**	3.60
Moonstone	2.57	**Spinel** (synthetic)	3.63
Chalcedony (agate, cornelian)	2.60	Hessonite	3.65
Quartz	2.65	**Pyrope**	3.7–3.8
Coral	2.68	**Chrysoberyl**	3.72
Beryl (aquamarine)	2.69	**Demantoid**	3.85
Beryl (yellow)	2.69	**Almandine**	3.9–4.20
Beryl (emerald)	2.71	**Corundum**	3.99
Beryl (pink)	2.80	**Zircon** (green)	4.0–4.5
Pearl (natural)	2.71	**Zircon** (blue, white, golden)	4.69
Pearl (cultured)	2.75	Pyrites	4.90
Turquoise (American)	2.7	Hematite	5.10
Turquoise (Iranian and Egyptian)	2.8	Silver	10.50
Lapis lazuli	2.8	Gold, 9 carat	11.40
Pink (conch) pearl	2.85	Gold, 14 carat	13.93
Nephrite	3.00	Gold, 18 carat	15.40
Tourmaline	3.05	Gold, 22 carat	17.70
Andalusite	3.15	Gold, Pure	19.30
Fluorspar	3.18	Platinum	21.50

Table 5.2 will certainly cover all the materials handled by the average jeweller. For the benefit of keen gemmologists or collectors who are interested in the rarer gemstones, a more comprehensive table is given in Appendix 4 at the end of the book.

6

The use of the microscope

To the gemmologist, the microscope is undoubtedly the most essential of all instruments. Though not so flexible as a pocket lens, its powers of discrimination are far wider. In everyday practice the jeweller is more often concerned with distinguishing natural stones from their synthetic counterparts or in detecting imitation stones than in separating natural stones of different species. In detecting all manner of counterfeits the microscope is supremely important, as we shall see in Chapter 9.

However, its importance does not stop there; its powers of distinguishing one stone from another are also very considerable, since the identity of a stone and even its place of origin can often be deduced from a study of its inclusions as revealed by the microscope. The nature and depth of flaws can also be determined, the presence of double refraction detected, and its extent approximately gauged, and refractive index also can be assessed with the aid of immersion liquids, or by 'direct measurement'.

The study of mineral inclusions began in the nineteenth century, but such work was sporadic and uncoordinated. During the past few decades it has been coming into its own as a new and practical means of identifying gemstones and of understanding something of the conditions under which they were formed in nature. This indeed is one of the most important contributions of gemmology to the parent sciences of mineralogy and geology. The major advances in inclusion study are chiefly due to the researches and enthusiastic advocacy of Dr Edward Gübelin of Lucerne, Switzerland. Many of his articles on the subject, illustrated by splendid photomicrographs, have appeared in the journals. In his most recent book, *Photoatlas of Inclusions in Gemstones*, he has collaborated with John Koivula of the Gemological Institute of America, who has specialized in the development of new methods of photomicrography. This new work is a feast of colour and scientific information on gemstone inclusions. The subject is a vast one, however, and plenty remains to be discovered by the patient and acute observer, and many problems still await solution.

Microscopes specially designed for use by jewellers and gemmologists have now been developed by many firms in the USA, West Germany, Japan,

Switzerland, and Britain, Prices range from a few hundred to several thousand pounds. Unfortunately, most of these are either expensive or not readily available; the special 'petrological' microscopes used by the mineralogist are also expensive luxuries.

Fortunately, most of the essential work needed can be carried out with quite a simple form of microscope, and we shall begin by describing a typical standard instrument (see Figure 6.1) and the manner in which it may be used in the study of gemstones.

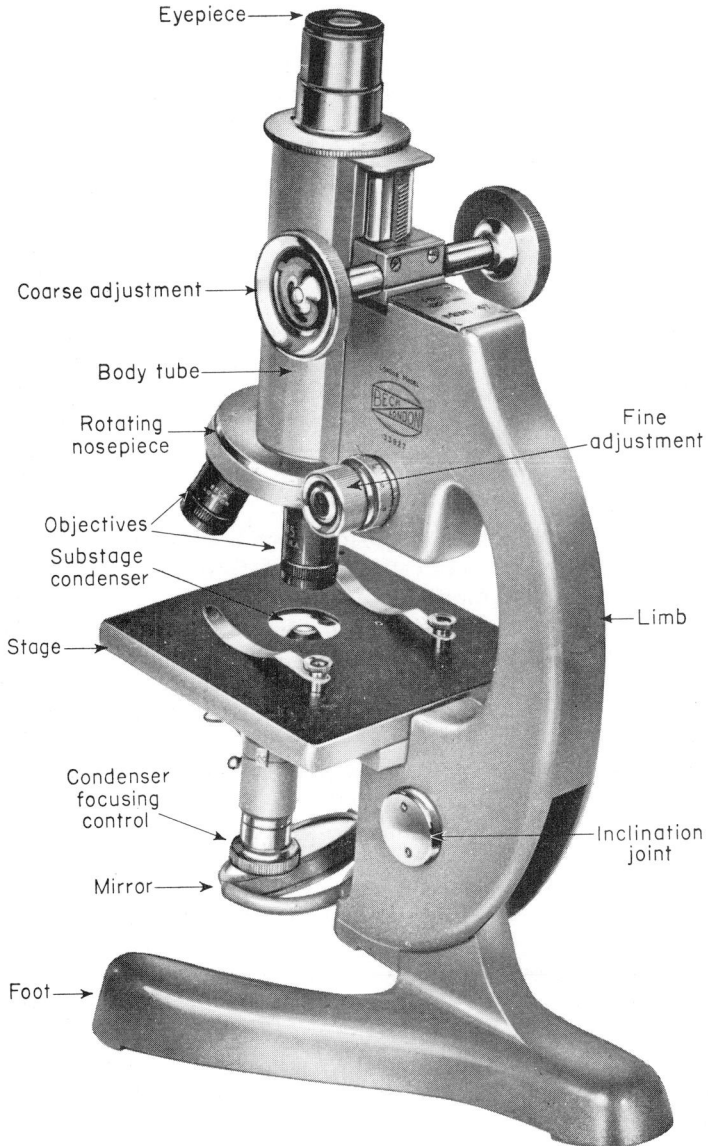

Eyepiece

Coarse adjustment

Body tube

Rotating nosepiece

Objectives

Substage condenser

Stage

Condenser focusing control

Mirror

Foot

Fine adjustment

Limb

Inclination joint

Figure 6.1 A typical standard microscope (courtesy: R. & J. Beck Ltd)

There are two lens systems producing the magnifying effect in a microscope, each being mounted and handled as a single piece. The lower group is known as the *objective* and the upper as the *eyepiece* or *ocular*. The objective is screwed into the lower end of the *body tube* of the instrument, and the eyepiece is slid into the upper end of the tube. The total magnification produced depends on the power of the objective and of the eyepiece, and the length of the microscope tube. The microscope *stand*, as the whole instrument (without the eyepiece or objective) is called, consists of a heavy metal base or *foot* to which is hinged a heavy metal *limb* by which the microscope is carried or lifted from its case and to which is attached a platform called the *stage* with a circular aperture in the centre through which the light reflected from the adjustable *mirror* of the microscope reaches the specimen to be examined. The body tube of the microscope (which often contains an inner *drawtube*) is attached to the limb and may be raised or lowered for focusing by a rack and pinion movement operated by two milled heads turning in the vertical plane on either side of the body tube.

This focusing arrangement, which has a play of several inches, is known as the *coarse adjustment* and is all that is needed to obtain sharp focus with low or medium powers. Better-class microscopes are also fitted with a *fine adjustment* focusing arrangement for making the final delicate focusing adjustments with higher power objectives, with which the focus is extremely critical.

In addition to the simple optical system of mirror, objective, and eyepiece, it is useful for our purpose to have a *substage condenser*; that is, a condensing lens fitted beneath the microscope stage, preferably combined with an iris diaphragm in the same fitting, housed below the microscope stage. This provides a convergent beam of light which greatly assists in illuminating a faceted specimen satisfactorily when it is not immersed in a liquid. It is a great convenience if this is adjustable by means of a rack and pinion.

The magnifying power of objectives is usually indicated by their focal length; the shorter this is, the greater the magnification. Medium- or low-power objectives, notably 16 mm (⅔ in), 25 mm (1 in), and 38 mm (1½ in), used in combination with a low-power eyepiece, will give all the magnifications needed for stones, that is, from 25 to 70 diameters.

The advantages of using fairly low powers are that (1) one can see more of the specimen in the field of the microscope at any one time; (2) the depth of the focus is greater; and (3) the working distance between the specimen and the front lens of the objective is also greater, so that one is less likely to damage the soft glass of the lens by contact with the hard edges or corners of the stone. Generally speaking, too, the definition is clearer with the lower powers and observation is much easier for the beginner.

Having thus very briefly detailed the main features of a standard microscope we can proceed to describe how it can best be used in examining the internal features of faceted stones. Incidentally, the microscope can, of course, also be used to examine the *external* surface of a stone if necessary; for instance, the precise nature of a chip or a flaw where it reaches the surface, the quality of the polish, and so on. Examination of a stone in its setting is often necessary and can usually be managed if the setting is an open one, but it is undoubtedly more difficult than with an unmounted gem, and the beginner will be well advised to practise first with the latter, as described in detail below.

To become accustomed to focusing the microscope the beginner should first start by examining simple objects such as scraps of blotting paper, cloth, leaves, etc. These will lie flat on a plain glass microscope slide placed on the stage of the microscope and can easily be focused, the specimen being illuminated by an adjustable microscope lamp or reading lamp placed near to one side and rather above the level of the stage.

It should be noted that in examining stones it is necessary to use the microscope with the body tube in the vertical position, that is, with the stage horizontal to prevent the stone, which is held in position only by gravity, from rolling off the stage. When observing an object mounted properly as a microscope slide, this is usually clipped into position on the stage or a stone held fast in a spring-loaded stone holder; the limb can then be tilted back towards the worker into a position which is more comfortable for prolonged scrutiny. The 'working distance' of any objective will be found to be rather less than the focal distance length by which the objective is known. A much greater 'working distance' is available when using a binocular microscope (Figure 6.2), described later.

Having become accustomed to the 'feel' of the microscope, the operator can now proceed to examine a stone. To make things easy to start with, take a plain glass microscope slide and a small piece of modelling clay or 'Blu-Tack'. Roll a small ball about the size of a pea (whether sweet pea or green pea must depend upon the size of the stone) and press this gently onto the slide near the centre. Then, having wiped the stone scrupulously clean, place it on the slide with the table facet uppermost and horizontal, the culet resting on the slide, and the stone supported by modelling clay near the girdle. Now place the slide on the stage of the microscope with the stone directly under the objective, which should then be racked down by means of the coarse adjustment until nearly touching the surface of the stone, while watching from the side to ensure that it is not allowed actually to touch. The operation of focusing the internal parts of the stone as seen through the table facet can now proceed with safety, since this will be accomplished by *raising* the focus, thus avoiding the very real risk of overshooting the mark in attempting to focus *downwards.*

In focusing upwards, as suggested, one should soon see the culet and the edges of the adjoining facets come into view; then, continuing to raise the focus slowly, the image of these should disappear, and before very long the surface of the table facet should be faintly visible by reason of line particles of dust settled upon it and sometimes slight imperfections of the surface such as polish marks, etc. If in the middle of the stone there are any flaws, inclusions, or bubbles, these will appear in sharp focus at some point between the correct focus for the culet and the table. If one is uncertain whether the table facet level has been reached (and this, admittedly, is not always easy to see) one can check on this by pushing the slide along a little, when the edge between the table facet and other crown facets should be clearly visible. The observer will find that moving the slide to the left will cause an apparent movement to the right as seen through the microscope.

By adjusting the distance of the condenser below the stage and tilting the mirror at the correct angle the best position can soon be found for obtaining optimum illumination of the interior of the stone. Stones can also be examined, of course, table facet down on slide; only one is then apt to be confused by the reflections and refractions caused by the many small facets. When one is completely at home with the microscope it will be found a simple matter merely to hold the stone table facet upward between thumb and forefinger of the left hand, resting the hand on the microscope stage to assist in steadying the stone. One can then focus the microscope down on to the stone safely, since the fingers will act as buffers to prevent the objective coming into contact with the specimen. Stones too small for this treatment can be held in tongs. Spring tongs with grooved lips are especially valuable for this work.

In the methods suggested above, it is essential for good results to have a substage condenser. Even so, the examination of the stone will be probably not quite complete, but will be chiefly confined to that part of the stone directly beneath the table facet. Small inclusions and other features near the girdle may therefore be missed. To make a more thorough examination of a 'difficult' (i.e. very clean) stone it will be advisable to immerse the stone in a glass or glass-bottomed cell containing a liquid of refractive index fairly near to that of the stone examined. Only just enough liquid to cover the stone completely should be added and the stone, the cell, and liquid should be clean to prevent particles floating at the surface of the liquid or adhering to the surface of the stone which will make observations more difficult. This can be done by filtering through folded filter paper in a glass funnel. A good immersion liquid to use is 1-bromonaphthalene, a liquid compounded of bromine and the naphthalene which is so well known in the form of 'moth balls' – the liquid itself has something of the 'moth balls' smell. Di-iodomethane is also often used, and has a refractive index (1.745) more nearly approaching the 1.76 of corundum than has 1-bromonaphthalene (1.66); but this close matching of the refractive index between stone and liquid makes the edges and facets of the former become practically invisible and this renders it difficult to 'find one's way about' in the stone. With 1-bromonaphthalene, refraction between the liquid and the stone is sufficiently small to allow the specimen to be easily examined throughout, but enough can be seen of the edges of the facets to enable the observer to know what part of the stone he or she is examining. Unfortunately, the fumes of most of these immersion fluids may prove troublesome to the eyes and mucous membranes of those liable to suffer from hay-fever. Benzyl benzoate, which has a refractive index of 1.56, does not have this disadvantage and has displaced 1-bromonaphthalene in many laboratories. However, it tends to 'stick' to surfaces and needs to be carefully removed using warm water and detergents. It must not be used near plastics. The best cells to use are those in which a ring of glass is fused or firmly cemented to a plain glass base. These are obtainable from the Gemmological Association, London.

When examining a stone in liquid it is normally convenient to place it so that it rests on its table facet. Care must be taken not to confuse air bubbles clinging to the surface with internal bubbles which may be found in synthetic stones. External bubbles usually are larger and look more luscious than those found in synthetics, and can be removed by wiping the stone, while internal bubbles, of course, will 'stay put'.

A few words as to the choice of a microscope may be helpful. Unless one is an experienced microscopist it is better to purchase an instrument from a reputable firm, whether it be new or second hand, as in this case the instruments so sold will be guaranteed to be optically sound. For occasional use a simple monocular stand of the type already described is quite adequate, but where stones are to be examined for prolonged periods some form of binocular instrument is preferable as it involves far less strain on the eyes. Examples of both inexpensive and advanced binocular microscopes are shown in Figures 6.2 and 6.3.

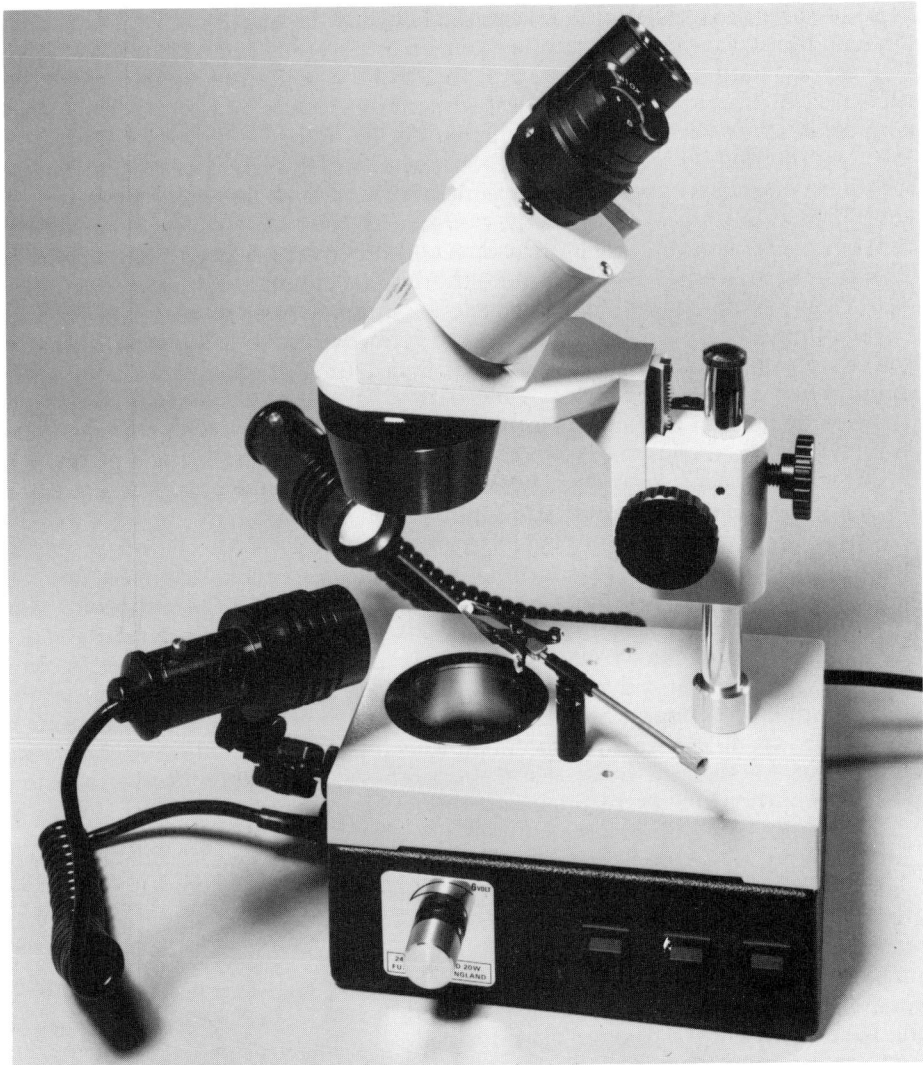

Figure 6.2 The 'Microgem' – a simple binocular microscope with ×10 and ×30 magnification and provision for transmitted and reflected lighting (courtesy: Gemmological Association)

The best microscopes for gemmology are undoubtedly the 'stereo-zoom' types with paired eyepieces and variable stereoscopic magnification built into the 'pod', which contains both eyepieces and objectives. Binocular microscopes without the zoom magnification are cheaper and are perfectly good for general use. Both types have a long working distance, considerable depth of focus, and give erect vision – that is, the orientation of the objects seen under the microscope is not reversed. This general type of instrument is made by many optical firms, some with and some without a stage, some with the usual mirror, others with built-in lighting. A substage condenser can be fitted as an extra, and is almost essential for the gemmologist; this should be adjustable, in order to

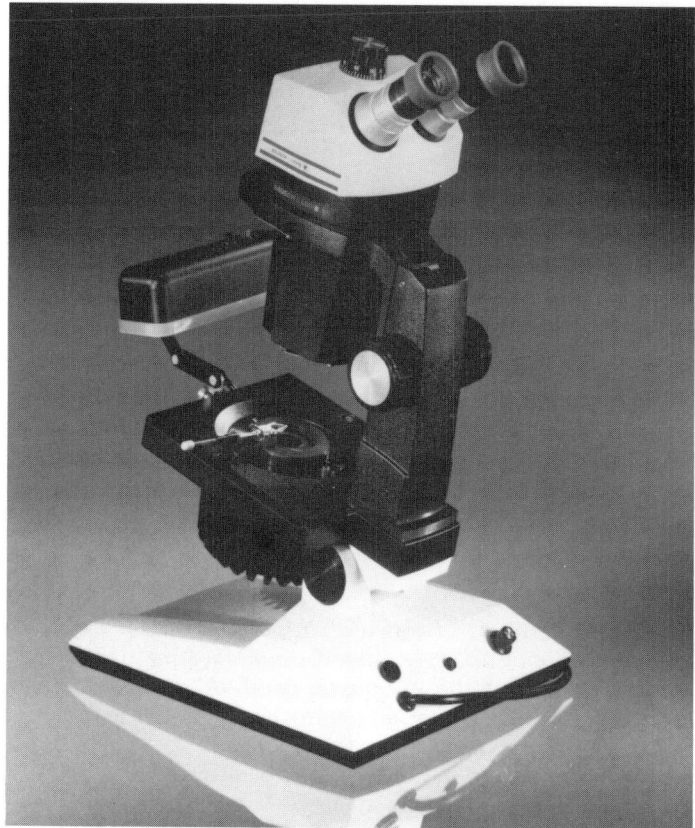

Figure 6.3 An advanced 'Gemolite' microscope providing ×10 to ×70 stereo-zoom magnification and transmitted, dark field, and overhead illumination (courtesy: The Gemological Institute of America)

allow for flexibility in the lighting of the stone. The fullest illumination is obtained when the condenser is raised, but fine details such as curved striae in a synthetic ruby show up much more clearly when the condenser is lowered. For prolonged work with the microscope, inclined eyepieces are a great advantage.

In recent years several new horizontal microscopes have appeared on the market. Expensive instruments of this type have been available from West Germany for some time, but more reasonably priced British instruments can now be obtained. They commonly use Russian-made optical 'pods' (objectives and eyepices with or without stereo-zoom facilities) but the stands are constructed in Britain. They are especially useful for examining immersed gemstones, which can be moved easily without upsetting the light path (see Figure 6.4).

It is a wise plan before purchasing either a new or second-hand microscope to try it out on a few stones containing inclusions, as often an instrument which will give excellent results with thin sections mounted on slides may prove unsuitable for three-dimensional examination of gemstones. Above all, it must be remembered that high-power objectives such as 3 mm (⅛ in) are useless in the

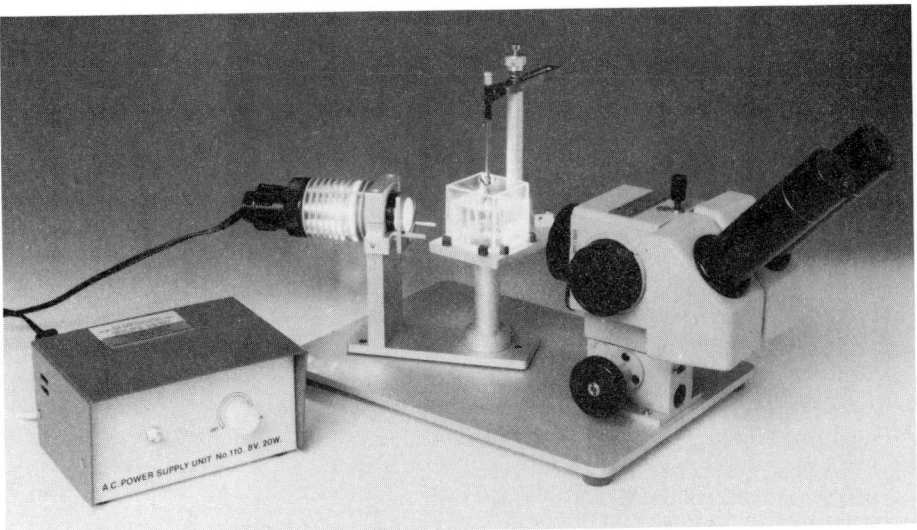

Figure 6.4 The Nelson horizontal immersion microscope (courtesy: Gemmological Association)

examination of gems; 25 mm (1 in) or 38 mm (1½ in) are the most generally useful, giving magnifications of about 20 to 30 diameters, according to the length of the body tube and the power of the eyepieces used. A double or triple nosepiece, enabling objectives of different power to be swung into position quickly, is a useful adjunct to the standard microscope.

So far, the question of illumination when using the microscope has been barely touched upon. Since daylight is such an uncertain and uncontrollable factor, it is usually better to employ artificial light. Many forms of special microscope lamps are on the market and in recent years many fibre-optic lights have become increasingly available and at reasonable cost. These fibre light guides can provide almost 'pinpoint' illumination and are easily adjustable to almost any position. However, an adjustable desk lamp will often serve the purpose very well, provided that this has an opaque or nearly opaque shade to cut out the glare produced when bright light, other than that from the microscope mirror, reaches the observer's eyes. An opal bulb of 40 or 60 watts is usually most suitable. 'Dark-field' illumination has many advantages, and is extensively used for the study of inclusions. The most usual arrangement is to have a small low-voltage lamp 'built-in' to a substage housing consisting essentially of a silvered bowl-shaped reflector. Direct light is prevented from entering the stone by a suitable stop, while oblique rays enter the specimen from all sides by reflection from the silvered reflector. Immersion in liquid is not necessary: the stone can be gripped in spring tongs attached to a pivot on the stage. Any feathers, flaws, or inclusions are clearly visible, being brightly lit against a dark background. This form of illumination is particularly suitable for diamond, through which it is difficult to transmit light in the ordinary way because of its high refractive index.

The interested reader is recommended to refer to an excellent set of diagrams which appears in *The Photoatlas of Inclusions in Gemstones* by Gübelin and Koivula. The diagrams illustrate the principles of construction in special gemmological microscopes, and the enthusiastic gemmologist or jeweller should be able to improvise many of these arrangements at a fraction of the cost. The technique of shadowing developed by Koivula and described in Chapter 2 should be read in conjunction with this chapter.

It is often necessary for the gemmologist to examine stones in their settings, and this can be an awkward business. A stone-set brooch can usually be placed on the stage and thus examined fairly easily. When examining stones in a ring it is often best to look through the back of the stones so that the particular stone under examination rests more or less with its table facet on a glass plate on the stage, only tilted sideways sufficiently to prevent the shank of the ring obscuring the view of the stone.

Ring stones are usually dusty at the back of the setting, and it is well worth cleaning them thoroughly with a small camel-hair or tooth brush and methylated spirit or warm water containing a little liquid detergent before attempting to examine them.

One can, of course, hold a ring with the stone uppermost so as to look through the table facet, but it is difficult in this position (unless a fibre-optic light is used) to obtain sufficient illumination. Most difficult of all to examine are backed stones, which can only be examined by overhead light. Even here, however, often enough detail can be seen within the stones to enable one to determine whether they are natural or synthetic.

If it is required to immerse mounted stones for their better examination it is advisable to use a glass dish of ample size, say, 7.5 cm (3 in) to 10 cm (4 in) diameter, in which the piece of jewellery can be placed and completely immersed. Some inexpensive and volatile liquid such as toluene should be used (in spite of its relatively low refractive index) to avoid waste of the more expensive liquids and also to avoid having afterwards to clean the jewellery to remove the oily and smelly remnants of the immersion medium. Stones with foiled backs or closed-back settings should not be immersed in any liquid.

In Chapter 3 it was explained how double refraction in gemstones (an important distinguishing factor) could be detected by observing the apparent doubling of the edges of the back facets when viewed with a lens through the front of the stone. In stones which have large double refraction such as rutile (0.287), sphene (0.12), zircon (0.059), peridot (0.036), and even in tourmaline (0.018) this should with practice be easily observed with a strong pocket lens even in small stones, but with species in which the double refraction is small, such as quartz (0.009), topaz (0.009), corundum (0.008), chrysoberyl (0.009), and beryl (0.006), the effect cannot be detected with a lens unless the specimen is large.

The microscope, however, enables one to extend this direct observation of double refraction even to those gems in which the effect is small, and, of course, makes it far more noticeable with those in which the double refraction is strong. In the 'petrological' type of microscope there are available two 'Nicol prisms' (or polarizers made of 'Polaroid' which fulfil the same purpose) which enable one to carry out the most sensitive of all tests for double refraction. This particular test between 'crossed Nicols', as it is called, was briefly explained in Chapter 3. It was there mentioned, also, that a number of isotropic (i.e. single refractive) substances, including glass, may show a certain amount of double refraction when under a condition of strain.

Such traces of anomalous double refraction, however, will never be large enough to cause the 'doubling' effect under the microscope, so that when this is seen, one may be sure that the stone is really doubly refractive and thus cannot be a paste or a mineral crystallizing in the cubic system. Such evidence is often of great practical importance and can be carried out at the same time as one is observing the internal inclusions of a stone. When, for instance, one is examining a reputed sapphire, which may be either a synthetic sapphire, a blue paste, or (less likely) a synthetic spinel, if the edges of the bubbles or of the back facets as viewed through the table are seen to be doubled then this establishes without further test that the stone is a synthetic sapphire, since glass or spinel would show no such evidence of double refraction.

As with all the other instruments mentioned in this book, good results can only be obtained by practice, but with a little perseverance any ordinarily intelligent person can handle a microscope with profit and pleasure.

Photomicrography

The keen student will sometimes wish to make a permanent record of some interesting feature or inclusion. Provided a good microscope is available, and the experimenter is sufficiently patient to arrange lighting to the best advantage, it is not a very difficult matter to procure a reasonably good photomicrograph.

The simplest techniques of all, involving no camera, is to remove the eyepiece and place a piece of finely ground glass on top of the body tube of the microscope, and, in a darkened room, to focus the microscope until the desired features can be seen sharply defined on the ground-glass surface (this surface, of course, should be the one in contact with the body tube). The microscope light should then be switched off, and, in darkness, a piece of slow 'line' film be placed, emulsion downwards, in place of the ground glass. This can be covered by a small square of cardboard or other light, opaque object. The lamp is then switched on for a sufficient length of time to make the required exposure, which may mean anything from about half a second to 30 seconds or so, according to the subject. When developed, the film should show a good sharp image, which can, of course, be enlarged at a later time. This method does not lend itself to masterpieces of photomicrography, but is cheap and simple.

If one of the forms of 35 mm camera is available, it is not difficult to take photomicrographs either in black and white or in colour. In this case the eyepice is not removed, but the desired features are carefully focused using the microscope in the ordinary way. The camera is then focused for infinity and the aperture set for a fairly wide stop (say, $f/4$). The camera can then usually be balanced with its lens resting on the eyepiece of the microscope, and an exposure be made either by fixing the shutter open and operating the microscope lamp for the required time, or by setting the shutter to 'bulb' and using a cable release. If the lighting is sufficiently strong to enable exposures of under a second to be made, the marked shutter speeds can be utilized with advantage, and the 'self-timer' clockwork device employed. This obviates risk of shifting the camera while operating the shutter. For steadiness and safety it may well be preferred to rig up some form of stand for the camera rather than simply to let it rest as suggested above.

Any of the modern single-lens reflex cameras which are so deservedly popular on account of their versatility can be adapted for use with the microscope. The camera, with its lens removed, is mounted over the eyepiece of the microscope by means of a special attachment available from the makers. The reflex image can be viewed and focused in the usual way, since the image seen and the picture on the film must always be the same.

The chief difficulty, apart from arranging the field so that all the required features are correctly lit and in focus, is in gauging the exposure time, especially where colour photographs are attempted. This, of course, is very largely a matter of experience. The ordinary hand-held photoelectric exposure meter may not be sufficiently sensitive to function properly under these conditions, but modern built-in exposure meters in 35 mm SLR cameras are very versatile.

In concluding this chapter it should be emphasized that a wide range of microscopes for gemmology is now available and is developing continuously. The interested student or jeweller is recommended to choose from the catalogues of the various suppliers (see Appendix 8) and to visit their showrooms if at all possible. Those to whom high cost is no barrier are advised to refer to the appropriate chapter in Peter G. Read's book *Gemmological Instruments* (Butterworths) where a full range of modern microscopes designed especially for the gemmologist is described and illustrated.

7

The manufacture of synthetic and imitation stones

Before describing the manufacture of the more important individual synthetic stones in some detail, a brief chronological survey may be attempted. For the reader requiring additional information the excellent and comprehensive book by Kurt Nassau, *Gems made by Man*, is thoroughly recommended.

Rubies were the first gem material to be made in quantity, and have been manufactured on an industrial scale by the remarkably successful Verneuil process since the early years of the twentieth century. Current annual production may be of the order of a thousand million carats. Much of this huge output is absorbed in the manufacture of the tiny 'jewels' used in the bearings of watches, meters, and aircraft instruments, but a good deal of overspill reaches the markets (particularly in the East) as spurious gem material. After a few years, sapphires followed rubies, not only in the traditional blue colour but in yellow, green, orange, pink, and colourless varieties. In the 1920s spinels, carrying an excess of alumina to enable the boules to grow properly, were made by the same fruitful process – mostly in pale shades or in colourless form; they were chiefly used to represent aquamarine quite plausibly, and diamond by skilled promotion. After World War II, the surprise production of star rubies and sapphires was achieved in West Germany and the USA, while later rutile and strontium titanate were made by modifying the Verneuil blowpipe to allow the presence of an excess of oxygen.

Synthetic emeralds were first produced in cuttable sizes by the I.G. Farbenindustrie firm in Germany shortly before World War II, but were first manufactured as a commercial proposition by Carroll Chatham in the USA some years later by crystallizing this most precious of beryls by an undisclosed process. He was followed, a good deal later, by the firm of Linde and, more notably, by Pierre Gilson in France, and others since.

Diamond, after a century of ingenious and persistent endeavour by a number of scientists, was first obtained by a reproducible process by a Swedish concern in 1953, but the first published account of success came from the General Electric Company in Schenectady, USA, in February 1955. In May 1970 scientists of the same great company announced the successful growth of diamonds of high gem quality up to one carat in size. The production of tiny crystals of diamond for industrial use has been a thriving industry for many years, and such material can compete in price with crushed boart. The cost of growing gem-quality crystals is believed, at present, to be relatively high. However, gem-quality yellow synthetic diamonds are grown commercially by the Japanese company, Sumitomo, for industrial purposes and some have been faceted into emerald-cut stones. De Beers have produced gem-quality synthetic diamond in relatively large sizes (10 ct+) for some years but only on an experimental basis.

Quartz is one of the commonest minerals in the world, but clear crystals free from twinning are comparatively rare, and it is these that are in great demand in industry for frequency control. For such purposes synthetic quartz crystals were produced extensively by both sides in World War II. These were of no consequence to gemmologists, but in recent years cobalt-blue, green, brown and, following irradiation, purple (amethyst) varieties of synthetic quartz have been produced in Russia (and elsewhere) and cut as gems, forming one more problem for the gemmologist. Other recent arrivals among man-made gemstones include fluorite, scheelite, alexandrite (including cat's-eyes), turquoise, and (most surprisingly) opal. In addition, there have appeared a whole series of transparent, hard, highly refractive crystals not found as minerals and not primarily intended to compete with natural gemstones. Among these have been double oxides of rare earths having a garnet structure, and for this reason referred to as 'garnets' by the physicists who made them. Best known of these has been the yttrium aluminate popularly known as 'YAG' (yttrium aluminium garnet), which had some success as a substitute for diamond in the mid-1970s. The latest 'diamond substitute' at the time of writing is a synthetically produced cubic form of zirconia, which is an unstable form of the oxide but can be 'stabilized' by the addition of calcium or other oxides. In nature zirconia occurs as the monoclinic mineral baddeleyite.

Of the wide range of synthetic crystals that are hard and transparent enough to make attractive gems, only those likely to be commercially exploited are described in any detail in the following pages, but the names and properties of others of lesser importance are tabulated to enable the gemmologist to identify them should they be encountered.

Before proceeding (in Chapter 9) to discuss the individual synthetic gems and how they may be recognized, it may be helpful to give a summary of the main methods which are now in use for growing materials of gem quality.

Growth from the melt

1. The Verneuil 'flame-fusion' method was first devised by Verneuil at the close of the nineteenth century. Ruby, followed by sapphire and other corundum gems, including star stones, spinels in various colours, rutile, and strontium titanate are made by this well-tried method.
2. The Czochralski method of crystal 'pulling' from a melt is successful for producing large single crystals of scheelite, fluorite, rare-earth 'garnets', etc. Rods of very pure ruby for laser work can also be prepared by this process.
3. The Bridgman–Stockbarger technique, in which a crucible of molten raw material is crystallized by lowering it very gradually into a cooler zone of the furnace, is also used for making scheelite and fluorite.

4. The 'zone-refining' technique is used for refining and crystal growth. A molten 'floating zone', induced by a radio-frequency heater, is passed along a sintered rod or melt and the impurities in the crystal or charge move ahead of the molten zone. Synthetic corundum, scheelite, fluorite, and alexandrite have been grown by this method.
5. A so-called 'skull' melting process, first devised by Russian scientists in 1973, can be brought into action where substances of exceptionally high melting points are involved. It has been used successfully to grow crystals of cubic zirconia by melting the material without the use of a conventional crucible by means of radio-frequency energy, and allowing it to cool slowly.

Growth from solution

1. Growth from a flux, i.e. from a molten solvent, is used successfully in growing euhedral crystals of emerald, ruby, spinel, alexandrite, and special 'garnets'. This technique is often known as the 'flux-fusion' or 'flux-melt' method.
2. Hydrothermal growth, from solution in alkaline and acid water at moderately high temperatures and pressures, is perhaps the nearest man-made approach to natural processes. It is used extensively for quartz, and also for ruby and emerald.
3. High-pressure growth is necessary for the production of diamond grit. High temperatures are also essential to speed the reaction.

Other methods

1. Sedimentation and compaction of artificially formed silica spheres are the processes by which synthetic opal has been made.
2. Ceramic type processes using precipitation, grinding, pressing, and possibly heating of very pure chemicals. The synthetic turquoise made by Gilson comes into this category.

The above summary is chiefly useful in making the gemmologist aware of the many possibilities. Since synthetic stones have essentially the same properties as their natural counterparts, in distinguishing between them great reliance has to be placed on the ability to recognize small internal differences which are usually to be found when the stones are carefully examined with lens or microscope. These differences are due to the very dissimilar conditions under which the crystals were formed. In nature a crystal grows slowly from hot aqueous solutions under pressure or from molten magmas (roughly equivalent to the hydrothermal and the flux-fusion categories listed above), but in the presence of many chemical ingredients which are capable, under the right conditions, of combining to form a number of different minerals. Also, in examining the crystal (or gemstone cut from such a crystal) one generally finds that it has acted as 'host' to tiny samples of other minerals which were being formed at the same time, or to traces of the surrounding fluid in which the crystal was formed. All three 'phases' or 'states' of matter – solid, liquid, and gaseous – are indeed often to be found in a natural crystal when diligently searched. Solid inclusions vary greatly and act as the best clue to the nature and origin of the host mineral. Liquid inclusions almost invariably consist of water containing salt in solution, and gaseous inclusions are commonly carbon dioxide. In the well-known 'three-phase' inclusions which are the hallmark of Colombian emeralds, flat cavities with spiky borders contain aqueous fluid in which a bubble of carbon dioxide and a tiny cube of rock salt can be seen. Under the conditions in which the fluid droplet was originally trapped it must certainly have been homogeneous. As the temperature dropped the several ingredients became immiscible and separated out, forming the three phases described.

Synthetic stones are grown under more chemically 'pure' conditions, and the foreign crystals enclosed are either those of similar chemical nature to the host crystal or are tiny crystals formed by solution from the crucible or container housing the reactive materials, and by subsequent recrystallization within the synthetic stone. Thus the beryllium silicate mineral phenakite is a common inclusion in synthetic emeralds, which are formed of beryllium aluminium silicate and crystals of platinum are commonly seen in the synthetic sapphires and rubies produced by Chatham in platinum-lined crucibles.

An exception to this rule must be admitted where the manufacturer has grown a crystal on the basis of a 'seed' of natural mineral in which the characteristic inclusions are still visible. Carroll Chatham has done this by crystallizing rubies on the basis of a Burma ruby seed, and Lechleitner has succeeded in coating a faceted seed of aquamarine or other gem beryl with a thin overgrowth of synthetic emerald. A Lechleitner overgrowth process has also been successfully used to grow a thin layer of ruby onto a natural corundum seed. Clever ideas of this kind naturally do not make things easier for jewellers or for gemmologists who are trying to protect jewellers, and through them the public.

In what follows the methods for the manufacture of all the main forms of synthetic gemstones now on the market are described in turn, but for convenience and keeping chapters shorter than in earlier editions the features which enable each to be distinguished from their natural counterpart are reserved for Chapter 9.

Early forms of synthetic ruby

The first artificially produced gemstones to reach the market were the rubies made by the French chemist Frémy, either working on his own or in collaboration with Feil and with Verneuil. In 1877 he succeeded in producing small, violet-red rhombohedral crystals of ruby, lining a clay crucible in which a mixture of alumina, potassium carbonate, barium fluoride, and some potassium dichromate had been fused together for some eight days. Tiny cut stones were, in fact, prepared from these, and some of the pretty little crystals were mounted as such for use in jewellery, thus anticipating a modern fashion.

A much more serious impact was made on the trade by rubies in sizes up to 1 or 2 carats, which made a sudden and mysterious appearance in 1885, and were first known as 'Geneva' rubies from their apparent place of origin. These had the chemical and physical properties of natural stones, and were first accepted as such. Later they were thought to have been made by fusing together small fragments of natural ruby, with the addition of potassium dichromate to improve the colour, and the term 'reconstructed' ruby came to be applied to these, and to similar products which appeared in several centres, including London. For those interested, a specific account of one such process is to be found on pp. 219–20 of Wodiska's *A Book of Precious Stones* (1909).

Small, button-like boules, some with a 'stalk' of natural Thai ruby (see Figure 9.1), and cut stones derived from these, exist today in some museums and in private collections, and lend colour to the 'reconstructed' story. Stones of this type, and the scanty available literature, have recently been critically examined by K. Nassau of the Bell Telephone Laboratories, and G. R. Crowningshield of the Gem Trade Laboratory in New York, and these workers are convinced that the 'reconstruction' process is not feasible with ruby, and that these early stones were produced by a primitive form of flame-fusion, using not ruby fragments but powdered alumina, though perhaps with natural ruby chips as seeds to induce crystallization.

The differences between these primitive stones and the present-day Verneuil synthetics which are described later are interesting and obvious to the keen gemmologist, but need not concern the jeweller or dealer, who can safely classify both types as synthetic rubies and can drop the unsatisfactory term 'reconstructed' for ever.

The Verneuil process

All previous methods for making synthetic rubies were superseded, and the era of commercial synthetic gemstones began, when the French scientist Verneuil designed his special furnace (Figure 7.1), which incorporated an inverted oxyhydrogen blowpipe or 'chalumeau'. The Verneuil process has since been successfully used not only in the preparation of ruby and sapphire corundums of many other colours but also of spinel, rutile, and strontium titanate, as will be described below.

Certain of the features, such as curved growth lines and included gas bubbles, which we rely upon for distinguishing Verneuil corundums from their natural counterparts, arise essentially from the nature of the process by which they are grown. It may therefore be useful to give here a brief description of the essentials of the Verneuil method.

Purity of the solid raw materials and of the oxygen and hydrogen used in the blowpipe flame is most important. Alumina, the essential constituent of corundum, is derived from ammonium alum, which is a double sulphate of ammonium and aluminium containing water of crystallization, crystallizing in octahedra. The alum is recrystallized to ensure purity, and later calcined in large crucibles at 1100°C. The alum decomposes with evolution of ammonia, sulphur dioxide, and water vapour, leaving a residue of pure alumina in the unstable 'gamma' form, as a very fine powder. When ruby is required, up to 8 per cent chromic oxide is added to the original batch of alum before calcination, and the calcined product then has a pale green colour. For blue sapphire, oxides of iron and titanium are added to the alum; for yellow sapphire, nickel oxide; for 'alexandrite'-coloured corundum, vanadium oxide is added, and so on. The powder is placed in a sieve at the top of the furnace and, after the gases have been lit, this is periodically tapped with a small hammer, which results in small controlled quantities of powder falling through the flame, where the powder melts and falls as white-hot droplets on to an inch-wide pipeclay pedestal beneath the flame, which is enclosed in a circular chamber. A conical mass of small corundum crystals is first formed. The flame is then so adjusted that a small single rod begins to grow in the centre of the mass, gradually swelling as

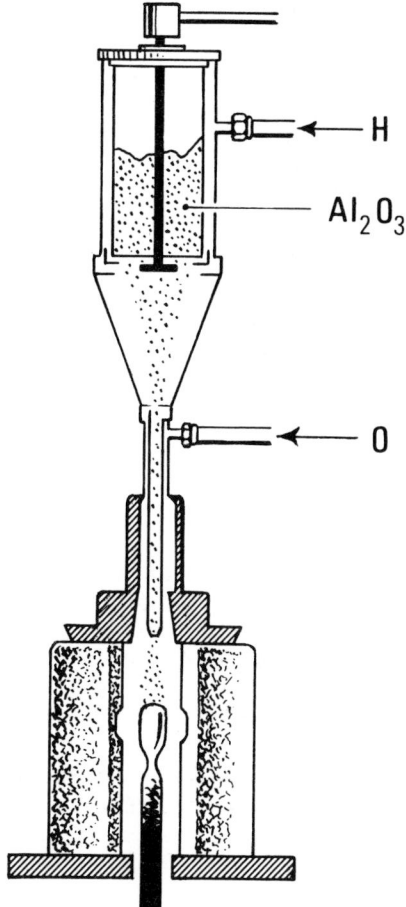

H

Al_2O_3

O

Figure 7.1 Diagram of a Verneuil furnace

the supply of powder is increased into the familiar boule form – a domed cylinder of some 190 mm (¾ in) width, tapering to a narrow base. Corundum boules have a strong tendency to split lengthways, and are always so split, if necessary by nipping them with pliers, to release the strain before they are cut or fashioned for use as gems or in industry.

Stones analogous to natural spinel are also grown in similar fashion, the raw materials here being alumina and magnesia, with cobalt oxide added for the popular shades of blue, manganese oxide for pale green stones, and iron oxide for those of pale pink shade. It has been found that where the ratio of alumina to magnesia used is 1:1, as in natural spinel, the boules do not grow successfully. The most favourable ratio is 3½ alumina to 1 magnesia. The resulting boule consists of a mixed crystal of spinel and 'gamma' alumina, with which it is isomorphous. The tendency of the cubic 'gamma' form of alumina to revert to the stable trigonal 'alpha' form (corundum) gives rise to strain within these synthetic spinels which, in consequence, always show an anomalous birefringence when examined between crossed polarizers.

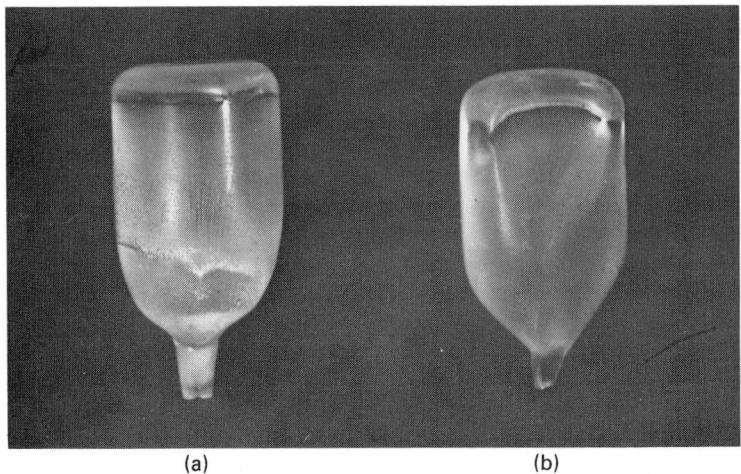

(a)   (b)

Figure 7.2 (a) Corundum and (b) spinel boules

Typical examples of corundum and spinel boules are shown in Figure 7.2. Whereas corundum boules do not commonly show any crystal form, spinel boules have flattened sides which correspond to crude crystal faces of the cubic system. The upper surfaces of the boules are frosted with tiny crystals formed during the relatively rapid cooling which takes place when the furnace gases are switched off. These, too, differ in the two cases. In corundum boules the surface, when examined with a lens, will be seen to consist of tiny overlapping rhombohedral platelets, rather resembling tiles on a roof, whereas on spinel boules are seen chains of linked octahedra intersecting at an angle of 90°.

Synthetic star corundum

In 1947 there was released a series of star sapphires and star rubies, which were made by preparing boules with a small proportion of titanium in addition to the usual colouring agents, and annealing these for periods of two to 72 hours at temperatures ranging from 1100 to 1500°C. Under this treatment the titania crystallized out as short, fine crystals of rutile, oriented at angles of 120°, in accordance with the structure of the corundum crystal lattice. Stones cut *en cabochon* with the axis of the cabochon coincident with the optic axis of the corundum showed a brilliant six-rayed star by reflected light. Thus the Verneuil process surprisingly entered a field where it was thought that Nature must remain unchallenged.

Synthetic spinel

As mentioned above, synthetic spinels of a kind are also made by the Verneuil process. These were originally made accidentally in the early years of the twentieth century in an endeavour to colour synthetic corundum blue by addition of cobalt, using magnesia as a flux. However, it was not until the mid-1920s that synthetic spinels were manufactured on a large scale. They are not made in colours associated with natural spinel, but usually in tints suitable for counterfeiting blue zircon, aquamarine, alexandrite, sapphire, or even diamond. All these stones have quite different properties from spinel, and thus, however convincing their appearance may be, distinction is easy by a number of orthodox tests, and the microscope is seldom called into play.

Synthetic rutile

Another product of the Verneuil flame-fusion method is that most spectacular of all synthetic stones – rutile. This caused a considerable stir on its first appearance in 1949, but has not been widely welcomed as a gem material, and has far less commercial importance than synthetic corundum, spinel, or even emerald. Rutile is well known to the gemmologist as needle-like crystals included in ruby, and as larger needles in quartz – the so-called 'flèches d'amour'. When found thus it has a typical reddish-brown colour, and crystals are seldom found in nature sufficiently transparent to warrant their being cut as gemstones.

Synthetic rutile boules are made in a special modification of the Verneuil furnace incorporating an additional outer jet of oxygen; this lessens the tendency to lose oxygen which this compound undergoes at high temperatures, with consequent darkening of colour. As it is, the boules are completely black and opaque when they leave the furnace, and have to be treated by heating to around 1000°C in a stream of oxygen before they have any ornamental value. During this treatment a variety of colours may be induced, rather unpredictably; but the makers' goal of a completely colourless rutile has not yet been achieved, a decidedly yellowish cast always remaining. An attempt to improve the 'whiteness' by adding strontium oxide has been partially successful, and stones which have been coated with a film of corundum to improve their surface hardness also seem to have a better appearance. Red, orange, blue, and brown rutiles have been produced by varying the aftertreatment of the boules, and provide stones of spectacular beauty far preferable, in the author's judgement, to the off-white types which seem to be the only ones marketed – presumably in a vain attempt to rival diamond.

Strontium titanate

Another product of the special Verneuil furnace in which an extra concentric oxygen pipe is added outside the hydrogen pipe is strontium titanate, only found in nature as a mineral (tausonite) in 1984. As with rutile, the boule is black when it leaves the furnace and must be annealed in oxygen to attain its desired colourless state.

Strontium titanate (often sold and referred to under the fancy name of 'fabulite') approaches more closely to diamond in its appearance and its optical properties than any other material, natural or synthetic.

Czochralski 'pulling' method

In this technique, which produces crystals of very high quality, the raw material, e.g. pure alumina (Al_2O_3) for colourless sapphire, is melted in a crucible made of high melting-point metal. Iridium is normally used for sapphire since platinum melts at 1774°C, which is below the melting point of corundum (Al_2O_3) at 2050°C.

However, platinum crucibles are used if the melting point of the material to be grown is below that of platinum. Materials grown by 'pulling' include ruby, sapphire, YAG, GGG, and other 'synthetic' garnets, lithium niobate ('Linobate'), and alexandrite. The melts are usually composed of oxides or mixtures of appropriate composition. The heating is provided by radio-frequency coils surrounding the crucible. The temperature has to be controlled very accurately at a point just above the melting point of the material to be grown. A small seed of sapphire (or other 'to be grown' material) on the end of a rod is touched onto the surface of the melt. If conditions are right more sapphire will solidify on the rotating seed which can be gradually raised from the melt – the seed is being 'pulled' from the melt. A simplified diagram of the apparatus is shown in Figure 7.3, but the process is usually more complex with heat shields around the growing crystal, with growth being conducted in an inert gas atmosphere.

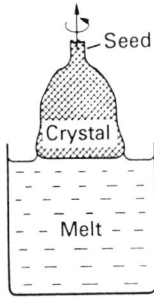

Figure 7.3 Diagram showing Czochralski growth or pulling from the melt

Zone-refining technique

The 'zone-refining' or 'zone-growth' method has been used to refine existing crystals or to grow new ones. A horizontal form of the apparatus uses a boat-shaped crucible which contains feed material of appropriate composition. A 'zone' is melted at one end, using radio-frequency coil heaters, and this is passed along the crucible by suitable mechanical means (see Figure 7.4). A single

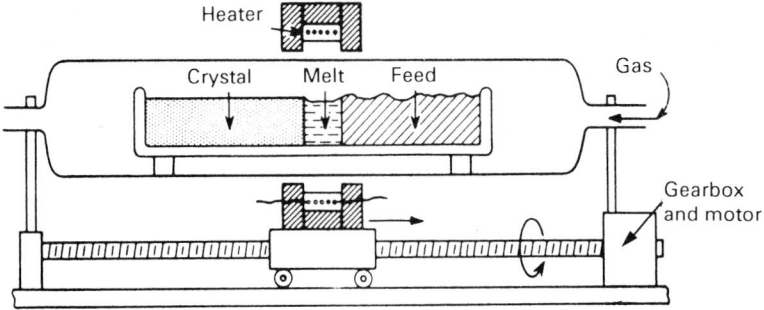

Figure 7.4 Diagram illustrating the 'zone-refining' technique (after Nassau)

crystal may be formed in this way. In the vertical form of the method a sintered rod is fixed top and bottom and a 'floating zone', induced by radio-frequency coils, is passed slowly along the rod. Impurities in the rod are carried along, in front of the melted zone, to the end of the rod and can be removed after cooling. Synthetic corundum, scheelite, fluorite and alexandrite have been grown or refined by this method.

Skull-melting process

At the time of writing (1989) the most favoured of the synthetically produced substitutes for diamond is the cubic form of zirconium oxide (zirconia). Since the melting point of cubic zirconia is high (2750°C) there are no refractory materials available in which it can be melted. In the early 1970s the Russians devised an ingenious method whereby the cubic zirconia is fused in a container or 'skull' of its own substance. An outer shell compound of separated copper fingers, through which water is passed, forms a cooled outer container (see Figure 7.5)

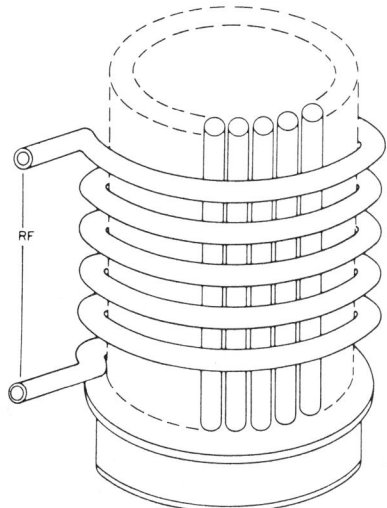

Figure 7.5 Diagram showing the skull-melting apparatus used for the growth of cubic zirconia (after Nassau)

into which zirconia powder is placed plus some zirconium metal to assist conductivity and help the initial melting. Radio frequency is used to heat the apparatus and the fused charge is retained by the 'skull' of solid zirconia in contact with the cooled copper fingers. After the initial melting more zirconia and a stabilizer is added and the charge held molten for some hours to ensure uniformity and to boil off impurities. The melt is cooled slowly and a series of columnar crystals grow upwards from the base. After annealing for some hours to remove strain the crystals are then ready for cutting.

Flux-melt growth of emerald

It may be noted that most of the synthetic gem materials described above are oxides. Generally, *silicate* minerals cannot be manufactured by the Verneuil process, since this demands the power to crystallize rapidly on cooling from the molten state, and most silicates when quickly cooled congeal into a non-crystalline glass.

For the record, it must be admitted that it has been proved possible to make emerald of a kind by the Verneuil process – but since all the synthetic emeralds used in commerce have been manufactured by processes involving slow crystallization from a melt or solution, it is important to clear the mind of all thought of the Verneuil synthetic features of curved growth lines and bubbles in considering how to differentiate between natural and synthetic emeralds.

The artificial production of emerald crystals in sizes large enough to be cut as gemstones was first achieved in the 1930s by Dr H. Espig and his colleagues. These scientists were working in the research laboratories of the giant German dyestuff and chemical concern, the I.G. Farbenindustrie.

The process, as later described by Dr Espig, had a basic similarity with that used in early experimental work by Hautefeuille and Perrey, in 1888. In Espig's process oxides of beryllium and aluminium are dissolved in fused acid lithium molybdate in a large platinum crucible. Floating on the molten mixture are slabs of silica glass, below which is a platinum sieve. Small seed crystals of synthetic emerald rest on the sieve, and act as nuclei for the growth of single crystals of emerald in the region where the beryllium and aluminium oxides react with the silica to form beryllium aluminium silicate (beryl). Some chromium salt must also be present to give the desired emerald-green colour.

The optimum temperature is 800°C, and the crucible must be maintained at that temperature for many months if crystals of useful size are to be grown. The melt must be replenished at intervals with further supplies of beryllium and aluminium oxides. This is done by means of a platinum funnel permanently fixed in the crucible and reaching nearly to its base (see Figure 7.6).

'Igmeralds', as these stones were known, were produced in some quantity but never reached the market: negotiations to that end were interrupted by the war.

Since the war the most serious producer of synthetic emeralds has been the San Francisco chemist, Carroll F. Chatham. His method is a secret one, and doubtless an improvement on the I.G. Farbenindustrie process, which had hardly proved economic. The internal features, appearance, and physical properties of Igmerald and Chatham emerald are so similar that their methods of growth must have much in common.

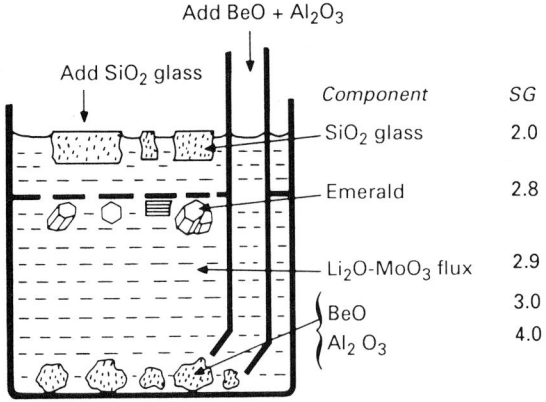

Add BeO + Al_2O_3

Add SiO_2 glass

Component	SG
SiO_2 glass	2.0
Emerald	2.8
Li_2O-MoO_3 flux	2.9
BeO	3.0
Al_2O_3	4.0

Figure 7.6 Diagram of the I. G. Farben process for the flux-melt growth of synthetic emerald

The emerald crystals made by C. F. Chatham show a simple, rather squat prismatic habit with basal plane, very similar to those found in nature, though there is a tendency for a number of parallel individuals to amalgamate, often leaving cavities of hexagonal outline in the direction of the main crystal axis. A large crystal made by Chatham, of 1275 carats, is in the Harvard Museum. This and several other large crystals are not of gem quality. Cut stones of over 6 carats are seldom seen, but specimens of 1–3 carats are now marketed in large numbers.

Of the other synthetic emeralds which have made their appearance, the most successful seems to have been the Gilson emerald, manufactured by a firm primarily concerned with ceramics. The process is clearly a flux-fusion one, large crystals being developed initially on the basis of crystal seeds. Some newly grown emerald is used subsequently as seed material. Gilson's process has now been taken over by Turgil SA of Geneva.

The fluxes used in these processes are probably composed of lithium molybdate possibly with lead or vanadium oxides. Photographs of Gilson synthetic emeralds being removed from the growth furnace indicate that the process may operate at a red heat. A schematic diagram (after Nassau) of the Gilson process is shown in Figure 7.7. Other flux-fusion processes are probably similar in principle.

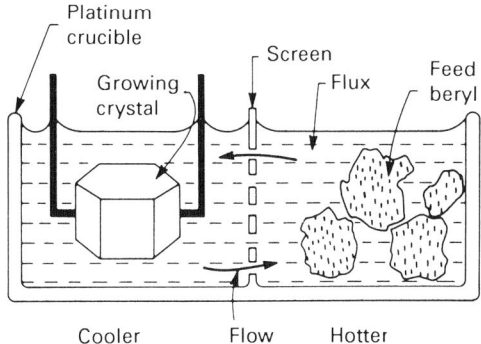

Figure 7.7 Diagram showing the flux-transport growth of synthetic emerald used by Gilson

Flux-melt emeralds have now been made by several manufacturers, including Inamori (Kyocera), Lennix, the Russians, and Zerfass; others will surely follow. Most of these have features in common, but there are some individual characteristics and these will be discussed in Chapter 9.

The flux-melt processes described above have been concerned with the manufacture of emerald, but other gem materials including ruby and sapphire, chrysoberyl and alexandrite, and various coloured spinels and rare-earth 'garnets' may also be made in a similar way. Properties of some of these stones will be discussed in Chapter 9.

Hydrothermal growth

Lechleitner synthetic emerald-coated beryls

Early in 1960 came news of an entirely different form of synthetic emerald, for which the name 'Emerita' was suggested. These stones were made in Innsbruck,

Austria, by a process perfected by Johann Lechleitner. In this, faceted or cabochon pieces of natural beryl too pale in colour to be valuable are placed in a hydrothermal bath containing the appropriate ingredients, and the beryl seeds become coated with a thin layer of synthetic emerald in crystallographic continuity with the underlying stone. All that remains is for the matt-coated facets to be lightly polished.

Synthetic emeralds of quite another type were produced by the ingenious Mr Leichleitner some years later (1964). These were made by inserting a thin plate of natural or synthetic colourless beryl into an autoclave and 'plating' this with thin layers of dark green synthetic emerald. The stone was then enlarged by placing it in another autoclave and growing further layers of colourless beryl (which is deposited much more rapidly than the green). The stone when finally faceted has the appearance of normal emerald when viewed from the front, but when immersed in liquid and looked at edge-on under the microscope it reveals its banded structure in a startling manner.

Around 1965 the Linde Company in the USA began to market hydrothermal emeralds of a very fine colour. Their success stemmed from the discovery that to keep the essential chromium in the growing emeralds it was necessary to use acid solutions instead of the alkaline ones which had been used in previous syntheses such as that of quartz. The whole operation takes place in an autoclave with temperatures in the 500-600°C range and pressures of $10\,000$–$20\,000\,\mathrm{lb/in^2}$ (700–1400 bar). The nutrients in the process might be gibbsite $Al(OH)_3$ to provide aluminium, beryllium hydroxide $Be(OH)_2$ for beryllium, quartz for SiO_2 with chromium chloride supplying the colouring agent, chromium. There may be a temperature difference of up to 25°C between the bottom of the autoclave and the top, which is cooler. The emerald is deposited on an initial seed of natural beryl. For economical growth rates it may be necessary to grow the crystals in several stages; the whole process may take weeks. The excellent schematic diagram (after Nassau) in Figure 7.8 compares the hydrothermal and flux-melt growth of emerald.

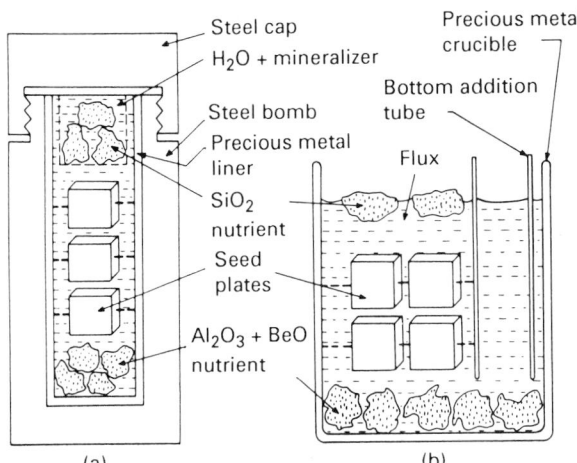

Figure 7.8 Schematic diagram comparing (a) hydrothermal growth and (b) flux-melt growth for synthetic emerald (after Nassau)

In 1978 the Linde process was sold to Vacuum Ventures Inc., who now market Regency-created emeralds. The Russians and Australians (Biron and Pool are the trade names) also produce hydrothermal emeralds.

Synthetic quartz

The so-called piezoelectric properties of quartz crystals enables thin, carefully oriented slices of the mineral to be used, for example, in the electronics industry for the exact control of broadcasting wavelengths, and in the accurate regulation of time-keeping devices.

Although quartz is the commonest crystalline mineral in nature, pure, untwinned crystals suitable for piezoelectric work are relatively scarce, so in many countries quartz crystals are being grown synthetically by a controlled hydrothermal process as an industrial necessity. In this process, a silica-rich alkaline aqueous solution is enclosed in a strong, steel-walled autoclave or 'bomb', which contains in its base chips of quartz as nutrient material, and in the upper part, racks from which are suspended untwinned slices of quartz to act as seeds on which crystal growth can proceed. These seeds are carefully oriented to produce optimum growth in the required directions. The base of the autoclave is heated electrically and held at 400°C, whereas at the top of the vessel the

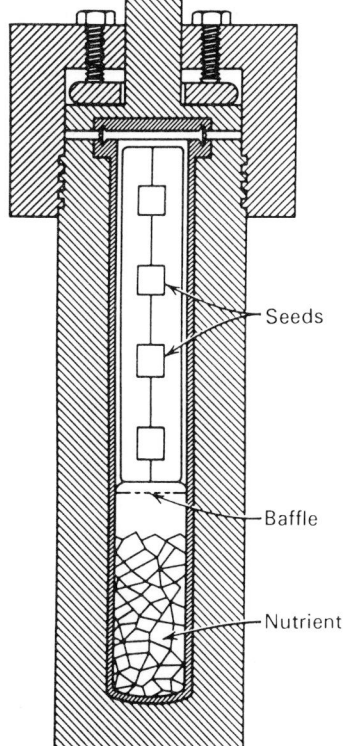

Seeds

Baffle

Nutrient

Figure 7.9 Diagram showing a small autoclave (about 35 cm long) used for the laboratory growth of synthetic quartz. Larger autoclaves are used in industry (after Nassau)

temperature is some 40°C cooler. The saturated solution rises slowly from the base, becomes supersaturated in the cooler portion, and thus crystallizes on the seeds provided (see Figure 7.9).

Colourless quartz is of small value for cut gemstones, and natural supplies are cheaper than factory-made crystals in this instance. In recent years, however, coloured synthetic quartzes have been manufactured in Russia and in the USA. First to appear were green and also blue types of colours not found in nature. The latter were coloured by cobalt and showed the usual cobalt spectrum of three broad absorption bands in the orange, yellow, and green – very much like those seen in blue cobalt glass. It is always worth remembering that except in rare instances a cobalt spectrum of the type seen in synthetic spinel and glass is not seen in any natural mineral used as a gemstone, and therefore signifies some sort of fake. Synthetic citrine and even amethyst have also been reported and the latter is now present on the gemstone market in large quantities. Small particles resembling breadcrumbs have sometimes been seen in synthetic quartzes and parallel banding near the flat seed plate around which they are grown; but suspicion is more likely to be roused by an absence in the synthetic stones of any of the many forms of natural inclusions to be found in the quartz minerals. For inclusion-free material tests have been devised to identify synthetic amethyst based upon the characteristic internal twinning of the natural stone.

Synthetic diamond

For more than a hundred years scientists had striven by all kinds of ingenious means to make diamonds from other forms of carbon before the first undoubted success could be claimed. The news was released in February 1955 by the General Electric Company of America: a team of workers in their research laboratories in Schenectady had subjected carbonaceous materials in a small pyrophyllite pressure chamber to pressures of over 100 000 atmospheres and temperatures of more than 2760°C, and under these conditions tiny crystals of diamond had formed very rapidly. The presence of nickel or other metal catalysts was found to aid the transition of graphite into diamond, and up to 0.2 per cent of nickel was found in early samples of GEC diamond grit, which was, as a result, distinctly magnetic.

Years of research on the physics and production of high pressures, notably by Professor P. W. Bridgman in the USA, had paved the way for this breakthrough in diamond synthesis. As often happens, once the possibility of an almost incredible feat has been demonstrated (climbing Everest, flying the Atlantic, the four-minute mile) the first success is rapidly followed by many others, thus making the whole adventure seem almost commonplace. Thus it is not perhaps surprising that the GEC triumph was but the first of many others in which, though similar principles were naturally involved, there was considerable variation in detail. Diamonds have now been produced artificially on a considerable scale, not only by the GEC and several other important research groups in America, but also by the large electrical concern ASEA in Sweden, Philips in Eindhoven, De Beers in Johannesburg and Shannon, the Shibaura Electric Co. in Tokyo, the USSR Academy of Sciences, and the National Physical Laboratory in the UK.

Much has been learned about the conditions under which carbon can be transformed into diamond, and the habit and even the colour of the tiny diamond crystals produced can to some extent be controlled. At first these were crude in shape and contained many skeletal forms, but now quite well-formed combinations of octahedra, cubes, and dodecahedra are to be observed. Nevertheless, the processes by which diamonds were formed in nature are still imperfectly understood.

Synthetic diamond are now produced commercially for industrial uses and can to some extent be 'tailor made' to suit certain purposes. Early in 1970 the General Electric Company of America announced that gem-quality diamond crystals of up to 1 carat in size had been produced in a variety of colours, subsequently De Beers and the Sumitomo company of Japan have grown yellow, green, and brown diamond crystals of gem quality and faceted stones have been cut from them. Those crystals so far manufactured have been found to have distinctive properties in electroconductivity, fluorescence, and infra-red spectroscopy.

Synthetic opal

Probably Pierre Gilson's most spectacular triumph has been in the production of synthetic opals of outstanding beauty. In 1964 the electron microscope revealed the structure of precious opal to be essentially closely packed spheres of amorphous silica forming a three-dimensional diffraction grating; when the spheres are of suitable dimensions, this gives rise to the iridescent colours for which opal is so famous. Since 1964 many attempts have been made to form deposits of silica of this nature. Some success was achieved, but the problem of stabilizing the structures and forming a stone that could be cut and used in jewellery was first solved by Gilson's genius, and since 1974 his products, which comprise both white and black opals, have been commercially available. Whenever a new synthetic gemstone appears on the market there follows a spate of papers in the gemmological journals describing the properties of the new material and comparing them with the natural mineral. These papers are essential reading for the keen gemmologist, as they are able to convey vital information a year or more in advance of any textbook on the subject. Moreover, ambitious manufacturers of synthetic gemstones are seldom content to repeat in every detail the process that brought them their initial success. Experiments continue in the hope of obtaining an improved product, perhaps one that will be less easy for the gemmologist to recognize as a man-made stone. Continual vigilance is therefore necessary, especially since 'plastic opals' have now arrived on the gemmological scene; these simulants are described in Chapter 9.

8

Gemstone enhancement

Before proceeding to the identification of the individual synthetic stones we must consider the various methods of treating natural (and synthetic) materials to improve their appearance. Some of these methods date back more than 4000 years, but since the turn of this century and particularly following the explosion in scientific technology associated with man's excursions into space, there has been a great expansion in methods for synthesizing and enhancing the appearance of gemstones. The principal enhancement methods are:

1. Heat treatments;
2. Irradiation treatments;
3. Chemical treatments, including dyeing, impregnating, bleaching, and surface modifications.

Heat treatment

Much, if not most, of the reddish chalcedony (carnelian) or agate encountered today has been heat treated, and this process dates back to around 2000 BC in India and elsewhere. Heat treatments known in the nineteeth century, or before, included the changing of greenish beryl to blue aquamarine, orange topaz to pink, and amethyst to yellow citrine. Sapphire, tourmaline, zircon, beryl, and other stones were also treated, sometimes with significant colour changes (e.g. reddish-brown to pale blue zircon) or in other cases from dull colours to brighter ones (e.g. bluish-green tourmaline to bright green). Precise temperatures were rarely recorded and stones were enclosed in crude crucibles sometimes enveloped in 'inert' powders such as sand or magnesia. Since the 1970s heat treatments have been increasingly reported in the literature, but there has probably been a gradual increase in treatments and their sophistication over many decades. Techniques for the improvement of the corundum family have made great strides, especially in the heating of oily-looking, greyish sapphires ('geuda') from Sri Lanka which had previously been discarded as waste. The

Thais discovered that this material could be heated at very high temperatures (±1800°C) to produce fine blue sapphires. It was some time before the Sri Lankans discovered the value of their 'waste' material, and they were not overjoyed to realize that much potential blue sapphire had been sold for very small sums.

Although the nature of the changes which take place on heating some stones are now reasonably well understood, there are still many processes where further research needs to be carried out. Many aspects of these changes are quite complex and it is not proposed to deal with them in this book, nor shall we discuss the precise details of the processes. The interested reader is referred to Kurt Nassau's comprehensive book *Gemstone Enhancement*. The results and tell-tale signs of heat treatment are, however, of great importance to the gemmologist, who needs to know what has and has not been treated. These signs (which are not always obvious) will be dealt with in some detail when individual gemstone species are described later.

During the 1940s scientists of the Linde Company in the USA developed methods of inducing asterism and colour into corundum by a diffusion process. The stones are heated for periods varying from 4 to 24 hours, embedded in powder composed of roughly one part of titania (TiO_2) to two parts of alumina (Al_2O_3). Diffusion of Ti into the corundum results in the formation of needles of TiO_2, which produces star stones. The addition of chromium compounds to the powder induces red colours and other compounds may enhance the blues of sapphires. The powders may be replaced by a painted-on slurry and the introduction of other elements results in the production of other colours. These colours are only 'skin deep', and therefore the treatment must be carried out on either preformed or pre-faceted material, and careful polishing is necessary to ensure that the coloured layer is not removed.

Diffusion at high temperatures may also be used to reduce the sharpness of the banding in corundums grown by the Verneuil method. The varying concentrations of the impurities, which form the colour bands, diffuse at these high temperatures and the banding is much more difficult to detect.

Amber is another material which is commonly modified by heating. There are many different shades of amber (in the colour sense), and to many people the darker-coloured ambers are the more desirable – deeper colours being equated with greater age in the eyes of many beholders. By heating amber in air or sand at temperatures below about 150–200°C it is possible to darken the skin of a specimen, to darken it throughout, or, on some occasions, to produce a darkened skin but a much paler (sometimes almost colourless) interior. These treatments need to be carried out with great care.

Much amber contains myriads of tiny bubbles which may cause cloudiness or even total opacity. By immersion in rapeseed oil (the classical liquid) or linseed, cottonseed, or even specially formulated lubricating oil, it is possible, by very gradual heating and cooling, to remove these bubbles, perhaps by a diffusion process of the air in the bubble to the outside. Such treatment may induce strain in the amber and produce 'strain discs' or 'sun spangles'.

By heating small fragments of poorer-quality amber at temperatures around 180°C and with pressures of 5000–120 000 lb/in² it is possible to fuse the pieces together. Forcing the fused melt through steel plates with narrow slits or holes in them can produce a homogeneous melt of clear amber – the so-called 'pressed amber'. Dyes may be introduced during this process and blue, green, and bright-red ambers are sometimes encountered which have been treated in this way.

Irradiation

The usual purpose of irradiation is to change the colour of a gemstone and thus to improve its visual appearance and value. The process is carried out by exposing the gem material to one of a variety of rays (visible or ultra-violet light, X-rays, or gamma rays) or energetic particles such as electrons, neutrons, protons, or alpha particles. Some stones may change colour simply by exposure to light after mining (e.g. some Burmese or Russian brownish topaz, which becomes almost colourless eventually). Other stones may change on exposure to ultra-violet light, but for most permanent colour changes more energetic radiation such as X- or gamma rays are necessary. X-rays are routinely produced in many everyday applications, but this energy is usually low and the penetration is only 'skin deep' or is rather localized. Colours induced by X-rays are usually short-lived – they fade back, sometimes quite slowly in normal daylight or very fast under sunlight or a desk lamp (e.g. brown topaz and some yellow sapphire). They are not, therefore, usually used for changing colours.

Gamma rays are commonly used and are usually produced from Co-60, an isotope of cobalt (normal cobalt, Co-59, has a nuclear mass of 59 units). Depending upon the energy and size of the treatment plant, it may take from half an hour to a few minutes to turn rock crystal into brown smoky quartz. Colourless topaz has been treated with Co-60, but the ultimate blue colours depend both on the initial radiation dose and the subsequent heat treatment.

Alpha and beta rays are energetic helium nuclei and energetic electrons or cathode rays, respectively. Electrons are produced in television tubes where they are accelerated by electric fields produced by high voltages. High-energy electrons are produced in linear accelerators which are very expensive to build and to maintain. Electrons are not ideal, since they are strongly absorbed and only produce surface coloration. They also result in localized heating, and it is necessary to cool stones in running water to prevent cracking. Other particles such as protons, deuterons, and alpha particles have been used for irradiation, but they may have high energy and can induce radioactivity. Around the turn of the twentieth century diamonds and sapphires were irradiated by alpha particles by burial in radium salts, but the decay products from the radium can be absorbed by the stones and some stones were left with hazardous residual radioactivity.

Neutrons, which are uncharged particles, are produced in nuclear reactors or 'atomic piles'. They may have considerable energy and are very penetrating. Colour changes induced by neutrons are, therefore, very uniform, but they may induce appreciable radioactivity and a 'cooling-off' period may be necessary before the stones are safe to wear. Nevertheless, some blue topaz is produced by neutron irradiation. From the various irradiation agents mentioned above, gamma rays are clearly to be preferred – they give uniform colouring, do not require electrical power, do not produce localized heating, and do not induce radioactivity.

Natural irradiation over periods of geological time has been responsible for the colouring of many natural minerals. Radioactive elements in the earth's crust (produced at the creation of the universe) have remained active because of their very long decay periods, and have provided the radiation for colour changes in many gemstones. Minerals such as Sri Lankan zircons, which may contain significant amounts of uranium and thorium, have been both coloured and become metamict by natural irradiation.

Research into the nature of colour centres is being actively pursued, and there is still much to be elucidated. However, most of the colour changes are already known to involve 'colour centres'. To explain the detailed nature of colour centres is beyond the scope of this book, and the interested reader is referred to Kurt Nassau's excellent book *Gemstone Enhancement*. However, to state the matter briefly, 'colour centres' involve the absence of one electron from a normally occupied position in the atomic structure of the mineral, giving rise to a 'hole colour centre'; or the presence of an extra electron which produces an 'electron colour centre'. By irradiating rock crystal to form smoky quartz, a 'hole colour centre' is formed, and this may be removed or destroyed by heating to produce rock crystal once again. Natural smoky quartz has resulted from the radioactivity of the surrounding rocks, which may be derived from elements such as uranium or thorium, but might also be from one of the isotopes of potassium, K-40, which emits beta and gamma rays, and although it forms only a very small proportion of all the potassium isotopes, the overall potassium content of rocks is large, and the effect of the K-40 proportion may be considerable in aggregate.

Chemical and other treatments

The most universal treatment of gemstones is the initial cleaning or preparation of the raw material for fashioning, and similar processes are likely to be repeated at intervals during the routine cleaning and repair of jewellery. It is important that the jeweller or gemmologist is aware of which processes are safe and of those which may incur the risk of damage. This section outlines many of the chemical treatments available and, it is hoped, will enable costly mistakes to be avoided.

Cleaning

This is a simple process which may involve nothing more than soap or detergent and water. If this does not produce the desired result there is the temptation to use dilute or even concentrated acids, but care must be exercised, since even vinegar (acetic acid in part) will mark polished coral or marble surfaces. Strong acids may severely etch some gem materials and trials should be carried out on an inconspicuous part or on a loose fragment, preferably under a lens or microscope using a fine-pointed dropper. A very small scraping in a small test tube will readily react to a drop of acid and no harm is done. Organic solvents such as alcohol or acetone are fine for cleaning rubies and sapphires, but could have disastrous effects upon plastic coatings or oil-impregnated emeralds. An awareness of the worst possibilities is very desirable. Ultrasonic cleaners can be very effective but can readily facilitate cleavage in gemstones such as topaz and tanzanite and are disastrously effective at removing oil from treated emeralds.

Bleaching

Household bleaching agents are seldom needed in connection with faceted gemstones, but they have been used on organic substances such as coral, pearl, and ivory, where the bleaching agent works on the organic content (conchiolin) of the material. In all cases, dilute solutions should be tried first and then on inconspicuous parts; strong solutions have been known to cause cracking in ivory. Oxalic acid solution is often used to 'bleach out' iron stains on mineral specimens; the solution may need to be warmed and the process may take days or even weeks. The solution, made up from crystals, is poisonous and needs to be treated with care.

Colourless impregnations

Various colourless oils, waxes, and plastics may be applied to the surfaces of gem materials; the depths of penetration and the effects are variable. The application of a well-known brand of 'baby-head' oil to recently mined rough opal gives an impression of high polish and hides the untreated surface; the oil also persists in hot climates where water would quickly evaporate. Smoothed lapis lazuli beads are often dipped in molten paraffin wax which readily penetrates between the constituent mineral grains, produces a deeper, richer colour, and dries to a 'polished' surface. Similar effects are produced with poor-quality turquoise, where the wax or resin improves the colour by reducing light scattering at the porous surface. Some turquoise and opal are improved (stabilized) by plastic impregnation, and in other instances the raw turquoise is ground up and the plastic used as a cement to reconstitute it. This often 'improves' the colour, and these treated turquoises will commonly give a much clearer or stronger absorption spectrum. Another reason for oiling or waxing objects is to hide or fill up cracks or other surface irregularities and to improve the polish. Waxing is commonly employed in this way for carvings in jade and its simulants, soapstone, and marble.

Cracks in crystalline materials which may penetrate quite deeply, as in emeralds, are commonly filled with a variety of liquid oils and resins. These fill up the cracks and greatly improve the optical continuity and, therefore, the transparency within the material; its value, to the casual buyer, is obviously enhanced. However, there may be many problems with oiled stones; even gentle heating may drive out the oil and household detergents can extract the oil if rings are not removed when placing the hands in water. The use of relatively low melting-point resins, such as Canada balsam and 'Opticon', provides a much more durable filling for flawed emeralds, and with the higher RI (1.53-4) almost matching that of emerald, it is much more difficult to detect than liquid oils. In addition, the balsam is less easily removed by ultrasonic cleaning – the great enemy of oiled emeralds.

Coloured impregnations

The colouring of oil used to impregnate emeralds and rubies has been known for over a thousand years. This type of treatment can provide very convincing optical continuity, and flawed stones are greatly, but fraudulently, improved in appearance. However, the dyes may fade with time and evaporation may unevenly concentrate the colouring. Gentle warming may cause the oils to 'bleed' to the surface and wiping with a cloth may reveal the coloration. Immersion in a suitable liquid may reveal the concentration of liquid in the cracks and, by suitable movement, the planes of the cracks can be revealed by obtaining specular reflections.

Dyeing

Many natural rocks and minerals, particularly granular and cryptocrystalline materials, are notably porous, and it is not difficult to deposit coloured solutions or dyes in these pores. For permanence it is desirable that the dyes are inorganic, since organic dyes often fade with time. In many cases, colouring compounds are deposited by means of chemical reactions taking place within the pores themselves. This process is facilitated if the material itself is resistant to heat, acids, and the chemicals utilized in the dyeing process. The cryptocrystalline quartzes (e.g. agates and chalcedonies) often contain iron within their pores, and this may be redistributed by acid treatment, or they may be immersed in a solution of an iron salt. After dyeing, heating will cause the deposition of iron oxides within the pores and the material takes on shades of brown and reddish-brown. Much greyish-brown and white-banded agate is treated in this manner, whereupon the coloured layer turns brownish-red but the white impervious bands are unchanged, thus producing sardonyx. The dyeing of agate has been practised in Idar Oberstein, West Germany, for hundreds of years and they have perfected techniques using inorganic reagents. Many greens are obtained by using compounds of chromium and some blues are produced with cobalt compounds. In the past, deep blues were obtained by depositing Prussian blue (Berlin and Turnbull's blues) using solutions of potassium ferricyanide followed by ferrous sulphate. Much onyx (black- and white-banded agate) is produced by soaking or boiling natural agate in sugar or honey solutions for up to three weeks, and then transferring the stones into concentrated sulphuric acid. The acid extracts water from the sugar ($C_6H_{12}O_6$) and black carbon remains in the porous bands. Marble, alabaster (gypsum), jade, and amber are other materials which have been extensively dyed. The identification of dyed material is sometimes possible by using a ×10 lens, but careful examination of coloured layers and surface cracks using a microscope will often reveal traces of dyestuffs.

Surface modifications

Some of the treatments described above, such as oiling, will affect outer surfaces as well as penetrating the interior, but there are other treatments which affect only the outer surfaces. These include waxing, painting, and covering with thin skins of plastic of various types. It is not uncommon to find so-called emeralds (particularly carved examples) to consist either of very pale natural emerald or other beryl coated with a layer of green gelatine. Close inspection with a lens will usually reveal a place where the coating has peeled away, and the stone has a tacky feel to the touch. In such cases, the spectroscope will show a broad absorption band in the red, due to the dyestuff, in place of the narrow chromium lines seen in emerald. Coated beads simulating amber are not uncommonly seen.

A relatively recent process for coating the upper surface of gemstones with a film of low refractive index to improve their brilliance and apparent depth of colour must be mentioned here. Such films reduce the surface reflections in a manner similar to the coating or 'bloom' on lenses in binoculars and cameras, and allows more light to enter the stone. There is, of course, a reduction of surface lustre. Such treatment is another form of 'faking' and, as such, should be condemned. Coated stones can be recognized by a slight bloom or iridescence on the surface. Refractive index readings on the refractometer may be unobtainable until the coating has been removed by rubbing with rouge or similar agent. These coloured films can be applied to stones (usually to the pavilion) to improve their colour (e.g. a blue film applied to yellowish diamonds to produce desirable 'colourless' diamonds).

Other surface treatments include cladding the back of a stone with metal foil or depositing silver from solution to form a mirror-like coating. These foils have often been coloured, and they may fade, sometimes unevenly. Such foiled stones should never be immersed in liquids since once the liquid has penetrated between stone and foil it is very difficult to remove, and it could redistribute colouring matter and any dirt trapped between the foil and stone.

The coating of stones with thin films of corundum has been known for some years, but it is now possible to deposit very durable films of diamond on various surfaces, including glass lenses. These coatings are produced by a combination of ion bombardment and the deposition of carbon from methane gas in a vacuum, the high energy of the bombardment reproducing conditions similar to the high temperatures and pressures in which natural diamond is formed. Such possible diamond-coated gems could present problems for the gemmologist, especially if the 'pre-form' is cubic zirconia. However, considerations of specific gravity, size, and weight should reveal their true nature, provided that the stones can be unset.

Surface overgrowths

Natural quartz, ruby, and emerald have been used as seeds for the growth of synthetic stones since the turn of the twentieth century for quartz and during the early 1920s for emerald. With a small seed in a large overgrowth (as with many modern synthetic growth techniques) the seed is not subsequently of great significance and can be cut away if the outer synthetic layers have sufficient volume. However, around 1960 Lechleitner in Austria produced an emerald which consisted of a pre-form of pale natural beryl coated with an overgrowth of dark emerald, the names 'Emerita' and 'Synemerald' being used to market the product. These thin surfaces tend to crack, and the characteristic crazing (see Chapter 9) makes recognition easy. Lechleitner has also coated natural and synthetic corundum with suitable corundum overgrowths.

Other stones with modified surfaces are the corundums subjected to high-temperature diffusion processes which have induced asterism or deeper and more desirable colour in the surface layers.

9

Detection of synthetic, imitation and composite stones

In earlier chapters it has been shown how it is possible by simple tests to assign any gemstone that is in common use in jewellery to its correct mineral species. If by refractive index or other tests we have determined a stone to be, say, a peridot or a zircon, that is the end of the matter as far as identification is concerned. Knowing certainly its nature, estimation of the stone's value can then be made on the basis of colour, purity, perfection of cutting, and weight. But where the stone tested turns out to be a ruby or a sapphire or an emerald – or indeed a growing list of other stones such as alexandrite, opal, and turquoise – then a further problem still awaits solution: is the stone natural or synthetic? In experienced hands the answer may often be easily arrived at. Sometimes, however, there are no readily apparent signs of distinction, and laboratory tests may be needed to settle the issue. Probably in no other branch of gem testing is there so great a disparity between the amateur gemmologist and the professional laboratory worker, with a skill, born of long experience, in interpreting correctly the small signs which differentiate between natural and synthetic stones. By careful study of the following pages the reader will at least know what to look for, and the accompanying photomicrographs should also prove helpful. But plenty of practice will be needed with actual specimens if any real skill is to be obtained.

It is internationally accepted among gemmologists and in the jewellery trade that any stone which has the composition, crystal structure, and properties of a gem mineral but which has been artificially produced (no matter how) should be known as a 'synthetic' stone. Strictly speaking, however, there is no warrant for using the term synthetic in this restricted sense only since any man-made compound can correctly be described as 'synthetic', whether it represents some natural substance or not. In fact, gemmologists themselves freely use the adjective in its more general sense when they refer to such gem substitutes as strontium titanate or yttrium aluminate as 'synthetic' stones.

To overcome the inconsistency inherent in this double standard the International Committee on Industrial Technology proposed in 1974 a newly coined word 'homocreate' (derived from the Greek *homos*, same, and the Latin *creare*, to make), to fit the restricted use of 'synthetic' for which the gemmologist feels the need. Their definitions were:

Synthetic (n.) A human-produced chemical compound or material formed by processes that combine separate elements or constituents so as to create a coherent whole; a product so formed.

Synthetic (adj.) Pertaining to, involving, or of the nature of synthesis; produced by synthesis; especially not of natural origin.

Homocreate (n.) A human-produced substance (solid, liquid, or gas) whose chemical and physical properties are within the range of those possessed by the specific variety of the natural substance that the *homocreate* is intended to duplicate.

Homocreate (adj.) Synthetic and possessing chemical and physical properties that are essentially the same as those of its natural counterpart; created the same as.

There are other difficulties in store for the gemmologist who is endeavouring to use the correct terms for the various materials used as substitutes for natural gemstones. The commonest, perhaps, relates to synthetic spinels; these are manufactured in large quantities by the Verneuil process in colours (or lack of colour) intended to represent not natural spinels but aquamarine or diamond, thus joining the ranks of imitation gems in which only the appearance links them with the natural stone represented. Moreover, neither in composition nor in properties are these Verneuil spinels identical with their natural namesake; only in their structure do they clearly have a family relationship.

Thus a substance such as emerald made in the laboratory is a homocreate. Its properties are designed to mimic those of the equivalent substance produced by nature. However, GGG and YAG are compounds made in the laboratory and have no natural counterparts. They are true synthetics by these definitions. Briefly, all homocreates are synthetic, but not all synthetics are homocreates.

These definitions were approved unanimously by the ICTT after three years of deliberation and have been adopted by most professional scientific societies. Whether gemmologists will agree to adopt the proposed word still remains to be seen, but some of the rules on nomenclature issued by CIBJO (The International Confederation of Jewellery, Silverware, Diamonds, Pearls and Precious Stones) in 1988 do not appear to fit in with the ICTT usage.

Synthetic corundum

The pre-Verneuil synthetics are still encountered surprisingly often in jewellery, at least in Britain (Figure 9.1). Logically, the date range of such jewellery should be fairly limited – say, between 1885 and about 1910, when the Verneuil synthetics were in commercial production. At first glance they are more natural-looking than the modern stones, being richer in colour and containing many bubbles and impurities which give an impression of 'silk' until more closely examined under a lens or microscope. They are usually oval or cushion-cut, and between 1 and 2 carats in weight. The microscope reveals strongly marked, almost crack-like curved lines of growth, which are often not strictly parallel, with finer growth lines between these major markings. Bubbles, often of considerable size, are common, and solid fragments and bleb-like

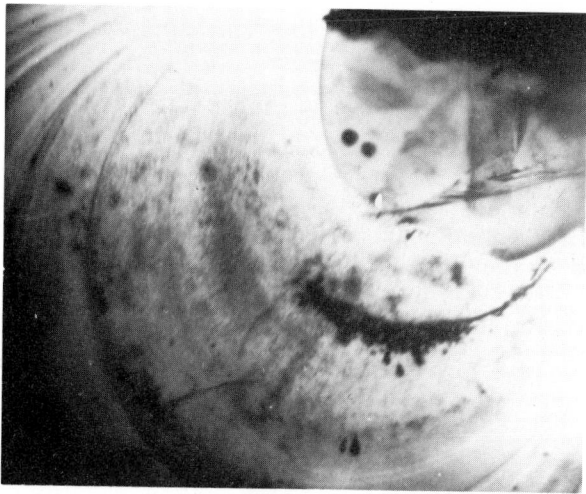

Figure 9.1 Part of a small 'boule' of early synthetic ruby. The top-right portion is a fragment of Thai ruby used as a seed during manufacture

unassimilated particles, which often seem to rest on one of the major lines of growth, are very typical. Where tailed bubbles are seen, these may show that the bubble was moving towards the *convex* side of the curved lines, whereas in Verneuil stones one finds, as would be expected in an upright boule, that signs of bubble movement are directed towards the *concave* surface. Another feature seen in the old stones is a curious 'frosted' appearance in the interior, reminiscent of the effect caused by exsolution of calcite in some Colombian emeralds.

In the normal Verneuil synthetic stones the curved lines of growth, which are commonly seen only in coloured synthetic corundums, are due to the intermittent fall of the droplets onto the boule's upper surface, and to the greater volatility of some of the colouring oxides compared with alumina, which causes minute differences in colour and refractive index as one layer is succeeded by another. The gas bubbles seen in so many synthetic corundums consist almost certainly of hydrogen, of which an excess is used in the furnace gases. The gas may be occluded in particles of the falling powder, and thus incorporated into the boule. In synthetic blue sapphires there are generally curved bands of colour broad enough to be observed by the naked eye – especially when the stone is immersed in a suitable liquid and observed against a white background. In synthetic ruby and other coloured synthetic corundums the growth lines are much finer and when magnified they resemble the lines on a gramophone record. The actual *growth lines* in blue sapphire are in fact as fine as this, as we have found by using the technique described below: normally, however, only the broad swathes of colour are visible. In all cases it should be realized that growth lines are only visible when the stone is viewed in the correct orientation (corresponding to directions at right angles to the length of the original boule) and under suitable lighting conditions lens inspection over white paper may prove to be best for sapphire. It can thus be understood that with mounted stones there may be great difficulty in detecting these valuable diagnostic features. The nature of the curved lines in synthetic ruby and sapphire are well shown in Figures 9.2 and 9.3, respectively.

Experiments in the author's laboratory have shown that 'immersion contact' photographs of synthetic stones may often reveal curved growth lines in synthetic corundums when these are invisible under lens or microscope when viewed in the same orientation. This fact may be of considerable practical value in certain cases. The general technique has already been described in Chapter 2 as a means for determining refractive index. It is essential that the beam of light passing through the specimen on the film below should be narrow and nearly parallel. Pure di-iodomethane serves very well for the immersion fluid, as it has very nearly the same index as corundum for blue and violet light, which are the

Figure 9.2 Curved lines and bubbles in synthetic ruby

Figure 9.3 Curved colour bands in synthetic sapphire

operative colours for the slow film used. If a photographic enlarger is available, light from this, with the lens stopped down to $f/22$, provides a very suitable illumination for the process. Under these conditions, curved growth striae have been observed even in colourless synthetic corundum, together, it must be admitted, with straight lines making an acute angle with these, the origin of which is not yet understood. Although curved lines may be made visible by this method at angles other than the optimum, the orientation must not be too far wrong. Large synthetic stones are normally cut with their table facet parallel to the length of the original boule, and as a consequence the lines should be visible at right angles to the table – a fact very convenient for the technique outlined above. Yellow synthetics usually fail to show these lines. The American

Figure 9.4 Zoned colour bands in natural sapphire

gemmologist William J. Jenkins, experimenting with this method of revealing curved striae, achieved surprisingly good results by using the narrow beam from a pen-type torch held some feet above the specimen, which rested unimmersed directly on the surface of a photographic film, but a simple and inexpensive technique reported by R. Hughes may be helpful. Placing a blue filter between the light source and the immersion cell provides the increased contrast necessary to resolve the colour banding (curved or straight) in difficult yellow/orange sapphires – synthetic or natural. Frosted blue glass filters are best and, up to a point, the deeper the blue, the better. The colour of the filter to be used should match the wavelengths of maximum absorption in the stone and for difficult rubies a yellowish-green filter might be tried.

Another useful technique in helping to distinguish natural from synthetic flawless corundums makes use of the so-called 'Plato effect'. The optic axis of the stone is first located by turning the stone between crossed polarizers using a polarizing microscope or a polariscope with a supplementary lens or 'conoscope' (see Chapter 2) until an interference figure (a black cross) is obtained – the optic axis of the uniaxial corundum is now aligned with that of the microscope or polariscope. The stone is then observed, retaining the same orientation, immersed in di-iodomethane between crossed polarizers. Turning the stone about the optic axis (by hand or by the microscope stage) may show two (or even three) sets of lines, reminiscent of twinning lines, at 60° or 120° to each other. Synthetic corundums may show this effect but it has not yet been seen in natural stones (see Figure 9.5).

Figure 9.5 The 'Plato effect': sets of lines at 60°/120° (reminiscent of axis twinning) seen in synthetic corundum looking down the optic axis

In contrast to the curved layers of growth seen in synthetic corundums, any lines or bands of colour seen in natural rubies or sapphires are rigidly straight, as they follow the outlines of the original crystal, which invariably has flat faces (see Figures 9.3 and 9.4). Moreover, such lines or bands in natural stones can frequently be seen to meet each other at angles of 120°, since they are parallel to the outlines of a hexagonal crystal. However, the flux-fusion growing techniques may also produce crystals within which may be sets of straight lines which may intersect at 120°. Thus synthetic sapphires and rubies of the types made by Chatham and others may show intersecting straight lines or bands and the former distinctions; curved bands = synthetic, straight bands = natural no longer holds, and one may need to look for other distinctive signs such as the tell-tale inclusion crystals or flakes of platinum crystallized from the vessel in which they were made. Gas bubbles of the type seen in synthetic corundums are never seen in natural stones, but, particularly following heat treatment, large and/or numerous gas bubbles have been observed in gas cavities in natural sapphires, which might be a little confusing if the observer looks no farther. On the other hand, small crystals of minerals associated with corundum in nature, such as zircon, rutile, spinel, mica, hematite, or even other crystals of corundum itself, are usually to be seen when the specimen is scrutinized under a low-power microscope. Tiny flat cavities containing liquid are also a common feature of natural stones. These are usually assembled together in the same plane, either flat or slightly curved, forming what is commonly known as a 'feather'. Rubies and sapphires from different localities each tend to have their own characteristic inclusions. Only in Burma rubies (Figure 9.6), for instance,

Figure 9.6 Rutile needles in Burma ruby

does one find the patches of slender, short, rutile needles, intersecting at angles of 120°, which, from their appearance in reflected light, are known as 'silk'. Also, only those corundums which come from Sri Lanka show included zircon crystals surrounded by crack-like halos. More will be said concerning these and other features in natural rubies and sapphires in the appropriate chapters. There are, however, certain further distinguishing features peculiar to synthetic corundums which should be given here.

Small, roughly parallel crack-like markings (see Figure 9.7 and Plate 9) are often seen at or near facet junctions in synthetic corundums. Lapidaries call these 'fire marks', and they are caused by local overheating during the polishing process. These are only seen on corundum, and while these have been observed on the surfaces of natural rubies and sapphires they are only indicative of a synthetic stone because in these less care is taken to avoid such blemishes. Another feature which only serves as an indication is the strong dichroism usually visible when a synthetic ruby is observed through a dichroscope at right angles to its table facet. In fashioning natural rubies the lapidary usually grinds a table facet at right angles to the so-called optic axis of the crystal, thus ensuring the best colour for the stone. When this is done, no dichroism will be visible through the table. Thus, strong dichroism in this direction is a suspicious sign, since it points to a synthetic ruby, which is randomly cut.

More stringent 'extra' tests for synthetic rubies are available for those with laboratory facilities. In the first place, it has been found by the author and others that synthetic rubies are decidedly more transparent to short-wave ultra-violet light than are natural rubies. The presence of traces of iron in the latter is

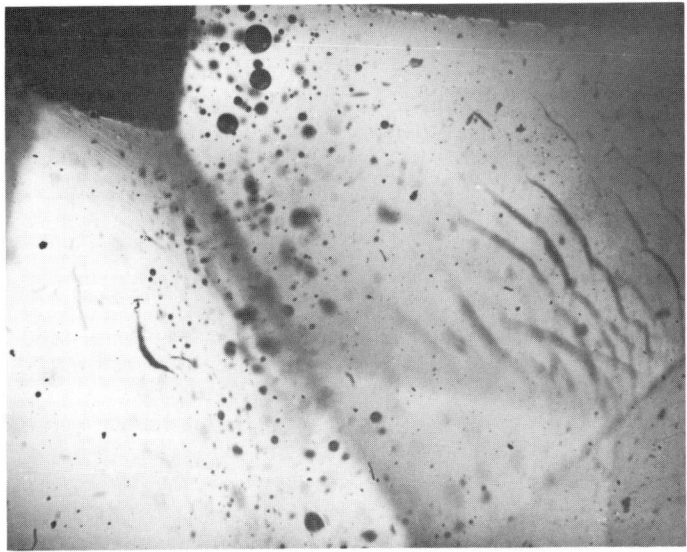

Figure 9.7 Bubbles and surface cracks in synthetic ruby

probably the main factor governing this difference. A determination on these lines is possible without the use of expensive apparatus if one calls immersion contact photography to one's aid, as first suggested by Mr Norman Day.

Stones to be tested are placed, table facet down, on a sheet of photographic paper which is immersed in a flat-bottomed dish of water to a depth sufficient to cover the rubies. This must be done, of course, either in a dark-room or in a room which is nearly dark. The stones are then exposed for a few seconds to light from a short-wave (253.7 nm) ultra-violet lamp held some eighteen inches or so above the dish. On developing the paper, provided the exposure time has been correct, the images of natural rubies will appear white (showing their relative opacity to the rays) whereas synthetic stones appear dark, except for a marginal white rim. An example of this rather spectacular test is seen in Figure 9.8. It has been found that 'Ilfospeed' printing paper gives good results in this

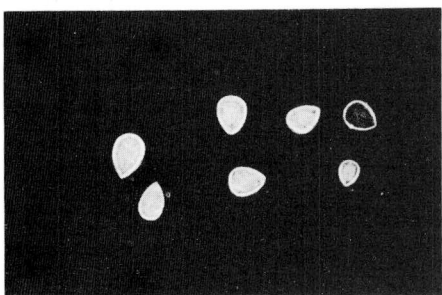

Figure 9.8 Immersion contact photograph in water of one synthetic and six natural rubies showing greater transparency of synthetic ruby to short-wave ultra-violet light

method. However, it must be made clear that this test for short-wave ultra-violet transparency is only good for stones that themselves are quite transparent. If a stone, natural or synthetic, has numerous feather-type inclusions the ultra-violet light may reflect from these and give the possibly false impression that the stone absorbs these wavelengths. In recent times tests have been devised based on this method which involve the measurement of the ultra-violet transmission area for each ruby with a UV/visible spectrophotometer, but such testing is laboratory based and goes beyond the scope of this book. Under a short-wave ultra-violet lamp, such as the 'Mineralight', synthetic rubies usually show a distinctly brighter red fluorescence than natural rubies. This enables doubtful stones to be picked from a large parcel for a microscope test. Chrome-rich synthetics show less fluorescence and too great reliance should not be placed on this test.

Short-wave ultra-violet light can also be of great assistance in separating synthetic from natural sapphires. Under these rays, synthetic sapphires show a bluish-white or greenish glow, which may be ascribed to the titanium in their make-up. Many heat-treated and some non-heat-treated sapphires from Sri Lanka may reveal a similar glow under the same conditions, but where an effect is seen, the test can be made more precise and certain by examination of the fluorescing stone under a lens when the curved striae typical of Verneuil sapphires will usually be plainly visible for the synthetic and angular zoning for the natural stone – more so than under ordinary lighting conditions. Furthermore, under long-wave ultra-violet radiation many blue sapphires from Sri Lanka, which contain a trace of chromium, will fluoresce orange or red.

Where a suitable X-ray plant is available another test may be used in those cases where the microscope does not easily yield the required information. Both natural and synthetic rubies show a bright red fluorescence under X-rays but when the radiation is switched off there is a pronounced afterglow with synthetic stones, whereas in natural rubies no afterglow can generally be detected. To be effective, this test should be carried out in a darkened room, and with the observer's eyes 'dark adapted', in which condition they are far more sensitive to the rather feeble red afterglow it is required to detect. A short but distinct afterglow has been seen with some natural rubies, and since some synthetic manufacturers produce a 'range' of 'colours' and qualities in the more modern synthetic ruby, the X-ray fluorescence and afterglow will vary from stone to stone. An example of this would be the Ramaura synthetic ruby in which some stones fluoresce and some are virtually inert, and some have an afterglow and some do not. It is worth noting that with Thai rubies there is virtually no fluorescence seen under X-rays, and the same seems to apply also to East African rubies, the chromium glow being 'killed' in both cases by the presence of small amounts of iron. This is useful knowledge to the gemmologist, since Thai rubies can on occasion be almost entirely devoid of inclusions and thus difficult to determine. It may also interest those who, for commercial reasons, like to categorize their rubies by place names instead of letting their value rest purely on colour, quality, and size. Rubies recently found in Kenya have, at their best, a colour near to those from Burma, and a similar irregular distribution of colour and bright red fluorescence under ultra-violet rays.

In the case of corundums of other colours, the spectroscope may often be of great assistance in distinguishing between natural and synthetic stones. Natural green sapphires, for instance, show a group of three absorption bands in the blue part of the spectrum, merging into one another to form what at first glance may appear to be a single broad band. No bands of this kind are seen in synthetic green corundum. What is far more important in practice is that in most natural *blue* sapphires the strongest at least of the three bands mentioned (wavelength 450 nm) can almost always be detected. Again, this is missing in all synthetic sapphires with the possible exception of those produced by Chatham, so that when seen it forms a reasonable test for natural sapphire. The same is true for yellow sapphires from Thailand, Australia, or Montana. In yellow sapphires from Sri Lanka there may not be enough iron present to produce any absorption bands, but in these there is a distinctive apricot-yellow fluorescence under long-wave ultra-violet rays. Yellow sapphires showing neither absorption bands nor fluorescence are almost certainly synthetic or heat treated. Synthetic sapphires of warm yellow or orange tint will show a red fluorescence under X-rays with a distinct afterglow when the rays are switched off. This phosphorescence gives a clear distinction from natural yellow stones. Finally, in those synthetic corundums coloured by vanadium which are intended to represent alexandrite, there is a very distinctive narrow line in the blue at 475 nm, while alexandrite chrysoberyl has an entirely different spectrum of its own (see Chapter 10).

A synthetic 'fancy' sapphire less frequently seen is presumably intended to represent amethyst. This can be annoying as, though palpably not a natural stone, it may be difficult to find evidence in the form of curved lines or bubbles. Robert Webster found that the use of long- and short-wave ultra-violet rays can be helpful in this case, since the long waves produce a red glow and short waves bluish-white fluorescence. This curious double effect is unique and entirely damning.

A word more ought to be said about the gas bubbles which have been merely mentioned as being typical for stones (particularly corundums) manufactured by the Verneuil flame-fusion method. Bubble shapes in synthetic stones and in glass are well worth study, since they vary a good deal with the nature of the medium (see Figure 9.9). It is also important to be able to distinguish the small rounded crystal particles found in many natural stones from bubbles of rather similar appearance. In synthetic corundums the smaller isolated bubbles are usually almost perfectly spherical, but where little groups or 'bubble clouds' are encountered (see Figure 9.10) some of the shapes are nearly always distorted, and there may be amalgamation into 'clots', which sometimes present an unexpectedly 'natural' appearance. Extreme elongation into hose-like forms is also possible, and confusing to the beginner. Larger individual bubbles, usually only encountered near the base of the growing boule, are often flask- or bomb-shaped in outline, the direction of apparent 'fall' of these bombs being at right angles to the curved layers of growth, as the bubbles are trapped on their way towards the surface. Bubbles in synthetic corundum are also sometimes enclosed in flat triangular cavities, which may present a misleadingly 'natural' appearance.

The experienced gemmologist learns to look out for useful indications which give warning that a stone is a Verneuil synthetic. The colour tends to be 'stark', since the stone is lacking in the traces of colouring elements other than the main one present. The cutting is, of necessity, of a 'mass-produced' type, with a tendency to use rather odd cuts, such as the 'scissors cut', which would never be chosen by the lapidary who is dealing with a valuable natural stone. The girdle in particular is often clumsily thick, and the diameter in round-cut stones, or the width and breadth in the case of step-cut stones, are regulated to exact millimetre dimensions, by which the stones are priced for the wholesale market.

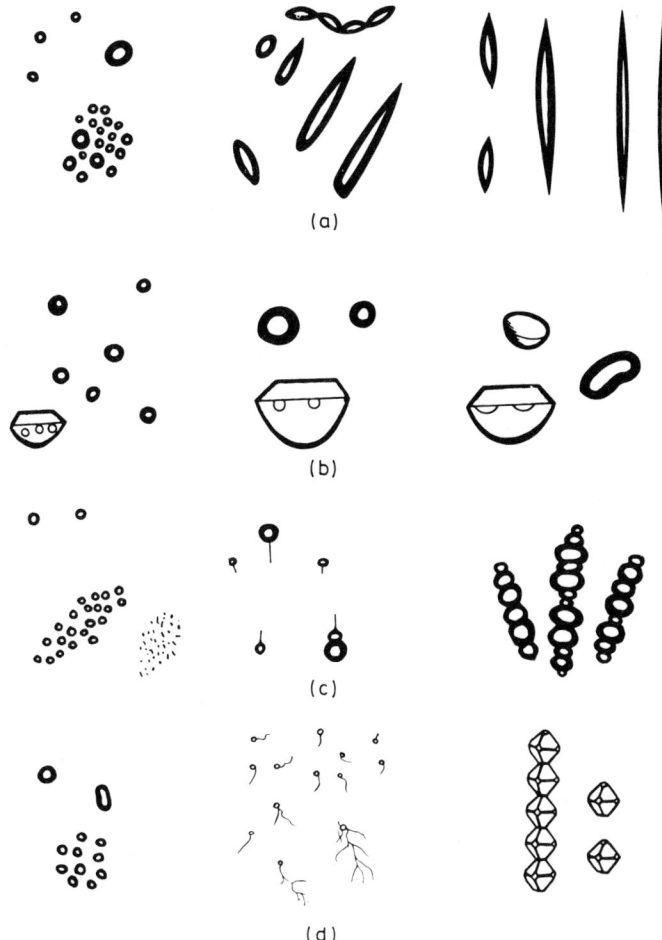

Figure 9.9 Typical bubble shapes as seen in (a) paste. (b) doublets, (c) synthetic corundum, (d) synthetic spinel (after D. Level)

In corundums, the frequent appearance of 'fire marks' in synthetics has already been mentioned, and in the case of synthetic spinels, soon to be described, the colours favoured by the manufacturers are not those found in the natural mineral.

Synthetic corundums made by other methods

So far, we have been describing exclusively those synthetic corundums made by the Verneuil process. Though these are greatly preponderant over all other types (particularly in the UK), we have in these days also to be on the lookout for synthetic corundums (particularly rubies) manufactured under conditions more akin to the processes of nature. Here the well-known 'hallmarks' of the Verneuil

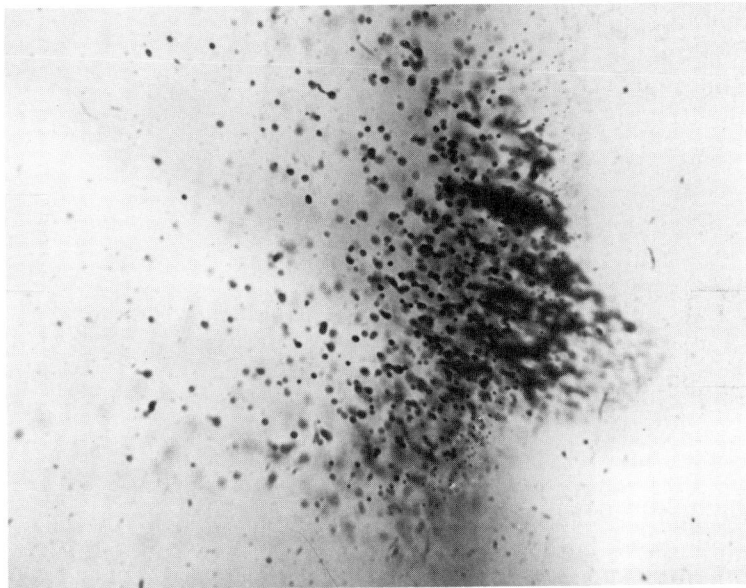

Figure 9.10 Typical bubble cloud in synthetic ruby

synthetics, such as curved striae and groups of small gas bubbles, are missing, and the gemmologist has to learn to rely on other features.

At present many firms, including Chatham, Kashan, Knischka, Kyocera (Inamori), Ramaura, and Lechleitner have succeeded in crystallizing corundum from a flux (see Plates 6–8). Chatham has concentrated on growing crystal groups. Other manufacturers have produced material from which significant stones can be faceted.

Some of the inclusions in these synthetics are reminiscent of the wisp-like 'feathers' familiar to gemmologists in many synthetic emeralds, these being a typical flux-fusion hallmark. In the Chatham crystals a natural 'seed' nucleus may be sometimes discerned, and the gemmologist has to be careful not to conclude from the inclusions in these nuclei that the stone has a Burmese origin. In Kashan stones the most distinctive signs are groups of strongly marked parallel flattened cavities, usually rather like elongated moccasin footprints but sometimes broken up into shorter lengths (see Figure 9.11). These flux droplets have been described by some writers as 'paint-splash' inclusions. In many flux-grown stones, inclusions of solid flux, often in the form of wispy feathers, are a characteristic feature. The fluorescence and other properties of the Kashan stones are too similar to those of natural rubies to act as helpful clues. The fluorescence, however, is too strong to correspond with that of rubies from Thailand, and the colour distribution is not that typically seen in Burma stones, nor is there any 'silk'. The transparency to short-wave ultra-violet rays is slightly greater than in natural rubies, but for this test the conditions have to be carefully controlled and comparison stones of known origin must be tested on the same photograph for any safe conclusions to be drawn and preferably the ultra-violet transmission measured with the aid of a spectrophotometer if this is available.

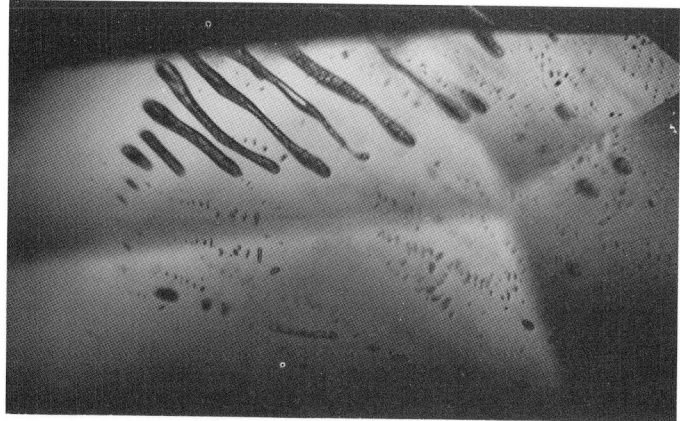

Figure 9.11 'Paint splash' or 'icicle' inclusions in Kashan synthetic ruby

However, observations made under the microscope still produce the most reliable evidence of natural or synthetic origin, irrespective of whether the synthetic is of modern manufacture or is somewhat older. The synthetic rubies produced by Chatham, apart from their wisp-like feathers, also often have needles or platelets of platinum. Kashan synthetic rubies often contain fine dust-like flux particles which may be arranged in 'comet' or 'hairpin'-like formations or just produce a general 'fog' in the stone. Large two-phase inclusions in which the 'bubble' is in high relief as well as platinum platelets are common in crystals of the Kashan synthetic ruby. Yellowish-orange flux inclusions, wispy feathers, swirled areas of colour similar to those in Burma ruby, and angular colour patches are seen in Ramaura synthetic ruby.

Synthetic star corundums

The synthetic star stones can usually be recognized at sight. The star effect is very brilliant, almost as though it were 'painted on' the outside of the stone (see Plate 2). The colour of the synthetic stones is also brighter than the natural star corundums, which seldom combine good colour with a sharply defined star. In addition to these fine coloured synthetic rubies and sapphires, white, brown, and other coloured star stones have been produced. The neatly ground-off base of the synthetics is distinctive, and there is an absence of the zoning seen in almost all natural star corundums. Bubbles are usually present in great numbers, as well as curved bands of colour. The stones vary from subtransparent to nearly opaque. In the almost opaque types one must look for bubbles immediately below the surface, the stone being illuminated from above.

Synthetic spinel

These synthetics are seldom made to imitate natural spinel, but are intended to simulate stones such as blue zircon, aquamarine, topaz, alexandrite, and sapphire. All these stones have quite different properties from natural spinel

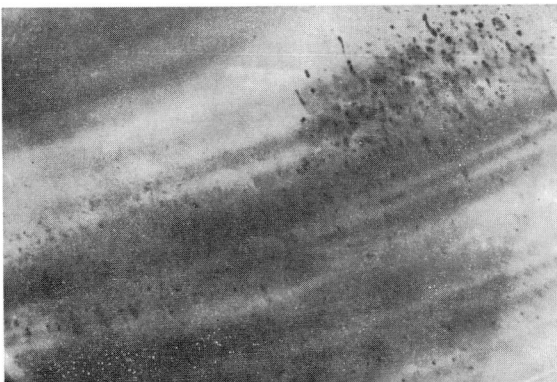

Figure 9.12 Curved striae and bubble clouds in synthetic red spinel

and distinction is easy by orthodox tests, the microscope seldom being called into play. This is just as well, as (except in the red variety (see Figure 9.12) which has occasionally been manufactured) curved structure lines are hardly ever to be seen, and the majority of stones are free from bubbles. Both these facts are due to the relatively slow growth of the boules: the same lack of growth lines and bubbles is noticeable in yellow synthetic corundums which (according to Professor W. F. Eppler) are also usually grown slowly. When bubbles do occur in synthetic spinels they may assume very bizarre forms, though tiny isolated spherical bubbles are sometimes encountered. Distortions include long, hose-like tubes, often oriented in parallel formation; large 'profilated' bubbles, like bulky furled umbrellas or long flasks made by an inexpert glass blower. These last have been shown by Dr E. Gübelin often to assume in part the shapes of negative crystals, roughly hexagonal when viewed end-on, and these may further be grouped in a hexagonal pattern in planes at right angles to the trigonal axis of the boule lattice, providing a very 'natural'-looking feature to the eye of a beginner.

A. J. Breebart has described two-phase inclusions which he was the first to notice in synthetic spinel. These consist of tiny flat cavities containing a bubble and either liquid or gas, often joined by a thin tube to a similar flat cavity parallel to and below the first. Tiny, emaciated, comma-shaped bubbles are also sometimes seen in synthetic spinels. All these variations in bubble shapes are extremely interesting to the keen gemmologist, but it is perhaps fortunate that so many other features in synthetic spinel provide an easy means of recognition.

As already noted, the composition of synthetic spinel is markedly different from the natural, in containing a considerable excess of alumina. This has an effect on its physical properties, raising the refractive index to a figure which seldom varies more than one unit in the third place of decimals from an average of 1.727. Natural spinels, unless they contain more than 2% ZnO or Cr_2O_3, lie within the range 1.715 to 1.720. The SG of synthetic stones is 3.63 or 3.64, as against 3.58 to 3.60 (again excepting the rare zinc-rich types of blue spinel in which both SG and refractive index may rise considerably). Another character induced by the excess alumina, under strain in its tendency to revert to corundum, is a marked anomalous birefringence, resulting in a very characteristic appearance between crossed polars for which the author coined

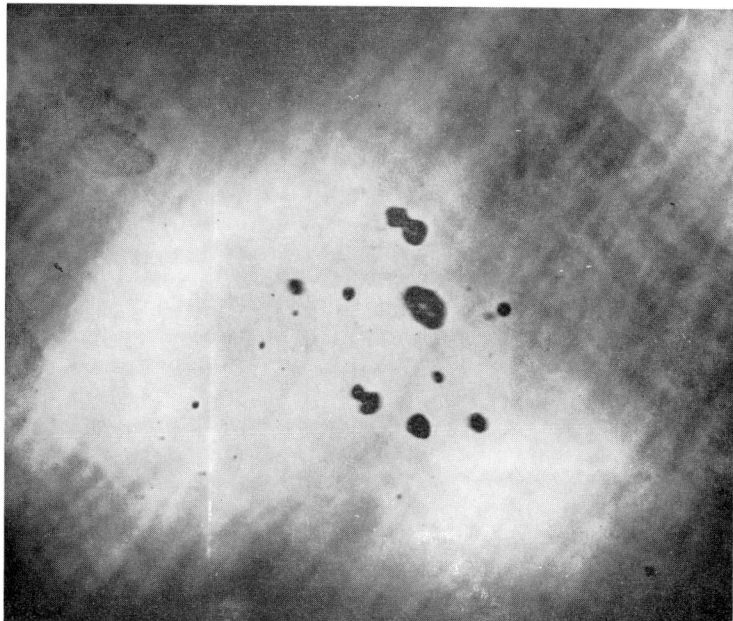

Figure 9.13 Synthetic spinel between crossed polarizers, showing 'tabby extinction' and profiled bubbles

the term 'tabby extinction'. This can be seen in Figure 9.13. It is well worth remembering this particular feature when struggling to identify small, colourless baguettes of synthetic spinel mounted, as they often are, as a substitute for diamond in pieces of jewellery. Immersion of the whole in di-iodomethane may reveal, by the lack of relief, that the stones are not diamond, but the tabby extinction will confirm them as synthetic spinel where they might otherwise be possibly synthetic sapphire or a highly refractive glass. While we began this section by stating that with the identification of synthetic spinel the microscope is seldom called into play we must add the warning that colourless synthetic spinel is often used to make composite stones, and if an opinion is expressed merely from the measurement of an RI on the table of a stone an error could be made.

The Chelsea colour filter and the spectroscope also play a useful part in the identification of synthetic spinel, especially the very popular blue types used to imitate aquamarine, etc. These are coloured with cobalt in almost all cases, and the absorption in the yellow-green and clear transmission in the deep red characteristic of cobalt blues result in the stone assuming a strong red colour through a Chelsea filter. Natural blue spinels may appear somewhat reddish through the filter, but are hardly likely to be confused with the synthetics in any case, as their daylight colour is so different a shade of blue.

With the spectroscope, blue synthetic spinels show three broad absorption bands due to cobalt. In the pale blue stones made to represent aquamarine these are quite faint, while in the deep blue stones resembling sapphire they are strong, and so broad as almost to amalgamate into a single wide band extending from the orange to the green. These dark blue stones give out red gleams by reflected light, which serve to reveal their nature to the naked eye. Usually, a little chromium is added to the stone to improve the colour, and this gives rise to a red fluorescence under ultra-violet or between crossed filters.

Other coloured varieties of synthetic spinel are used less frequently. An attractive pink type resembling morganite beryl is produced by addition of iron and a pale yellowish-green type by using manganese as colouring agent. These latter have a strong green fluorescence under ultra-violet light and two absorption bands in the violet. Synthetic spinel resembling alexandrite in colour and showing a perceptible colour change in artificial light can be made, but is very seldom seen. It has, of course, no dichroism and this fact, together with its lower, single refractive index, make distinction from alexandrite chrysoberyl easy. However, as far as appearance goes, this variety of synthetic spinel is far more deceptive than the synthetic sapphires coloured with vanadium which are so commonly encountered.

Colourless synthetic spinel is very clear and bright, and utterly devoid of colour. Although in figures its dispersion of 0.020 between the B and G lines of the solar spectrum is very little greater than the 0.018 of corundum, there is perceptibly more 'fire' than in the latter, and it has often been used to represent diamond, particularly in baguette form, where its inferiority in lustre and fire to diamond are not so readily apparent as when cut as a brilliant. Immersion in di-iodomethane and examination between crossed polars has already been suggested above as a simple means of identification. The strong bluish-white fluorescence usually shown by synthetic colourless spinels under short-wave ultra-violet light is another aid to identification.

Although synthetic red spinels have been made by the Verneuil method, those which have appeared on the market were quite small and production does not seem to have been continued. In these the ratio of magnesia to alumina must be nearly the same as in natural spinels. They are chrome-rich, entailing a saturated red colour and rather high refractivity (1.725) with SG 3.60. Under the microscope curious curved swathes of colour are easily seen, which are distinctly different from those seen in Verneuil rubies, and also clouds of small gas bubbles. The stones are, as one would expect, red-fluorescent under crossed filters or long-wave ultra-violet, but, strangely enough, do not display the same 'organ pipe' group of fluorescence lines so characteristically shown by natural red spinel; the light emitted being mainly concentrated in a line at 685 nm.

The same is true of handsome red groups of synthetic spinel crystals which have been grown by a flux-fusion process. Not only red but blue, green, yellow, and colourless spinel crystals have been grown in this way, for sale as pretty crystal groups for mounting as such in the modern manner rather than as material for the lapidary. Such groups can be distinguished at sight from natural spinels by their habit and bright crude colours. Further, in common with all man-made crystals, they feel peculiarly harsh to the touch. This last observation was made by Mlle D. Levei of the Paris Laboratory.

Synthetic rutile

This material caused a stir when it first appeared in the late 1940s, but it was superseded by strontium titanate which may be colourless, whereas synthetic rutile never lost its yellow tinge despite strenuous efforts to counter this defect by adding various dopants.

Here, in brief, are the properties of synthetic rutile: hardness, 6; SG, 4.25; refractive indices, ordinary 2.61 and extraordinary 2.90; and uniaxial positive, with birefringence 0.287. The dispersion (i.e. the difference in refractive index) between wavelengths 670.8 nm and 435.8 nm was found to be 0.2851 for the ordinary ray, which corresponds to approximately 0.300 for the standard range between B and G lines of the solar spectrum. The dispersion is thus some seven times greater than for diamond even for this ray, and must be higher still for the extraordinary ray. From the figure given above it will be realized that the double refraction, too, is enormous, being nearly five times that of zircon, and far greater even than that for Iceland spar, in which the double refraction of 0.172 is, by normal standards, reckoned very strong indeed.

Considering these properties, it will be realized that synthetic rutile can quite easily be identified without any elaborate tests. The stupendous 'fire' makes it as full of colours as a water opal; the high refractive indices give it a high surface lustre, but with a 'soft' look compared with diamond, while the double refraction causes a tremendous doubling of the edges of the back facets when viewed through the front of the stone. Clever lapidaries have learned to orient the stone so that the optic axis (direction of single refraction) is at right angles to the table facet, thus mitigating the 'out-of-focus' appearance caused by the strong double refraction when viewing the stone from directly above.

However, when viewing the stone with a lens the slightest tilt will reveal the 'doubling' in great strength: far stronger than in the case of sphene, for instance, which is perhaps the only natural stone with which this synthetic could be confused.

One other feature of synthetic rutile may be mentioned in conclusion: there is a strong absorption band in the violet, beginning at about 425 nm. The presence of this band probably accounts for the yellowish colour of the stone, and also for its high dispersion, since refractive index always rises steeply when a region of intense absorption is approached.

Strontium titanate

This Verneuil product, with its less extreme dispersion, rapidly superseded synthetic rutile when it appeared in the mid-1950s. It is practically colourless, it is cubic, and therefore singly refracting, and its refractive index is 2.41, and therefore practically equal to that of diamond. The dispersion is approximately four times that of diamond, giving rise to a play of colour noticeably in excess of that shown by diamond, but not so extravagant as that seen in rutile. The SG is high (5.13) and the hardness low (about 5½); this latter fact provides, one must admit, one of the best ways of distinguishing the stone from diamond when mounted in a setting. A steel needle, which is one of the most easily controlled 'hardness points' readily available and one not likely to do much damage, will be found to mark the surface of strontium titanate, while, of course, it leaves diamond entirely unscathed. Under a lens the facet edges of this synthetic stone are seen to be slightly rounded, unlike the sharp junctions of a diamond.

Tiny inclusions, reminiscent of a centipede or a ladder with bent rungs and no uprights, have been noticed by the author in several specimens of strontium titanate. No absorption bands have ben observed, but complete absorption begins at about 415 nm in the violet.

For those with the necessary equipment, the lack of fluorescence under ultra-violet rays or X-rays and the opacity to X-rays will be useful signs distinguishing strontium titanate from diamond. The jeweller without equipment must be on guard against any 'diamond' which shows an unusual degree of fire, and may resort to a careful test with a needle, as suggested above, or to testing the stone itself carefully against a piece of polished corundum, which only diamond will mark. If the stone should be free from its setting its high SG will be revealed by its surprisingly high weight compared with its spread as measured on a diamond gauge.

Provided a clean, flat facet is available, a strontium titanate specimen can be clearly differentiated from diamond by testing on one of the new reflectivity meters. On account of its high dispersion, it gives a much *lower* reading than diamond for the infra-red beam that is operative in these instruments.

The production of strontium titanate is now greatly diminished and synthetic rutile is only made for scientific purposes. Nevertheless, large quantities of both synthetics were made and cut stones may still be encountered.

Synthetic rare-earth 'garnets'

During the past two decades a whole series of compounds incorporating various rare-earth metals and having a garnet-type crystal structure has been manufactured by the 'pulling' method and also by flux-fusion. These compounds are primarily prepared for laser and other technical purposes, and are somewhat costly to produce, but since the resultant crystals are hard, transparent, and lustrous it was inevitable that some of them should enter the field as gemstones. It is thus wise for the gemmologist to become acquainted with their properties.

The two best-known members of the tribe are conveniently nicknamed YAG (yttrium aluminium garnet) and YIG (yttrium iron garnet), and the first-named is the only type so far to have been deliberately used as a gem substitute, and has been marketed, with suitable trumpetings, as 'Diamonair'. When pure, YAG is colourless, with SG 4.57, hardness 8, refractive index 1.832, and dispersion 0.028 – properties which make it a reasonably effective diamond substitute. It is an yttrium aluminate, with a chemical formula $Y_3Al_5O_{12}$ (Table 9.1). However, this

Table 9.1   Properties of some synthetic rare-earth aluminium garnets having the general formula $Y_3Al_5O_{12}$

Element	Colour	SG	RI
Yttrium	Colourless	4.57	1.832
Yttrium (doped)	Green	4.60	1.834
Terbium	Pale yellow	6.06	1.873
Dysprosium	Yellowish-green	6.20	1.85
Holmium	Golden yellow	6.30	1.863
Erbium	Yellowish-pink	6.43	1.853
Thulium	Pale green	6.48	1.854
Ytterbium	Pale yellow	6.62	1.848
Lutecium	Pale yellow	6.69	1.842

material becomes even more dangerous when coloured green with chromium. One such, in the form of a well-cut gem, resembled a fine demantoid so closely as to be bought and sold as such among experts without arousing suspicion until its absorption spectrum happened to be studied. The stone was doped not only with chromium but with neodymium, and this gave it a startlingly rich spectrum of fine lines, betraying at once its artificial origin. Addition of manganese yields a red type, cobalt a blue, and titanium a yellow form.

Another of the man-made family of so-called 'garnets' that has had some success as a diamond simulant is nicknamed 'GGG' (gadolinium gallium garnet). This has a very brilliant lustre, but usually has a slight brownish tinge. Its refractive index is 2.03 and dispersion 0.038 (similar to that of white zircon), while its density is extremely high (7.05) matching that of cassiterite. Its hardness is only 6½ on the Mohs scale – rather too low to withstand daily wear as a ring stone.

Cubic zirconia

At present (1989) the most favoured of all the synthetically produced substitutes for gem diamond is a cubic form of zirconium oxide. This form of ZrO_2 is unstable unless combined with some other oxide, such as CaO or Y_2O_3. In nature ZrO_2 is found as the monoclinic mineral baddeleyite in the gem gravels of Sri Lanka. The properties of the stabilized zirconia naturally vary somewhat according to the choice of the stabilizer, but not to an important degree. When CaO is the stabilizer its refractive index is given as 2.17, dispersion 0.060, and hardness 8½ on the Mohs scale – figures that, when combined with clarity and freedom from colour, point to the makings of an effective diamond substitute. As with all these materials embodying heavy atoms, the density is high (5.65) and even higher (near 6) if yttrium oxide is the chosen stabilizer. Some specimens have suffered from lack of clarity and tinges of colour, but one supplier has turned even this to advantage by offering such stones at a lower price and indeed grading the material in much the same manner as diamond itself is graded.

At least three firms are currently producing cubic zirconia, in Russia, Switzerland, and the USA, the most prominent being the famous Swiss firm of Djevahirdjian, known since the beginning of the twentieth century as the largest producer of synthetic corundum by the Verneuil process. Predictably, they have coined the name 'Djevalite' for the new synthetic, adding yet another fancy name to the list that the unfortunate gemmologist and jeweller now have to remember in order to ascribe to it its true nature and relevant properties.

The special technique devised by Russian scientists to enable this very high melting-point material to be fused and allowed to crystallize by controlled slow cooling has been mentioned in Chapter 7.

For the reader's convenience, a list of the properties of the synthetic materials liable to be used as diamond substitutes is given in Table 12.1 in Chapter 12, which deals with the identification of diamond. Fortunately for the jeweller, the only real concern is to be able to say with confidence 'this stone is (or is not) a diamond'. Gemmologists, at least if they work in a gem-testing laboratory, are expected to put a name to the product.

Synthetic diamond

In 1970 General Electric of the USA announced the production of gem-quality synthetic diamonds in various colours, but stated that they had no plans to produce them commercially. In 1985 Sumitomo of Japan produced gem-quality synthetic diamonds and this was followed in 1987 by the De Beers synthesis of very similar material. As yet these synthetic diamonds are crystals of yellow to yellowish-green and yellowish-brown colours, and cut stones of up to 0.2 carat appeared on the New York market in the late 1980s. Other sizes and colours will surely follow. The problems for the jeweller/gemmologist lie in the distinction of these synthetic diamonds from natural coloured diamonds and treated (irradiated) diamonds, and these are described in some detail in Chapter 12. However, it may be useful to summarize the main distinguishing points for the Sumitomo and De Beers material. They do not show sharp absorption bands, but display a general and increasing absorption towards the violet end of the spectrum. Under the microscope many synthetics show an 'hourglass' or 'bow-tie' structure which often appears as areas of deeper yellow. Under short-wave ultra-violet radiation they may show somewhat similar geometric zoned areas fluorescing in 'yellows' with intervening inert areas. Inclusions of metallic flux, derived from the high-pressure flux method of manufacture, are not uncommon, and some synthetics are distinctly magnetic. Any yellow to greenish-yellow or brownish-yellow diamonds should be tested with care.

Synthetic emerald

Emeralds, both natural and synthetic, are increasingly posing problems for the gemmologist. Because of changed government policies the output of Colombian emeralds has risen markedly in the last decade or two and an increasing number of sources of good emeralds are being discovered in Brazil. To these must be added several new localities in Africa and others from the 'Indian' subcontinent. To complicate the situation further, synthetic emeralds of convincing appearance are being produced by both hydrothermal and flux-melt methods in various parts of the world. Let the jeweller and gemmologist beware, especially of clean, fine-coloured stones.

In appearance, many of the synthetics have a rather characteristic saturated bluish-green colour, which is, to some degree, distinctive, though Colombian emeralds of matching tint are sometimes encountered. Through the Chelsea filter many appear intensely red, and the same is true of their appearance between crossed filters, due to fluorescence. However, some natural stones, especially those from Colombia, may exhibit similar properties and not all synthetics behave in the same way.

The SG of most of the flux-melt synthetics (including Chatham, Gilson, Inamori, Lennix, Russian, Seiko, and Zerfass) are generally low, almost exactly matching that of quartz (2.651), so that diluted bromoform in which quartz (and, say, a Chatham stone) is suspended as an indicator forms a very useful and rapid check for suspect synthetic emeralds. However, it should be mentioned that Gilson did produce (but has now phased out) his 'N' type synthetic with extra iron which increases the SG to 2.68–70 and inhibits fluorescence. In contrast, most natural emeralds have SGs above 2.70, whereas hydrothermal synthetics are usually in the range of 2.68–70.

The RIs of the flux-melt synthetics are also low, usually around the range 1.561–1.564 with a birefringence of 0.003–4. There is slight variation, and the birefringence of Zerfass synthetics may rise to 0.006. Hydrothermal synthetics usually have RIs in the range of 1.566–1.580 with a birefringence of 0.005–7, the Russian product having slightly higher constants. The lowest index in natural emeralds is usually from 1.570 upwards with a birefringence of 0.005–6 and above. A Rayner spinel refractometer (if available) is particularly valuable in enabling accurate readings in this region to be made easily in ordinary light and for this test, of course, the stones do not normally need to be unmounted. Some synthetic stones (except Gilson 'N' types) fluoresce strongly in varying shades of red under long-wave ultra-violet light, and some natural stones, especially Colombian, behave similarly, but in general the synthetics show brighter reds. Thus the test is one indicator only and not conclusive. Seiko stones are reported to appear green under ultra-violet stimulation.

The above features are valuable as testing aids but it is always possible that the manufacturers may so vary their ingredients as to raise both the RI and SG of their products. The inclusions, which are the most diagnostic feature, could not be so easily altered or controlled. In flux-melt synthetics common inclusions are veil- or lace-like feathers, usually curved (see Figure 9.14), which are quite unlike any feathers seen in natural emeralds. These veils, which resemble liquid or gas fingerprints, are usually filled with residual flux, but may contain gas as well (see Figure 9.15). Hydrothermal synthetics commonly contain spiky or 'nail-head' inclusions containing liquid and gas and capped at the broad end with phenakite crystals (see Figure 9.16). Isolated phenakite crystals may occur in both types (see Figure 9.17). Portions of seed crystals are not uncommon, and wave growth disturbances at seed/synthetic boundaries are seen in some types, as are other characteristic growth features. Some of these inclusions are illustrated in Chapter 15 and in the colour plates.

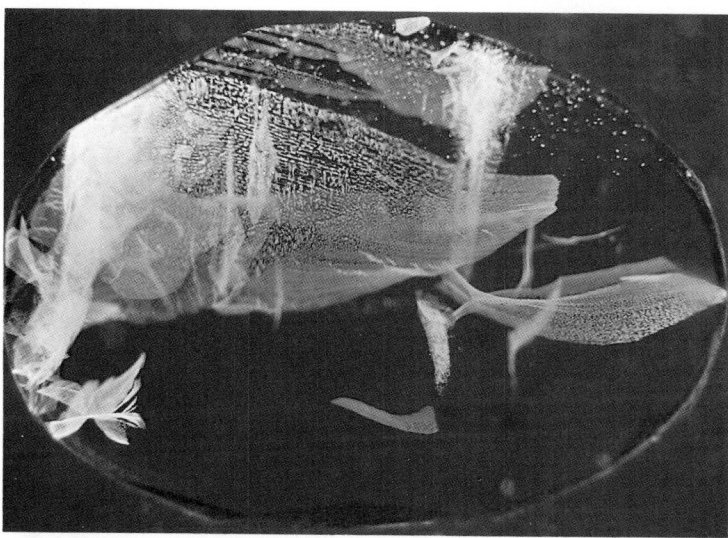

Figure 9.14 Twisted veils in Gilson synthetic emerald

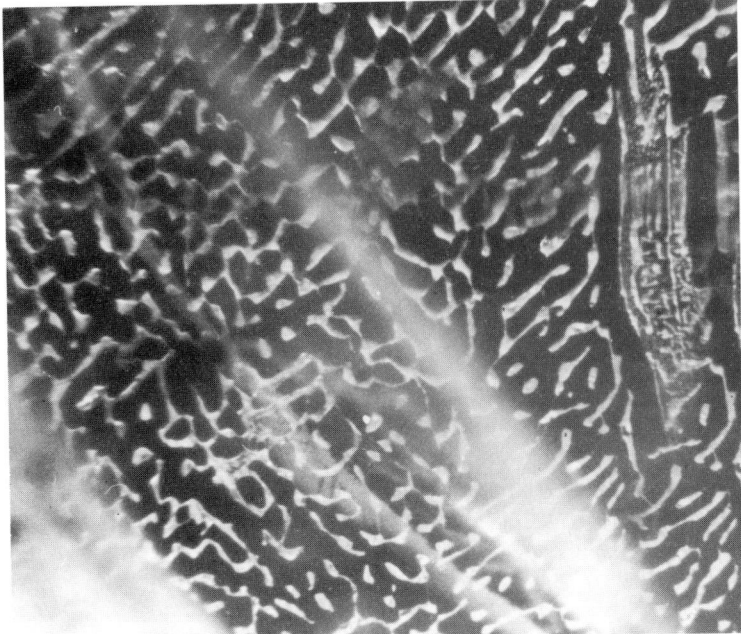

Figure 9.15 Typical pattern of liquid wisp-like 'feather' in Chatham synthetic emerald (photo: E. Gübelin)

Figure 9.16 Needles capped with phenakite crystals in Linde synthetic emerald

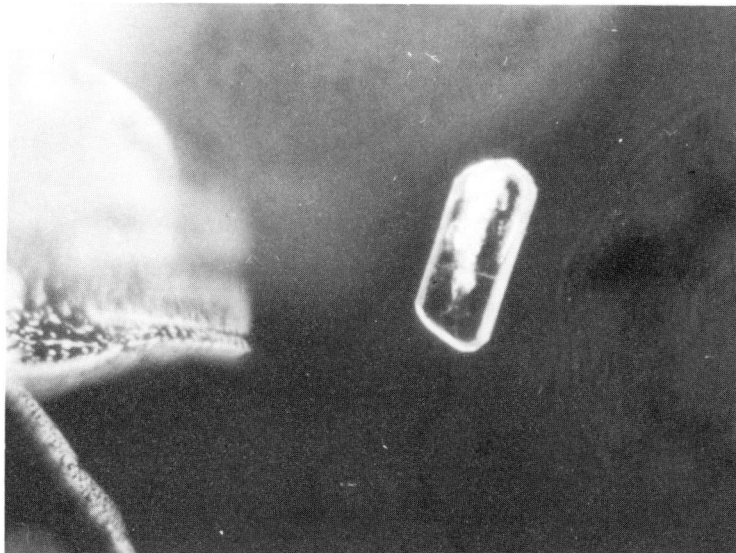

Figure 9.17 Phenakite crystal in synthetic emerald (photo: E. Gübelin)

One other marked difference between synthetic and natural emeralds may be mentioned here as it provides a valuable laboratory test for distinguishing between them. It was discovered by the author (1953) that whereas emeralds from all known natural sources are opaque to ultra-violet rays of wavelength shorter than 300 nm, synthetic (Chatham) emeralds transmit short-wave light freely down to about 230 nm. These results were first established by means of a small quartz spectrograph, but can be demonstrated much more cheaply and simply by the method proposed by Mr Norman Day, which has already been described as a means of distinguishing between natural and synthetic ruby.

In common with some other hydrothermally grown synthetic emeralds and unlike flux-grown synthetic emeralds, the Lechleitner overgrowth stones cannot be identified by their physical constants.

The refractive indices are usually about 1.575–1.582 or even higher, with a constant birefringence of 0.007. This, of course, represents the figures for the *coating*. Some Lechleitner stones reported by Professor H. Bank showed remarkably high refractive index readings, ranging from 1.583 to 1.605 for the ordinary ray. In some cases the table facet may have been polished free from its coating and will then give refractometer readings for the underlying natural beryl.

The SG of the stones, on the other hand, depends almost entirely upon the type of beryl chosen for the faceted nucleus. Where this is pale aquamarine, the SG will probably be near 2.69, which is lower than that for any natural emerald; but if pale varieties of morganite (pink beryl) are utilized the density may be much higher.

Detection of these coated beryls is not difficult, however, once suspicion has been aroused by their inclusions (which, though typical of natural beryls, are *not* typical of emerald) and general appearance. Parts of the rear facets, for instance, are often deliberately left unpolished in order not to weaken the colour, and the

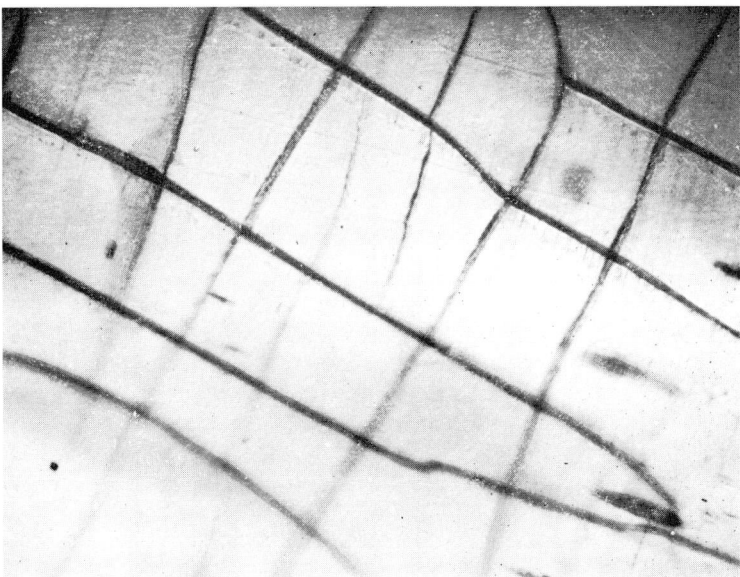

Figure 9.18 Crack-like markings in a surface layer of Lechleitner synthetic emerald-coated beryl

Figure 9.19 Hydrothermally grown Lechleitner synthetic emerald showing central seed with two symmetrically arranged growth layers above and below, each layer showing several sub-layers (photo: K. Schmetzer)

distinctive appearance of the crystallized overgrowth can then be recognized by careful examination with a lens or low-powered microscope. The most general and easily noted feature, however, is a series of parallel fissures resembling the 'crazing' on pottery, which are confined to the hydrothermal coating. Sometimes two intersecting sets of cracks are seen forming a net pattern (see Figure 9.18 and Plate 16). These fissures can be seen under the microscope to stop abruptly at the surface of the beryl 'core'. It will further be observed that the cracks are filled with liquid droplets as in a typical 'healing fissure'.

If there remains any doubt, and the stone can be removed from its setting, or has edges unobscured by the setting, the stone should be immersed in a liquid of matching refractive index such as bromobenzene or benzyl benzoate and examined under the microscope. In a Lechleitner synthetic the dark rim of emerald-coloured beryl can then be seen.

Synthetic emeralds of quite another type were produced by the ingenious Mr Lechleitner some years later (1964). These were made by inserting a thin plate of natural or synthetic colourless beryl into an autoclave and 'plating' this with thin layers of dark green synthetic emerald. The stone was then enlarged by placing it in another autoclave and growing further layers of colourless beryl (which is deposited much more rapidly than the green). The stone when finally faceted has the appearance of normal emerald when viewed from the front, but when immersed in liquid and looked at edge-on under the microscope it reveals its banded structure in a startling manner (see Figure 9.19). The refractive indices and density of such 'sandwich' stones are much the same as for some natural emeralds, so neither RI nor density is of much assistance in their identification.

One interesting type of synthetic green beryl has been produced in which the colour is not due to chromium, as in true emerald, but to vanadium. This synthesis was initiated by Dr A. M. Taylor, working for Crystals Research Co. of Melbourne. The element vanadium is responsible in part for the colour seen in natural emeralds, to which chromium gives the necessary richness and brilliance which have made the stone so famous and so desirable. These vanadium 'emeralds' are grown hydrothermally and have refractive indices between 1.571 and 1.575 for the ordinary ray and 1.566 and 1.570 for the extraordinary, with double refraction about 0.005. When immersed, stones show a banded structure indicating the different periods of growth. There are no chromium lines, of course, but there is a broad absorption band in the orange-red centred near 610 nm. Synthetic beryls doped with cobalt and of a reddish colour have also been produced.

Synthetic quartz

It has been mentioned in Chapter 7 that quartz is grown hydrothermally in Russia, Japan, and elsewhere and cut stones are now available in colourless, yellow, green, brown, amethyst, and blue colours. Since this blue has not been reported in nature and since it is coloured by cobalt, it is easily distinguished as it appears red under the Chelsea filter. However, the distinction of natural from synthetic amethyst has posed a problem, especially since some dealers have reported that, after testing, up to 25 per cent of their stock was found to be synthetic. Fortunately, the test first suggested by K. Nassau, developed by K. Schmetzer and refined by R. Crowningshield, C. Hurlbut, and C. W. Fryer has made separation easier. The test depends upon the fact that almost all natural amethyst is twinned on the Brazil Law, whereas virtually all synthetic material is purposely grown untwinned. Figure 9.20 shows the nature of Brazil twinning and its relationship to the major and minor rhombohedral faces of a natural crystal. The cut stone, held in tweezers or the fingers, is placed between the crossed polars of a polariscope, microscope, or two sheets of suitably fixed

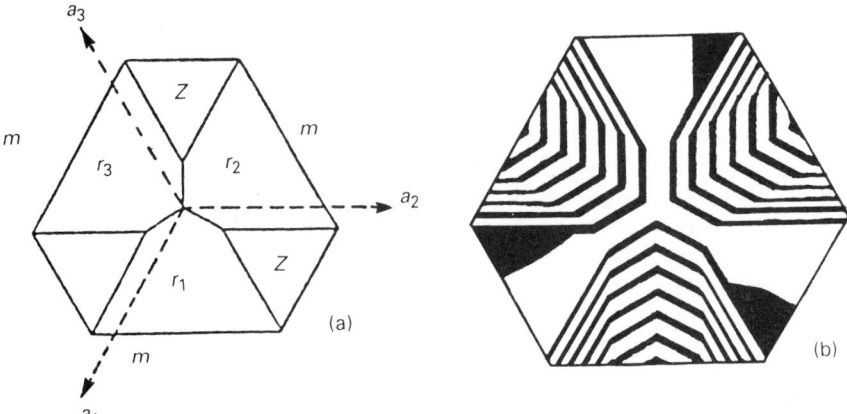

Figure 9.20 (a) Diagrammatic view down the optic axis of an amethyst crystal showing the major (*r*) and minor (*Z*) rhombohedral faces; (b) a similar view between crossed polars showing polysynthetic twinning under the major (*r*) rhombohedra with only single twin boundaries entering the minor (*Z*) sectors

polaroid and manoeuvred so that the optic axis of the stone is coincident with that of the polariscope or other instrument. This operation requires skill, patience, and persistence, but given all three, the fine bands of the polysynthetic twinning may appear. However, if the stone is cut from under the minor rhombohedron part of a crystal, twinning may not be obvious and careful search may be necessary to find twinning in a part of the stone cut from the major rhombohedron. The operation requires practice with a series of stones.

It should be remembered that much natural material displays 'tiger stripes' and may show a finely rippled surface on natural fractures. Natural stones may exhibit typical colour zoning where the darkest colour appears under the major rhombohedron. Synthetic stones may show 'breadcrumb' inclusions and some Japanese material has been reported with 'feathers'. Twinning in the form of an acutely angled or flame-like pattern is associated with some synthetic stones. However, by observing twinning, colour zoning, and inclusions it should be possible to identify all stones. Infra-red spectrometry promises to provide positive proof of origin, but many people would see this expensive technique as using a 'sledgehammer to crack a nut', and a relatively inexpensive nut at that.

Having said all this, it must be mentioned that in the relatively unlikely event of commercial synthetic material being deliberately grown from a twinned seed then the synthetic quartz produced may well reveal the Brazil twinning described above.

Synthetic opal

The synthetic opals so far examined (including white, black, and fire 'opals', all with a play of colour) have a density slightly lower than natural opals of comparable appearance (about 2.05 as against 2.10), and their hardness is also slightly lower (4½ as against 5½ on Mohs' scale). But these small differences are not of much practical value. The phosphorescence after exposure to long-wave ultra-violet is much shorter in duration in synthetic than in natural stones, but a more practical means of distinction is provided by a close study of the structure of the iridescent scales and patches under a lens or low-power microscope. In natural opal the patches of colour have a silky or satiny sheen and are striated along one direction, whereas in many of the Gilson opals the colour patches have a puckered seam across them – resembling, as K. Scarratt has observed, the skin surrounding a healing scar – while in others there is a scaly 'lizard skin' surface to the coloured area (Figure 9.21 and Plate 25). Some stones display a

Figure 9.21 'Lizard skin' effect in Gilson synthetic white opal (photo: W. F. Eppler)

distinct columnar structure (see Plates 24 and 26). Some black Gilson opals have long ribbon-like streamers of colour, not at all like those seen in nature. The overall display of colours (including a generous amount of red) outbids, one must reluctantly admit, all but the very finest natural black opals. In the earlier Gilson black opals the stones were far more translucent than true black opals, but in some seen more recently there is a greater degree of opacity. Many Gilson opals are slightly porous, and these tend to stick to the tongue when it is applied to the surface. Gem-quality natural opals show no porosity.

Some Gilson synthetic opals examined by K. Schmetzer have slight compositional differences from natural opal and in some very small amounts of ZrO_2 have been detected; this author has questioned whether the Gilson material should be designated as imitation opal.

'Plastic opals' have now been manufactured and they are very convincing at first sight, but their nature is immediately apparent on picking them up – the density may be as low as 1.0. The 'plastic' may be composed of spheres of a co-polymer of styrene impregnated with methyl methacrylate as a matrix and of different refractive index. A typical RI for the material is 1.465–1.480 with a SG of 1.17.

Synthetic turquoise

There have been many so-called 'synthetic turquoises' produced during the past half-century, some of which have had the chemical composition of true turquoise, while others have merely had a plausible appearance. In 1972, however, Pierre Gilson succeeded in manufacturing a true synthetic turquoise, having not only the composition but also the crystal structure of the natural material, as proved by X-ray powder photography. In appearance this new production is rather pale in tint, and the density is rather lower than in top-quality Iranian or Egyptian material – near 2.7 rather than 2.8. The most revealing sign by which the Gilson stones can be recognized is the surface structure under a magnification of some ×50, which reveals angular blue grains embedded in a whitish matrix. The absorption band at 432 nm in the violet, which can usually be detected by reflected blue light in natural turquoise, has not been observed in the Gilson version.

'Synthetic' lapis lazuli

A Gilson version of the decorative rock, lapis lazuli, has not been so successful in reproducing nature. It has a fine colour, much the same composition as true lapis, and shows a predominantly lazurite structure by powder photography, but it is notably softer than lapis, lower in density (around 2.38 as against an average value of about 2.8) and is decidedly *porous*. Tests on the new material by A. E. Farn also revealed that a strong reaction (with effervescence and production of sulphuretted hydrogen, with its strong smell of rotten eggs) occurred when a spot of hydrochloric acid was applied to an inconspicuous surface. True lapis gave a negative result when tested in this manner. To add verisimilitude to the manufactured material, fragments of iron pyrites are incorporated, but these show no crystal outlines and are friable when the point of a needle is applied to them. Possibly the simplest non-destructive test is one that can be applied to either an unmounted or a mounted stone, which is to weigh the stone or jewel very carefully and note the weight, then immerse the specimen in water for 15 minutes, dry the surface carefully, and reweigh. A marked increase in weight will signal the fact that the stone is not true lapis lazuli. Another point to mention is that the Gilson imitation is opaque, whereas genuine lapis is slightly translucent. The Gilson product has been examined by K. Schmetzer, who finds that a significant proportion of hydrous zinc phosphate is present, and maintains that it should be described as an 'imitation' rather than a synthetic.

Some other synthetics

In the years since the publication of the ninth edition of this book some surprising successes have been achieved in the field of synthetic gemstones, creating still further problems for the gemmologist to solve. It is hoped that the following descriptions, which include some earlier syntheses, will help materially in enabling correct identifications to be made.

Synthetic alexandrite

For the would-be manufacturer of precious stones, the enormous care and skill and patience needed to produce a saleable product can only be rewarded if the stones concerned are among the rarest and most coveted of those found in nature. Of these, alexandrite is undoubtedly a prime example – due, the author feels, more to the extreme rarity of this peculiar variety of chrysoberyl than to its intrinsic beauty. With the discovery of very fine material at Hematita in Brazil in 1986, alexandrite has become more easily available (at a price) and so has the incentive to manufacture convincing synthetics. Since 1970 the firm of Creative Crystals Inc. in California has succeeded in crystallizing this chrome-bearing form of chrysoberyl by the flux-melt method. The stones show a more emphatic red than Sri Lankan stones in electric light and are more fluorescent. They also show the type of 'feathers' typical of this form of crystal growth, with which we have become familiar in so many of the synthetic emeralds grown by Chatham and by Gilson – veil-like patterns of interconnected canals (Figure 9.22). More

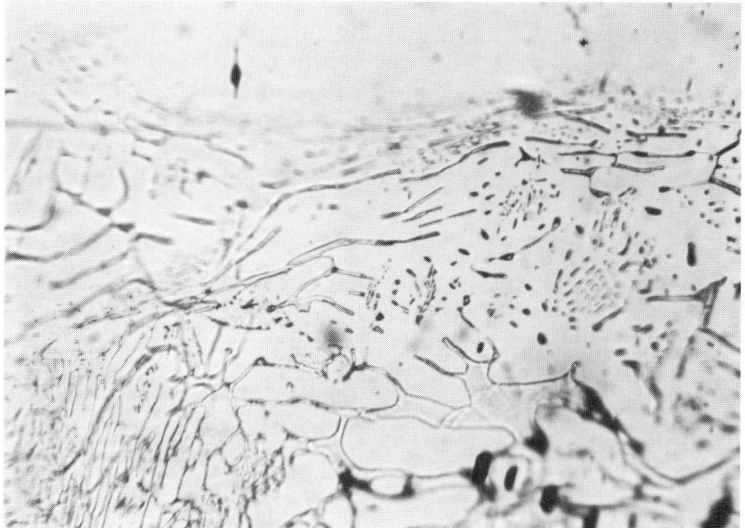

Figure 9.22 Typical liquid 'healing feather' in flux-fusion synthetic alexandrite (photo: W. F. Eppler)

recently, alexandrites have been successfully grown in Japan and the USA by the Czochralski 'pulling' technique. These show curved striae of a sort, and occasionally gas bubbles. For those familiar with natural alexandrites the synthetic stone will at once appear unusually chrome-rich and opulent in its night appearance, and its fluorescence is unusually strong. R. T. Liddicoat reports the occasional appearance of small metallic hexagonal platelets in these synthetics.

Seiko in Japan has now produced synthetic alexandrite by the floating-zone method, the Russians have perfected hydrothermal techniques, and Kyocera (Inamori) have recently produced synthetic alexandrite cat's-eyes, probably by the 'pulling' method. The synthetics are often brighter in colour than the natural stones and may show better colour changes. Most of the synthetics show strong reds through colour filters with strong fluorescence, but the cat's-eyes may show a chalky yellow surface fluorescence with an underlying weak orange-red fluorescence. Examination of the internal structures under the microscope is usually the best method of distinction, especially since the natural stones often contain solid crystals and ultra-fine tubes or needles in the case of natural cat's-eyes. The synthetic cat's-eyes contain fine whitish dust particles oriented in parallel planes, and this gives rise to the chatoyancy.

Synthetic jadeite

This was prepared in 1984 by General Electric of America using pressures in the 30–50 kilobar range and temperatures of 1200–1400°C. Best results were obtained using a prepared glass of jadeite composition. The colours range from white to greens, mauve, and even black, and they may be stronger than in nature and mottling may be more intense. Hardness may be up to 7½–8 in contrast to the 6½–7 for natural material. There are no marked differences in refractive indices, absorption spectra, specific gravity, or fluorescence. GE do not envisage marketing synthetic jadeite, but the process is straightforward, although expensive, and a Japanese firm has also synthesized jadeite. Given that jadeite is readily available commercially, the expense of synthesis would not seem worthwhile, unless transparent emerald-green 'Imperial jade' was produced; GE did not synthesize such material.

Synthetic scheelite

Gem-quality scheelite is so rare a commodity that cut stones of this calcium tungstate mineral are to be regarded as prizes for collectors rather than as serious material for use in jewellery, but it so happens that it has been extensively manufactured in large, clear pieces by the Czochralski 'pulling' technique, and synthetic material has been fraudulently offered for sale at high prices. The gemmologist may thus be faced with the need to distinguish synthetic from natural material – not an easy task.

The properties of pure scheelite are as follows: SG 6.1, refractive indices 1.918 for the ordinary and 1.934 for the extraordinary ray, giving a double refraction of 0.016 – perceptible to a lens. The hardness is only 5 on Mohs' scale, but when well cut its dispersion of 0.026 makes it a very attractive gem. The colour varies from pale yellow to orange or brown.

The most distinctive property of scheelite, for which it is well known, is its strong fluorescence under short-wave ultra-violet light. This is usually bluish-white, but tends to become green when molybdenum is present, as it frequently is in nature.

Natural scheelite nearly always reveals a faint group of lines, due to 'didymium' rare earths, in the yellow near 585 nm. In the artificial scheelite these lines will either be absent altogether or will be present with many other lines due to neodymium, having been purposely 'doped'. Another sign of synthesis can be the presence of curved lines, very much as seen in a Verneuil synthetic, and clouds of very minute bubbles. The natural scheelites are likely on their part not to be entirely faultless, and to contain such revealing signs of natural growth as liquid 'feathers' and minute crystals.

Lithium niobate is one such synthetic and has been manufactured both in Europe and the USA, where it is marketed under the trade name 'linobate'. Intrinsically colourless, linobate has been made in bright green (chromium), yellow (nickel), blue (cobalt), and red (ferrous iron) colours by addition of the trace elements indicated. It is manufactured by the 'pulling' method from a melt of the niobate fused in a platinum crucible, the melting point being 1250°C. Its SG is 4.64 and hardness 5½. The refractive indices are high, being 2.30 for the ordinary and 2.21 for the extraordinary index, with a correspondingly high dispersion of 0.120 for the B–G range. When it is freshly cut, linobate, with its high lustre and fire, makes an attractive gemstone. The easily discernible double refraction 0.09 might cause it to be confused with zircon (double refraction 0.059), but serves to distinguish it readily from diamond or demantoid garnet.

Silicon carbide, popularly known under its trade name of **carborundum**, is one of the oldest synthetic products, being manufactured on a large scale for use as an abrasive by fusing sand and coke together in an electric furnace. At the surface of the resultant mass it forms lustrous, platy hexagonal crystals of bluish-green colour, and has occasionally been cut as a gem, when it may resemble a green irradiated diamond. Its high refractive indices of 2.65 and 2.69 give it an almost metallic lustre; its SG is 3.17. The extreme hardness of carborundum (9½ on Mohs' scale) gives it an importance next to diamond as an abrasive agent.

Periclase, the crystalline form of magnesium oxide, is rare as a mineral, but for many years has been available in sizeable, clear, colourless pieces as a synthetic material. It crystallizes in the cubic system and has a perfect cubic cleavage. The refractive index is 1.734, SG 3.58, and hardness 5½. It has nothing in particular to commend it. It is worth noting, however, that the properties (except for hardness) are very similar to those of spinel.

Bromellite (beryllium oxide) is another rare mineral which has been manufactured to some extent. It forms colourless hexagonal crystals with refractive indices 1.719 and 1.733. The SG is 3.02, and the hardness (9 on Mohs' scale) is exceptionally high.

Glass imitations

Although the abundance of synthetic corundums and spinels manufactured by the Verneuil process has provided stones of greatly superior hardness to any glass imitations, the latter are still widely used, and are not always easy to detect at sight. As far as colour is concerned, glass can be made which matches most of the gemstones represented very exactly, and those stones which have a rather low refractive index have decidedly a vitreous (i.e. glassy) lustre, though their greater hardness enables them to take and to retain a higher polish than is possible with glass.

Jewellers were formerly accustomed to rely on a test with a hard file as an effective method for discriminating between real stones and glass imitations, but unless great care is taken damage may easily be done even to genuine specimens by the use of this implement.

Better and safer methods are in any case available for detecting pastes, and some of these can now be listed.

1. *Warmth to the touch.* Glass conducts heat less readily than most crystals, and thus feels warmer to the touch. The tongue, being thin-skinned and always practically at blood heat, is the most sensitive gauge of this property. The specimen should be cleaned and handled with corn tongs (unless it is in a setting) to prevent it being warmed by the hand, which would vitiate the test. Direct comparison with a known crystalline gem makes a decision more easy. When properly carried out this is a very useful test.

2. *Conchoidal fracture.* Glass is brittle and has a marked conchoidal fracture – that is, it breaks or chips in curved shell-like pieces, leaving shell-like concavities on the specimen. Examination with a lens of the edges of the specimen, especially where the claws of the setting (if any) bear upon it, will usually reveal some of these typical fractures. It must be said, however, that many gemstones also have a conchoidal fracture, though with these it is not so readily produced, nor on such a large scale.

3. *Bubbles.* The presence of one or more comparatively large bubbles is a frequent feature in glass imitations, and, when seen, is a sure sign that the specimen containing them is not genuine.

   With experience, one learns to distinguish between the bubbles seen in glass and those seen in synthetics of the Verneuil type. There are, of course, plenty of other distinguishing features between pastes and synthetics, but it is as well to be able to gain as much information as possible even from such humble things as bubbles. In pastes, bubbles are most often spherical and not grouped with others, but a pointed ellipsoid is very often seen, and this particular shape seems peculiar to glass alone (see Figure 9.23). Sometimes chains or sheets of bubbles are found in paste and these in imitations of emerald help to give the illusion of natural flaws and feathers until they are closely examined (see Figure 9.24).

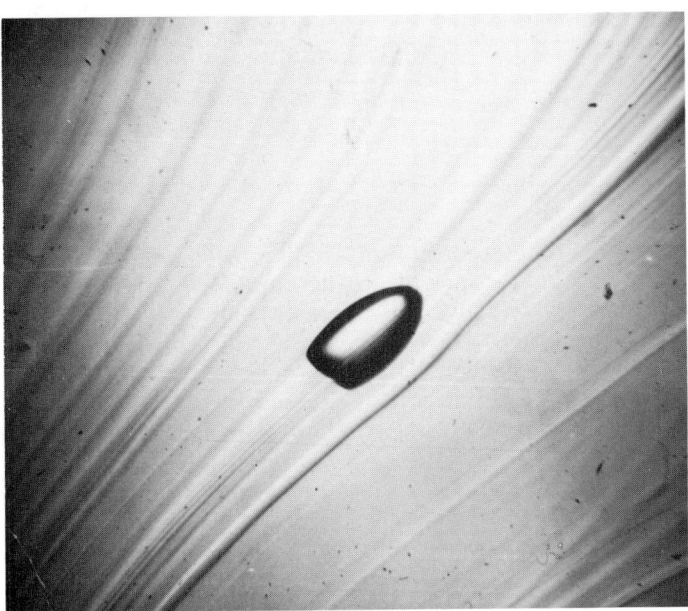

Figure 9.23 Bubble and swirl marks in paste

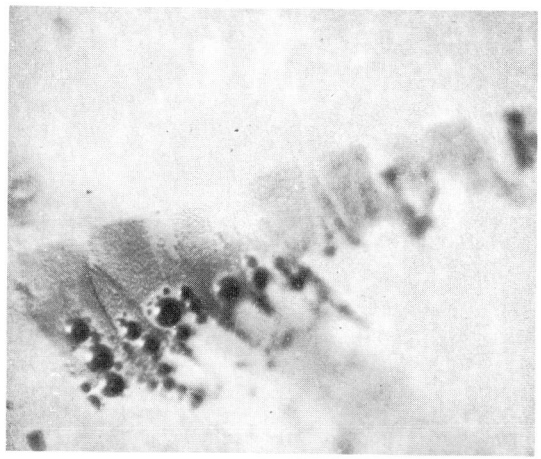

Figure 9.24 Bubble 'feather' in glass imitation emerald

Even in the translucent and semi-opaque glasses used to simulate jade, amber, or turquoise one can often detect bubbles just below the surface or circular cavities on the surface where bubbles have been partly polished away.

4. *Striae.* Glass often contains curved striae or 'swirl marks', due to imperfect mixing of the ingredients. These striae are not so regular as the curved growth lines in synthetic corundum. The effect is rather like that seen when sugar syrup is poured into water, and indeed is due to a similar cause – contiguous streams or layers of slightly different refractive index.

5. *Single refraction.* Glass, being amorphous (non-crystalline), is always singly refracting, and thus shows only a single shadow-edge on the refractometer. Between crossed polarizers, however, glass may show irregular birefringence due to strain (see Plate 58), and in rare cases may even show extinction positions at more or less regular intervals in the manner of crystalline gems, which is very disconcerting. Glass never shows dichroism, and since most coloured gems are pleochroic this is often a simple means of distinction.

6. *Refractive index and SG.* In a glass imitation these constants are seldom the same as those of the stone represented. Some typical values for imitation gemstones will be found in Table 9.2.

7. *Interference figure.* Between crossed polars in convergent light, many pastes show a crude version of a uniaxial interference figure, resembling a letter 'x' in which the two arms are curved and do not quite meet. When seen, this is a very revealing phenomenon (see Plate 58). It is probably caused by a crude orientation of the glass molecules produced by pressure when moulding the stone in the shape of a faceted gem.

As can be seen, the properties of glass vary widely according to its ingredients. Of the various types which have been used to imitate gems, by far the commonest are the calcium or 'crown' glasses (consisting of silica, potash or soda, and lime) and the lead or 'flint' glasses, in which lead oxide replaces the lime in the above to a greater or lesser extent. The more lead that is present, the greater the refractivity, dispersion, and SG, but at the same time the hardness decreases progressively. In the borosilicate crown glasses, some of the silica is replaced by boric oxide. Borosilicate glass can be very hard (nearly 7 as against

Table 9.2   Properties of some coloured glasses

Colour	SG	RI	Type of glass
Colourless	2.30	1.47	Borosilicate crown
Pale blue	2.35	1.50	Borosilicate crown
Pale blue	2.37	1.51	Borosilicate crown
Yellow	2.43	1.498	Calcium crown
Pale blue	2.46	1.51	Calcium crown
Blue	2.44	1.515	Fused beryl
Emerald-green	2.49	1.516	Fused beryl
Pale blue	2.70	1.59	Calcium-iron
Colourless	2.87	1.54	Light flint (lead)
Pale blue	3.19	1.575	Flint (lead)
Yellow	3.627	1.633	Flint (lead)
Ruby-red	3.69	1.63	Flint (lead)
'Jade'	3.73	–	Flint (lead)
Colourless	3.74	1.635	Flint (lead)
Pale blue	3.84	1.642	Flint (lead)
Orange-brown	4.12	1.68	Flint (lead)
Red	4.16	1.683	Flint (lead)
Emerald-green	4.25	1.70	Flint (lead)
Yellow	4.98	1.77	Flint (lead)
Yellow	5.12	1.775	Extra dense flint
Refractometer	6.33	1.962	Extra dense flint

5½ for 'crown' glass), and this is the type used in the preparation of the so-called 'mass-aqua' imitations of which the second and third specimens in Table 9.2 are examples. The lowest figures for SG and refractive index are found in the 'opal' glasses, containing fluorine and certain other additions to the normal crown-glass melt. With these, the SG may be as low as 2.07, with refractive index 1.44. At the other end of the scale, 1.69 may be taken as practically the limit for the refractive index of glass imitations. The only exceptions to this rule encountered by the writer were the remarkable pastes listed in Table 9.2. Such heavy lead glasses are extremely soft and susceptible to tarnish. Last in the table comes the heaviest type of lead glass, used in some refractometers, and quite unsuitable for other purposes; it is included here merely as a matter of interest.

As far as values of SG and refractive index are concerned, the only glasses liable to cause confusion with actual gemstones are the calcium–iron glasses with constants near those for beryl, and those of the lead glasses having values close to those for topaz. On account of their higher dispersion (more nearly matching that of the refractometer glass), the lead glasses with refractive index over 1.60 give a sharper shadow edge in white light on an ordinary refractometer than any gem of similar refractivity. With a spinel refractometer the reverse is true: while gems give a sharp shadow edge, lead-glass imitations show a colour fringe which is very distinctive.

The fused beryl glasses listed in Table 9.2 have the same composition as the beryl from which they are made, but they are non-crystalline and thus have lower constants. As another example of the lower constants of a glass compared with a crystal of the same composition one may cite fused quartz, which has SG 2.21 and refractive index 1.46 against 2.65 and 1.54–1.55 for the crystalline mineral. The beryl glasses are very hard (about 7) and can be made in various colours, such as blue by addition of cobalt and green by addition of chromium oxide. Usually they contain numerous large bubbles.

Glass which is coloured by cobalt has a characteristic absorption spectrum containing three broad dark bands centred at about 655 nm in the red, 580 in the yellow, and 535 in the green. They appear a rich red under the Chelsea colour filter. There are other blue glasses having no distinctive absorption spectrum, which appear a dirty green under the filter. Red glasses coloured by selenium have a single broad band in the green of their absorption spectrum, not unlike that seen in red tourmaline. Pink or red glasses containing rare earths and showing a strong didymium absorption spectrum are sometimes seen.

Glass imitations are often moulded, or very badly cut, showing the ridge marks left by the mill – another indication of their nature. Where the table facet is not truly plane there may be difficulty in getting a good reading on the refractometer.

'Slocum Stone'

This is a convenient term to describe an extraordinary form of glass that has been developed by John Slocum in the USA to vie with the most precious forms of natural opal (see Plate 28). These 'stones' are the result of many years of painstaking and expensive experiment with silica materials, first with the idea of making synthetic opals and later of concentrating on a virtually anhydrous form of silica glass that would not only equal opal in iridescence but would avoid its vulnerability. The results of Slocum's process are unpredictable from batch to batch, some producing 'pinfire' iridescent samples, other with fiery flashes from a dark background, some virtually clear or milky with iridescent spangles. With the naked eye it is difficult to distinguish some of these imitations from natural opal, but under a low-power microscope the structure of the iridescent patches is seen to be entirely different. Sometimes there is an effect of crumpled coloured tinsel, sometimes there are triangular patches of iridescent colour, and in no case does one see (in the limited number examined) the completely flat planes with striated satiny finish that are the glory of true precious opals. The properties of these remarkable imitations are sufficiently different from opal to distinguish them either by refractive index or density tests, though one must admit that with a mounted cabochon specimen, where a density test is precluded, one must be well practised in the 'distant vision' method on the refractometer to make plain the rather small difference in RI. The density of the Slocum glass is usually between 2.40 and 2.50, and the refractive index between 1.49 and 1.52. The hardness is near 6 on Mohs' scale. In composition the glass appears to be a sodium-rich silicate, containing some calcium and magnesium. In recent promotion leaflets the Slocum product is referred to under the unacceptable fancy name of 'Opal Essence'.

Quite recently, electron micrographs (taken by P. J. Darragh and J. V. Sanders) of the surface of Slocum Stones have revealed the physical basis for the production of the iridescent colours displayed. Under a magnification of some 18 000, zones of parallel, extremely thin layers could be seen to occur at intervals in the otherwise structureless glassy matrix. These layers are of equal thickness and of the correct dimensions for producing a display of diffraction colours by reflected light.

Natural glasses

In addition to man-made glasses there are several types of glass found in nature which may be conveniently mentioned here.

Silica-glass, consisting of almost pure silica (SG 2.21; refractive index 1.46), was found lying on the surface of the sands of the Libyan desert, from which it had apparently been derived. It has a pale yellowish-green colour, and is slightly opalescent. The glass formed from fusion of sand in the New Mexico desert as a result of the first atomic bomb retained considerable radioactivity.

Obsidian is a volcanic glass, produced by the rapid cooling of molten rock material which would have produced granite if allowed to cool slowly. It is usually black or very dark brown in colour, though iridescent, and mottled brown and black types have been found, and make rather attractive beads. The SG is near 2.4 and the refractive index is 1.50, hardness being about 5 on Mohs' scale. Ordinary clear green bottle-glass such as may be picked up on the beach, or even blue artificial glass, has been fashioned and sold as 'obsidian' to the ignorant public at seaside resorts. Under the microscope obsidian shows signs of incipient crystallization not seen in artificial glass.

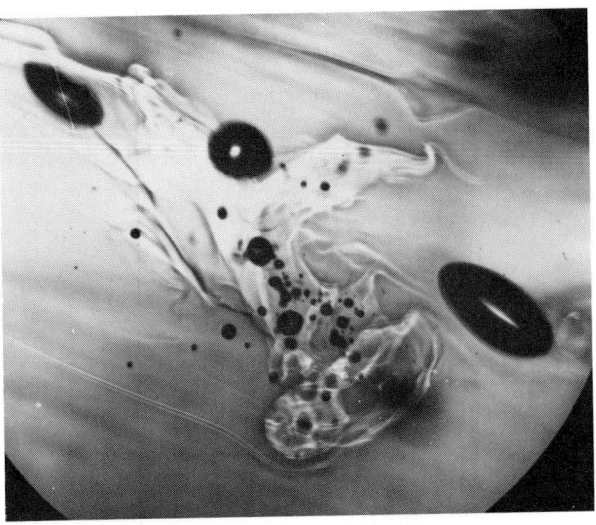

Figure 9.25 Bubbles and swirl marks in moldavite

Moldavite is the name given to clear green or brownish-green varieties of small glassy bodies sometimes classed with tektites. They have probably been caused by the fall of a huge meteorite. The energy of the impact fused the surrounding country rocks which were then 'splashed' outwards for considerable distances. Passage of the still-plastic melt through the atmosphere caused their characteristic shape. The properties of moldavite are practically the same as those of obsidian, but the internal characters are quite different. Moldavite contains large and numerous bubbles and swirl marks (see Figure 9.25), and lacks the incipient crystals which are seen in obsidian.

Composite stones

While the jeweller–gemmologist must view with some alarm the ever-increasing variety of synthetic gemstones now on the market, the ingenuity recently exercised in the art of making composite stones may prove an equally serious menace.

At the beginning of the twentieth century the most common doublets consisted of the garnet-on-glass type, in which a thin sliver of almandine was fused onto a coloured glass base and the whole was then faceted and used to represent ruby or sapphire. The junction between garnet and glass was usually above the girdle and could be observed with a lens, chiefly by the sharp change in lustre on either side of the junction line (Figure 9.26 and Plate 35). Many

Figure 9.26 Garnet-topped doublet showing change of lustre at the junction line between the garnet and the glass base (photo: R. Webster)

millions of such doublets must have poured out of workshops at one time, and they are quite often encountered in old jewellery to this day. The almandine top frequently contains rutile needles, and typical squashed gas bubbles can be seen through the table where the garnet meets the glass (Figure 9.27 and Plate 34). Less frequently encountered was the so-called soudé (soldered) emerald, in which two pieces of pale beryl, or more often rock-crystal quartz, were cemented together with a dense green compound that provided the necessary emerald-like colour, the junction being at the girdle of the finished stone and thus hopefully concealed by the setting. Other classic types were the opal-on-opal matrix, the opal-on-onyx doublet, and the diamond doublet in which the crown consisted of diamond and the base usually of synthetic white sapphire or other suitable material. These will be discussed in more detail later.

Detection of these faked specimens is usually easy once suspicion has been aroused – particularly when the stone can be examined out of its setting – but can be dangerously deceptive and puzzling when mounted in jewellery.

IN recent years doublet making has become quite a fashion, and the task of assembling such stones has been made easy by the availability of epoxy resins

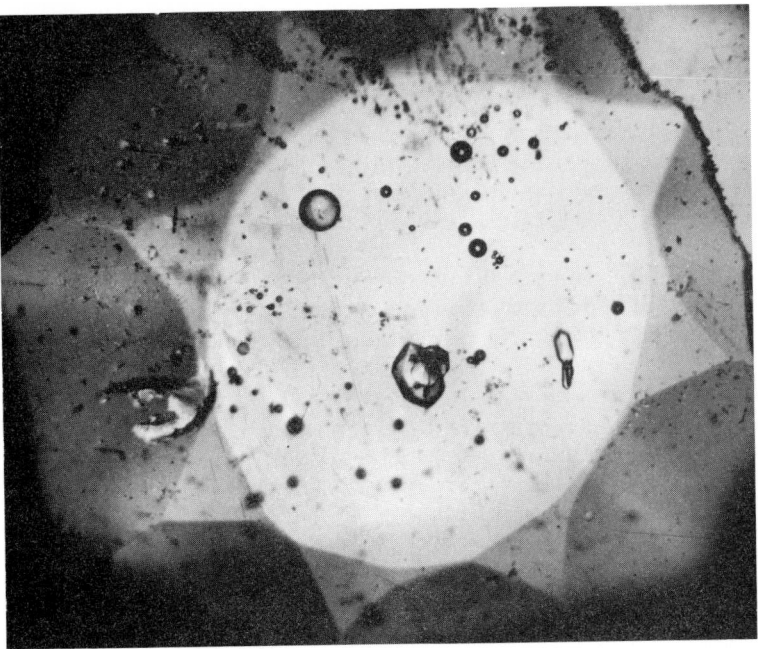

Figure 9.27 Photomicrograph of doublet, showing bubbles in the junction plane and crystals in the overlying garnet

such as Araldite, which enable a powerful and hardly detectable bonding layer to form a junction between the individual parts.

Some of the most effective fakes have made clever use of thin pieces of poor-quality Australian sapphire to form the crown of a composite stone in which the base consists either of blue synthetic sapphire or synthetic ruby. With the synthetic-on-natural sapphire doublets the junction between the two parts was exactly at the girdle plane, and so skilfully done that parcels of such stones were unsuspectingly changing hands between dealers at a 'bargain' price of £60 per carat. The silk, zoning, and other natural features in the crown of the stone gave a strong impression to the eye even under lens examination that the stones were natural, and tests with a refractometer and even the spectroscope would tend to confirm this. The stones backed with synthetic ruby were the basis of several daring sales of fraudulent jewellery involving thousands of pounds before the fraud was detected by tests carried out in the Gem Testing Laboratory in London. In this case the 'rubies' were mounted in collets to form a most impressive necklace, and offered for sale to a West End firm by an attractive young woman, at the 'bargain' price of £10 000.

Many of the recent doublets have made use of the attractive properties of some of the diamond substitutes while attempting to disguise their weaknesses. Strontium titanate, one of the most successful of the diamond simulants, has the drawback of being too soft for extensive wear in jewellery, and moreover has a dispersion more than four times that of diamond, which gives it away to the practised observer. These disadvantages are at least partly mitigated by the construction of doublets having a crown of synthetic white spinel or synthetic white sapphire and a base of strontium titanate. The junction plane is frequently placed well below the girdle, thus avoiding excessive fire in the finished article.

Opal doublets come into a rather different category from other composite stones, in that their beauty is authentically due to opal itself, which, however, is present in too thin a layer to be cut and mounted as a gemstone without reinforcement. Hence, it is cemented to a substantial base either of black onyx, or of dark-coloured 'potch' opal. If it is offered as a doublet, it can form an attractive and commercially legitimate gem material which can be used by opal lovers who cannot afford the very high price of solid black opal of high quality. Often, however, these doublets are mounted in such a way as to conceal the junction between the precious opal layer and the base. If, from the back of the jewel, this can be seen to be black onyx, there can be no doubt that the stone is a composite one. When the base is opal matrix or potch, an attempt must be made to tackle the problem. A diligent search round the perimeter will often permit a glimpse of the junction line to be seen. Alternatively, looking through the nearly transparent top of the stone under a good light a layer of flattened bubbles on the junction plane may give the show away. This frequently encountered problem is fully discussed in Chapter 21.

Mention should be made of an attractive form of opal triplet in which the flatness inherent in the thin top layer of an opal doublet is masked by cementing a dome of clear rock crystal to the iridescent layer. Such triplets should not deceive a careful observer, since the top of the stone can be seen to be free from any iridescence – and a quartz reading can be obtained on the refractometer by the distant vision technique.

The modern doublet maker has exercised considerable ingenuity in the opal field, and extreme caution must be used by the gemmologist before affirming that a stone is entirely natural. A convincingly genuine sample of opal growing naturally on its matrix may be found to consist of Coober Pedy opal cut deliberately with an irregular base, which has then been backed with a mixture of powdered opal, ironstone, and epoxy resin.

In all cases where composite stones are concerned the gemmologist's best weapons are a suspicious mind and a practised use of a ×10 lens. Where the stone is unmounted or can be taken from its setting, the classic procedure is to immerse the suspected stone in a liquid and view it from a sideways position. This will almost always reveal differences in colour or lustre of the component parts. A porcelain dish containing water is usually effective, but in cases such as natural-on-synthetic sapphire doublets it may be necessary to immerse the stone in a cell of di-iodomethane and study it under a low-powered microscope to discover the tell-tale inclusions or striae. Ultra-violet lamps can be useful: different parts of a composite stone usually have a difference in fluorescence, and the typical whitish fluorescence of synthetic sapphire can be revealing under short-wave ultra-violet rays. More information about doublets is given in the chapters dealing with the individual gem varieties.

Coated stones

A warning note on a recent process for coating the upper surface of gemstones with a film of low refractive index to improve their brilliancy and apparent depth of colour may be given here. Such a film reduces the surface reflections in similar manner to the coating or 'bloom' on lenses in cameras and binoculars, and allows more light to enter the stone. There is, of course, a reduction of surface lustre. Such treatment is a form of 'faking', and as such, should be condemned.

Coated stones can be recognized by a slight bloom or iridescence of the surface. Refractive index readings on the refractometer may be unobtainable until the coating has been removed by rubbing with rouge or similar polishing agent.

It is not uncommon to find so-called emeralds (particularly carved examples) to consist either of very pale natural emerald or other beryl coated with a layer of green gelatine. Close inspection with a lens will usually reveal a place where the coating has peeled away, and the stone has a tacky feel to the touch. In such cases the spectroscope will show a broad adsorption band in the red, due to the dyestuff, in place of the narrow chromium lines seen in emerald. Plausible emerald fakes (as 'rough' or 'crystals') composed of glass or quartz coated with green plastic have been encountered for some time. The more sophisticated examples have various additional coatings, including crushed mica-schist, detrital brown quartz, tourmaline crystals, and other natural-looking, but fabricated, matrices (see Plate 20). Even experienced gem dealers have been deceived by these clever fakes.

In recent years diamond technology has advanced to the point where ultra-thin diamond coatings can be applied to many different surfaces, including metal and glass. It is only a matter of time, therefore, before such coatings are applied to cut gemstones. This will, of course, enhance the surface lustre and greatly improve the durability of a 'base' pre-form. A diamond-coated cubic zirconia is an obvious possibility and a determination with a reflectivity/thermal tester might be misleading. This would need to be followed by a careful check of specific gravity, which would, of course, be far too high for a solid diamond.

Glass infills

Rubies and sapphires with glass infills have been encountered for several years and similar infills in other stones are obviously possible. The glass, which is fused in, is used to fill up cavities in the surface of a valuable gem when it would be uneconomic to grind away further expensive material to obtain a perfect plane facet. Infills are usually revealed by the lower lustre of the glass, consequent upon the lower refractive index, and by the presence of bubbles. Careful observation is needed since the infills are intended to deceive and are easily missed.

As this edition was going to press diamond-grading laboratories were encountering yet another type of diamond treatment. This involves the use of **solder glasses** to infill cleavages and other surface imperfections in cut diamonds. The glasses used have a very high lead content, sometimes with bismuth, and are not of the usual silicate type. The refractive indices may approach 2.4 and, therefore, they blend well with the diamond host. The exact method of treatment is not yet clear, but may involve vapour deposition in a vacuum. Detection of the treatment is not easy and is probably best left to professional laboratories. Monochromatic flash colours changing from blue to orange as the stone is moved (or single colour flashes), flow lines, and trapped

(a)

(b)

Figure 9.28 Cut diamond with a naat with an extending cleavage reaching the surface. (a) Before treatment with solder glass; (b) after treatment

gas bubbles may be seen in some cases under normal lighting with a microscope. A cut diamond before and after treatment with solder glass is shown in Figure 9.28.

10

The use of the spectroscope

The spectroscope is the third leg of the tripod of instruments on which modern determinative gemmology rests secure. The other two instruments forming the 'tripod' are, of course, the refractometer and the microscope.

We have seen how the refractometer enables us to determine the species of most faceted gemstones, while the microscope tells us of their origin. But the refractometer cannot be applied to rough stones, and cannot give readings with stones of high refractive index; and under the microscope not every stone displays revealing features. The spectroscope, it is true, also has its limitations, but often it can fill the gaps left by the other two standard instruments, and for a number of gems it provides the most rapid positive test available. It can be applied as easily to rough stones as to polished gems, and as easily to stones of high index as to those within the range of the refractometer. In many cases the spectroscope can determine the natural or synthetic origin of a gem, and detect cases of artificial coloration, whether by staining as in jadeite or by irradiation as in diamond.

All this it does merely by analysing the light transmitted by a gem or scattered from its surface. We have already seen in Chapter 4 that white light consists of a mixture of all the colours of the rainbow: red, orange, yellow, green, blue, and violet. Of these, red has the longest wavelength and violet the shortest. Formerly the Ångstrom was the accepted unit for the measurement of wavelengths in light, but with the advance of metrication and with the recommendation of the Royal Society in 1973 for the introduction of the Système International (SI, for short), the nanometre has become the accepted unit. The nanometre is one millionth of a millimetre. Fortunately, the conversion from one unit to the other is simple: 1 nanometre = 10 Ångstroms. The range of visible light is approximately from 700 to 400 nm or 7000 to 4000 Å in the older units. The wavelength of sodium light is now 589.3 nm instead of 5893 Å. Beyond the red end of the spectrum are invisible so-called infra-red rays which merge into the heat waves of still longer wavelength. Beyond the violet end of the spectrum are the invisible ultra-violet rays, which occupy the range 400 to 200 nm. X-rays, which have the same essential nature as light rays, have wavelengths of the order of 0.1 nm only.

Another consequence of the introduction of SI units is the demise of the term 'micron', so extensively used in the sizing of diamond particles to mean one thousandth of a millimetre. The new unit name is the micrometre. Clearly, there will be much digging in of heels on the part of gemmologists brought up in the old traditions that suited them so well.

The manner in which the spectroscope analyses light into its component wavelengths is, in principle, very simple: it depends upon the different degree of refraction which rays of each colour (wavelength) undergo when they are made to pass through a prism made of glass or other transparent substance. Thus a narrow parallel beam of white light, after passing through a prism, is spread out into a ribbon of rainbow colours – the visible spectrum. A spectrum can also be formed by passing light through a grating of lines spaced very closely at regular intervals. Such a grating is known as a 'diffraction grating' and is used in many spectroscopes. Both prism and grating spectroscopes have certain advantages and disadvantages. Prism spectroscopes produce a brighter spectrum, but the 'spread' of colours becomes greater and greater towards the violet end, in accordance with the increasing dispersion of the glass or other medium used for the prisms. With a diffraction grating there is an even spread of colours throughout the whole spectrum range, but since a series of spectra are produced on either side of the incident ray, far less light reaches the eye in the one spectrum which is observed. Since the brightness of the spectrum seen is of prime importance, the author recommends the use of a small prism spectroscope for 'spotting' purposes and for the rapid diagnosis of gemstones by means of their absorption spectra. Such a spectroscope consists essentially of a metal tube, at one end of which is fitted an adjustable slit through which the light to be examined is admitted. Beyond this is a lens to render rays from the slit parallel: then a series of three or five glass prisms which are in contact and with their refracting edges in opposing directions. Glass of a different degree of dispersion is also used in alternate prisms of the group, with the result that rays in the middle of the spectrum emerge from the eyepiece very little deviated. This constitutes a 'direct vision' spectroscope. Ideally, the total dispersion should be about 10°, which just enables the observer to see the whole of the visible spectrum. The lines in different parts of the spectrum can be brought sharply into focus by means of the drawtube of the instrument. For the red end of the spectrum it will be found necessary to pull out the drawtube about 60 mm (¼ in). For blue and violet rays the focus is sharp when the drawtube is pushed right home.

If the slit of a small pocket spectroscope be directed towards an electric light bulb a rectangular ribbon of the spectrum colours will be seen. When turned towards the sun or a bright sky the same continuous band of colours can be observed, but if the slit width is narrow enough and the drawtube of the instrument is correctly adjusted for sharp focus a series of fine dark lines will be discerned crossing the bright ribbon of colour. The lines referred to are at right angles to the length of the spectrum: any dark lines or streaks parallel to the *length* of the spectrum are due to dirt on the slit, and can usually be cleared by opening the slit very slightly. The dark lines crossing the spectrum are known as the Fraunhofer lines of the solar spectrum after the German physicist who first described and mapped them.

Fraunhofer called the more prominent of the lines A, B, C, etc., starting from the deep red end (see Appendix 1). We now know that these are *absorption* lines corresponding exactly in position (that is, in wavelength) with the bright lines emitted by the glowing vapour of metallic elements. Two lines in the yellow, for instance, so close together that they appear as one line in a small spectroscope, correspond exactly with the yellow lines emitted by glowing sodium vapour.

Putting the matter as briefly as possible, this *absorption spectrum* of the sun is produced because certain wavelengths of the bright continuous spectrum emitted by its glowing solid core are absorbed by metallic atoms in the cooler gaseous atmosphere which surrounds the star. Each kind of atom in the gaseous state has the power to absorb those same wavelengths of light which it emits when heated. But in solids the absorptive power of the atoms is much more restricted than in gases, and any absorption bands are much broader, more vague in outline and variable in position than the Fraunhofer lines described above. However, the fact that the absorption bands vary in position according to the mineral in which the colouring metal occurs is in itself a distinct advantage, as it enables us in many cases to identify the mineral.

There is a small group of metals which, when present in a solution or glass or mineral, tend to absorb certain wavelengths from white light and thus exert a colouring action on the substances which contain them. The most important as far as gemstones is concerned is chromium, which produces the splendid reds of ruby, spinel, and pyrope, and the rich and brilliant greens of emerald and jade. Iron, nature's most common colouring agent, gives rise to less brilliant greens as well as red and yellow and occasionally blue. Green sapphire, peridot, almandine, and blue spinel are examples of these. The influence of copper is seen in turquoise, and in the ornamental stones malachite and azurite. Manganese provides a peculiar rosy pink or orange in the rare spessartine garnet and the translucent rhodochrosite and rhodonite. Nickel earns credit for the green of true chrysoprase. The well-known blue of cobalt is seldom found in natural minerals, but is common in blue glass and in the blue synthetic spinels which are so prevalent today.

Titanium undoubtedly is responsible for the blue colour of synthetic sapphire, while in natural sapphire the colour is influenced also by iron. The role of vanadium, the eighth and last of these transition elements, is varied and rather obscure. In beryl this element produces an attractive green tint, and most emeralds contain some vanadium in addition to the chromium, which alone can provide that touch of richness and brilliance that characterizes the true emerald-green. The magnificent violet-blue of the gem-quality zoisite discovered (1967) in Tanzania has also been ascribed to the presence of 0.02 per cent vanadium which analysis showed to be present. Finally may be mentioned the influence of vanadium in producing the curious mauve to purple form of synthetic sapphire which is intended to imitate alexandrite chrysoberyl.

This forms a convenient point to return to the subject in hand, which is to explain how a spectroscope enables one to analyse the colour of a gemstone and hence, in many cases, to determine the nature of the stone. A piece of blue cobalt glass is not difficult to obtain, and provides perhaps the easiest demonstration of what we mean by absorption bands and an absorption spectrum. If the spectroscope is directed to a bright continuous source of light, such as an incandescent bulb, or the sun, and the cobalt glass is then placed in front of the slit, instead of the complete series of spectrum colours formerly visible the observer sees virtually only two colours – a patch of blue, as one might expect, and a patch of deep red, which may seem somewhat surprising in a blue material. The remainder of the spectrum is blotted out by three broad dark bands centred in the orange, yellow, and green. This, then, is the typical absorption spectrum of cobalt, and is seen, as indicated above, not only in cobalt glass but also in synthetic blue spinel, though the position of the bands differs slightly in the two cases.

When the specimen to be examined is not a flat sheet of glass but a faceted and often mounted gemstone, it will be realized that it is not such an easy matter to pass light through the stone onto the spectroscope slit. Before describing the more important gem spectra, therefore, it will be wise to give some advice on the methods of going to work which have proved most satisfactory, and to consider what types of spectroscope and what sources of light are most suitable.

The Beck prism spectroscope formerly recommended is no longer being produced, but an inexpensive Japanese-made prism spectroscope is available, and has a slightly higher dispersion, which is an advantage at the cramped red end of the spectrum. This has a narrow fixed slit protected by a dust-proof glass cover. It gives a beautifully clear spectrum, but is more suited to the observation of the narrow bright lines of an emission spectrum than of absorption bands in gemstones, the observation of which often requires the increased illumination obtainable from a widened slit. The slit width is indeed an important factor in obtaining good results with the spectroscope, and will be discussed later.

Many modern spectroscopes are of the diffraction grating type, and range from the inexpensive OPL model with a fixed slit to others with adjustable slits and movable drawtubes. Several types are available from the Gemmological Association and other firms (Figure 10.1). Most suppliers will provide opportunities for testing instruments and intending buyers are strongly recommended to avail themselves of these facilities.

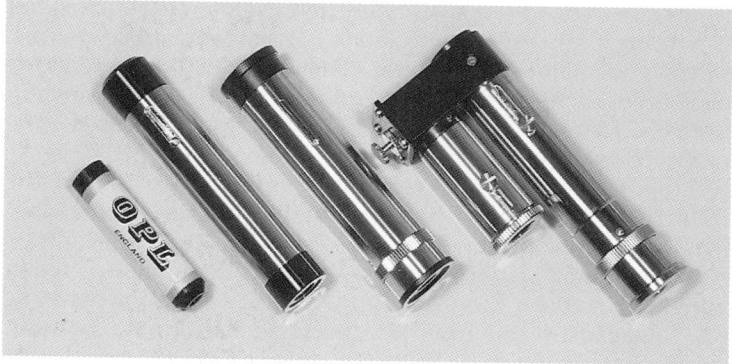

Figure 10.1 A series of spectroscopes. (*From left to right*) Simple diffraction grating type with fixed slit and eyepiece; prism type with fixed slit and adjustable eyepiece; prism type with adjustable slit and eyepiece; prism type with adjustable slit and eyepiece and wavelength scale

For those who require the reassurance of a wavelength scale, instruments are available that are essentially similar to those just mentioned but have an accessory tube containing a scale, marked in nanometres, which, when suitably illuminated, can be seen superimposed above the spectrum. The Japanese make such a spectroscope, and so do Zeiss, Krüss, and other Continental manufacturers. Dr E. Gübelin has designed a table instrument incorporating this type of spectroscope with built-in lighting and an illuminated scale (both controllable as to brightness), a holder for the stone, diaphragms, filters, and other arrangements to make observations easy.

While these superimposed scales make it simple for the student to recognize a band or line already accurately listed and measured in the literature, they are not accurate enough, particularly in the cramped red end of the spectrum, to enable a new or unknown band or line to be measured with any certainty. For such admittedly specialized work a recording spectrophotometer or an instrument such as the Hartridge reversion spectroscope, made by Beck, is needed. In this, absorption bands seen in contiguous reversed images of the spectrum are made to coincide by turning a calibrated drum.

However, it has always been the author's contention that wavelength measurements of absorption bands are very seldom necessary for diagnostic purposes. After a little practice the nature and grouping of the absorption bands which characterize the main gem species are quite sufficiently distinctive to enable them to be recognized with certainty simply by inspection with a small spectroscope. After all, our ability to recognize friends when we meet them in the street is not dependent upon our knowing the exact length of their noses or the distance between their eyes. Spectra which cannot be recognized at sight even by an experienced observer are of little value for purposes of identification, even when the wavelengths of the bands are measured.

An adequate source of light must be available to obtain satisfactory spectra. Quite a lot can be done with a 60- or 100-watt frosted bulb in a microscope lamp or a desk lamp, but the beam from a low-voltage 'intensity' lamp is better, and the author prefers still more to use a 250- or 500-watt projection lamp housed in a fire-proof box to avoid any glare reaching the eyes except through the specimen.

One of the most convenient means of concentrating light on to the specimen and ensuring that the transmitted light adequately and evenly covers the slit of the spectroscope is to bring a microscope into service. A simple microscope with a substage condenser is ideal for the purpose. The stone to be examined is placed on a glass slide on the stage of the microscope and the eyepiece removed. A low-power objective, 25 mm (1 in) or 37 mm (1½ in), should be used. The mirror and condenser should be adjusted and the position of the stone so arranged as to give the greatest possible amount of transmitted light in the centre of the field. The focus of the microscope should then be raised somewhat, so that the field of view is filled with a uniform 'out-of-focus' glare of transmitted light. The spectroscope can next be lightly rested on or inside the body tube in the position normally occupied by the eyepiece and, if the adjustments have been properly made, a spectrum free from horizontal streaks should be seen, evenly illuminated save for such vertical absorption bands as the stone may be causing (see Figure 10.2).

Light from a projection lamp contains a large proportion of heat rays, and, when concentrated by lenses as described, may heat the specimen unduly. Though normally this does no harm, it may be wiser when employing so powerful a lamp to filter the rays through a flask of water placed in front of the lamp (see Figure 10.3). Suitably arranged, the flask of water can itself serve as a condensing lens, and cuts out the unwanted heat rays very effectively, while passing almost all the visible light. A 600- or 750-ml flat-bottomed flask is ideal for the purpose. In addition to one of these filled with pure water it is well worth having handy a similar one filled with a concentrated solution of copper

Figure 10.2 B. W. Anderson observing a spectrum using a Beck wavelength spectroscope placed in a monocular microscope

Figure 10.3 Illumination for study of absorption spectra. Light from a 500-watt projection lamp condensed and cooled through a water flask

sulphate. This yields a fine blue liquid of the kind which used to be seen in large pear-shaped glass containers in chemists' shop windows. If this is made with tap water it will be necessary to filter carefully to ensure a completely clear solution and perhaps add a little dilute sulphuric acid. By using the copper sulphate flask as a condenser and filter it will be found that every vestige of red and orange light has been absorbed from the spectrum. This enables one to see fluorescence lines in the red (such as those emitted by ruby or red spinel, to be described later) very clearly against a dark background. It also enables one to view more clearly absorption bands in the blue and violet part of the spectrum. Naturally, one must only use the blue flask for such special purposes, since for observation of the complete absorption spectrum of any gemstone a white light source is essential.

To avoid being dazzled by the glare while adjusting the specimen and light to best advantage, it is wise to rest a piece of frosted glass on the top of the body tube of the microscope; a slightly enlarged image of the illuminated stone can be seen on the frosted glass and brought to focus.

When a gemstone is too small to fill the field it should be placed on a metal diaphragm of slightly smaller diameter to ensure that no unabsorbed white light reaches the slit of the spectroscope, as this can seriously weaken the strength of the absorption bands seen. All this has been written with an unmounted stone in mind, but with a little skill and practice it is usually possible to make similar arrangements with single-stone or even five-stone rings, taking one stone at a time.

Some workers prefer to dispense with the microscope and to concentrate the rays from their light source onto the specimen by means of a suitable condenser. A water flask serves here also very well, and is considerably cheaper, though for extreme concentration a lens may be used in addition. The specimen is placed table facet down on a strip of black velvet and the light reflected from the interior of the specimen is examined through the spectroscope with the slit a few inches from the stone (see Figure 10.4). Weak absorption bands in pale-coloured specimens appear more strongly by this method, and the bright fluorescence lines seen in ruby and spinel also show to good advantage. It is not so good as the transmitted light method for deep-coloured stones or for bands in the blue and violet. When using a spectroscope with attached scale sufficient light to

Figure 10.4 Examining absorption spectra of crystals by reflected light

illuminate the scale can be obtained by placing the stone to be tested on a ground-glass plate.

It must be admitted that the 250- and 500-watt projection bulbs recommended above (which are still being successfully used in the London Gem Testing Laboratory) have for some years been largely replaced by much smaller quartz-iodine bulbs, which give an intense and concentrated beam of light but have to be fan cooled to prevent overheating. These are ideal for absorption spectroscopy, particularly if used in conjunction with light-conducting fibres housed in a flexible metal-clad tube (fibre-optic lights).

Light sources of this kind, adjustable in intensity, are now commercially available, and though expensive are well worth consideration by the serious gemmologist. The 'piped' light provides intense illumination to be applied to the specimen, whether it be a cut gemstone or something unwieldy such as a jade carving, and although the source is hot the fibre-conducted beam is cool (see Figure 10.5).

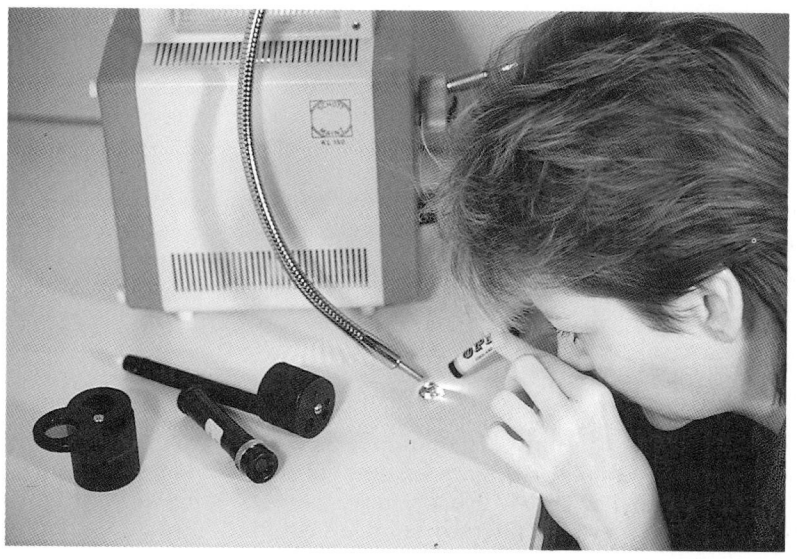

Figure 10.5 Examining absorption spectra of gemstones using a fibre-optic light

A custom-built 'Kaltlicht' spectroscope assembly manufactured by the firm of Eickhorst of Hamburg makes good use of fibre guides, one below and one above the specimen to be tested, which is placed on a rotatable mount. The prism spectroscope employed can be moved in an arc with the specimen as focal point, so that the spectrum of the stone can be examined by either transmitted or reflected light. The spectroscope itself is a Zeiss prism instrument of fine quality incorporating an illuminated scale of wavelengths, the red end of the spectrum being on the left of the observer. The bright image of the scale is superimposed on the spectrum itself, and would interfere with the observation of faint absorption lines were it not for a dimming device.

Modern pen-light torches or free-standing types with exposed bulbs can also be used for spectroscopy. If the specimen is attached (usually table facet downwards) to the 'lens end' of the bulb by using modelling clay or 'Blue-Tack' the intensity of the light provided by the two small batteries (AA type) is perfectly adequate for the illumination of most gems or transparent fragments. By using a small spectroscope such as the OPL or other compact instruments in the hand, identification can be effected anywhere (see Figure 10.6(a)). The torch, spectroscope, and the Blu-Tack can all be carried in the pocket or handbag (see Figure 10.5). Alternatively, one can use the Rayner Scopelight which provides holes of various sizes in which a range of gems can be rested; a penlight torch is incorporated for illumination (see Figure 10.6(b)).

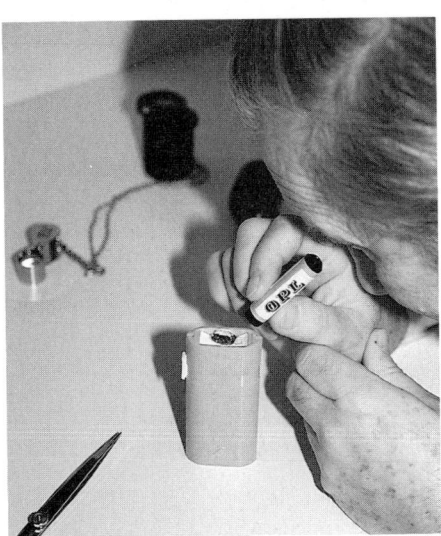 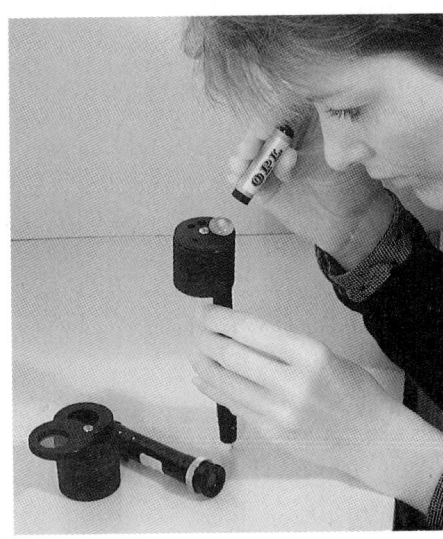

Figure 10.6 Using (*left*) a pocket torch or (*right*) a Rayner Scopelight for easy spectroscopy

In contrast to the simple equipment and methods just described a new spectroscope designed by the Gemological Institute of America was launched as this edition was going to press. Their new 'Discan' (digital scanning) spectroscope provides fibre-optic reflected light, transmitted light and a digital readout window which provides instant measurement in nanometres of absorption or emission lines or bands (see Figure 10.7). Such sophistication cannot be cheap!

In an assembly designed and used successfully in the USA a Beck spectroscope was used, in which the wavelength scale reads with the red end of the spectrum on the right, and for this reason Robert Crowningshield's splendidly lifelike series of drawings of absorption spectra in earlier editions of Liddicoat's *Gem Identification* are portrayed in this manner. When one is accustomed to view the spectrum in one direction it is surprisingly difficult to recognize the pattern of bands when viewed in the opposite sense. Hence in these pages drawings according to each convention are now included (Figures 10.9 to 10.12). It has always been the author's contention that for purposes of spectrum recognition a wavelength scale is in most cases an unnecessary distraction, since the distribution, width, relative strength, and general nature of the bands are the essential features by which the majority of absorption spectra

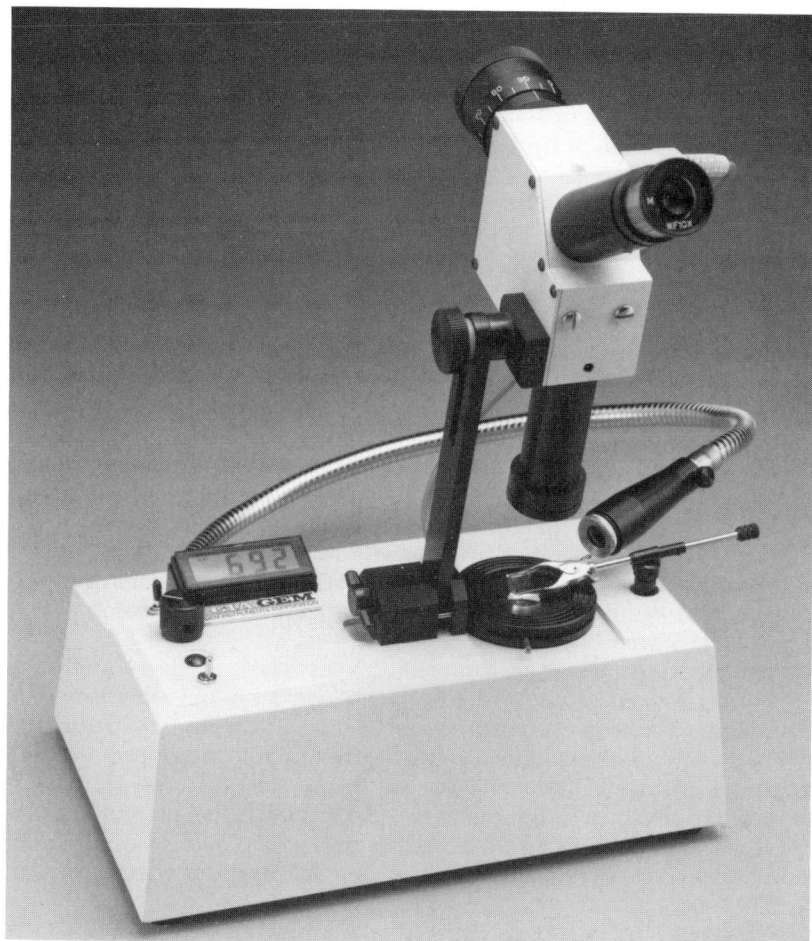

Figure 10.7 'Discan' (digital scanning) spectroscope (courtesy: GIA)

can be recognized at a glance. Moreover, spectroscopes without a scale are far less expensive. Where there is a need for some check on the position of broad and not very distinctive bands, such as those shown by pyrope and red spinel, a little ingenuity will get over the difficulty – by introducing, for instance, a dilute solution of potassium permanganate into the light train as detailed in Chapter 22. The permanganate provides five evenly spaced absorption bands, centred at 570, 545, 524, 504 and 487 nm, which can act as a 'built-in' wavelength scale. In similar fashion, the familiar bright-line mercury spectrum, which can be seen in light from an ordinary fluorescent lighting tube as used in offices or factories, provides a yellow doublet at 579 and 576.9 nm, a green line at 546 nm and one in the violet at 435.8 nm, which can be introduced into the spectrum train by means of a well-placed mirror. Sunlight also, with its tremendous array of measured Fraunhofer lines, can be pressed into use. But such admittedly tiresome expedients are very seldom necessary.

What has just been stated does not mean that accurate wavelength measurements of all absorption bands should not be made and recorded for reference, and the author and his colleague C. J. Payne were at great pains to carry out this work as well as possible in the early days of their researches on absorption spectra. It could fairly be said that the published data in all the world's textbooks of gemmology are based on these observations.

Measurements were in the main carried out with the aid of excellent instruments made by Beck. One was the 'Wavelength' spectroscope – a dog-leg instrument incorporating a diffraction grating, in which the cross-hairs seen through the eyepiece could be adjusted to a given line or band by means of a knurled drum marked in wavelengths; a modified instrument of this type is still available, but is very expensive. Where a well-defined line or band was concerned, an accuracy to ±1 nm could be achieved with this splendid instrument. The other instrument was the 'Hartridge' spectroscope, in which by an ingenious optical arrangement two images of the absorption spectrum in reverse position could be seen in contact, one above the other. Movement of one relative to the other was achieved by means of a drum calibrated in wavelengths, the correct figure being obtained when the reversed images of the absorption band in question were brought into coincidence. The accuracy here was similar to that obtained with the 'Wavelength' instrument. In each case the greatest difficulty arose where the band measured was not only broad but of different density throughout its breadth.

In a case where sharp line spectra were observable – such as with the chromium lines in ruby, emerald, and alexandrite, or in suitable specimens of zircon and of diamond – attempts to achieve greater accuracy were made using a table spectrometer and comparison with known emission spectra provided by chosen elements introduced into a carbon arc.

Reverting to the techniques necessary to obtain the best results from the usual small spectroscope when observing the absorption spectra of gemstones (having concerned ourselves so far chiefly with the provision of an adequate light source), the next essentials for good results are the correct adjustment of the slit and the sharp focus of the part of the spectrum examined.

Many inexperienced examinees have found themselves unable to obtain good results with spectroscopes available because careless previous users have left the slits either wide open or tight shut! The aim should always be to have the slit width as narrow as possible, consistent with allowing sufficient light to be transmitted to enable any bands present to be seen against the background of the continuous spectrum. When the slit is nearly closed, dark horizontal lines will appear along the whole length of the spectrum. These are caused by dust particles bridging the jaws of the slit at certain points. Where these are very obtrusive careful cleaning with a tapered matchstick or soft camel hair brush should help to remove them. In some instruments the slit is protected from dust by being enclosed inside a glass window. In one Rayner prism spectroscope fixed slits of varying widths are provided, which can be used as required. Makers admit that the provision of an accurately made slit with strictly parallel jaws is one of their most difficult engineering problems.

Focusing the spectrum is carried out by pushing home or pulling out the drawtube slightly. The furthest-out position is that for correct focus of the red end; furthest-in for the blue and violet. Either the bright-line spectrum of mercury vapour from a handy fluorescent light tube or the Fraunhofer lines of the sun can be used to practise this focusing manoeuvre, since in each case the lines in the spectrum are almost infinitely narrow and should therefore show a beautifully sharp image when correctly focused. When studying the absorption spectrum of a gemstone it may well be necessary to adjust both the slit width and the focus when examining different parts of the same spectrum.

The above directions have occupied a lot of space, but it is hoped that the instructions given may enable this most pleasurable form of gem testing to be carried out with greater success by those to whom the arts of observation seemed unduly difficult.

The detailed adjustments as described above will not, of course, be applicable to small spectroscopes with fixed slits and no drawtubes. However, these instruments are usually adjusted for optimum performance and will be found adequate for most purposes.

Easy spectra with which to practise are those seen in brown or greenish Sri Lankan zircons and in almandine garnet (see Figure 10.8). These happen to be 'classic' examples, since it was in these gemstones that Church, in 1866, first observed absorption bands. Synthetic ruby also shows a well-defined spectrum for practice. Details of these spectra will be found below in their appropriate places.

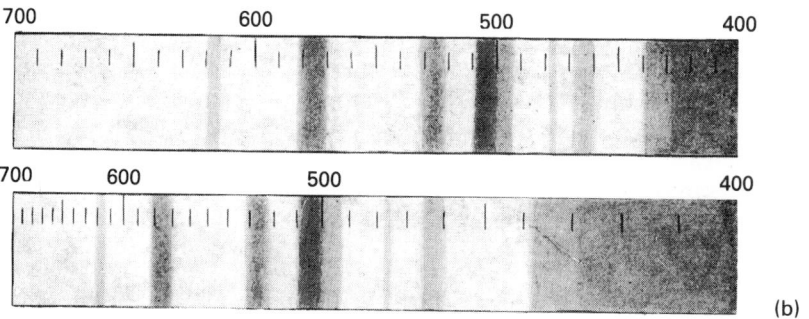

Figure 10.8 Distribution of absorption bands in almandine garnet as seen through (a) a grating and (b) a prism spectroscope (compare Figure 10.9(5))

To decide in which order the absorption spectra of individual gemstones should be described is by no means an easy matter. From a scientific and systematic viewpoint it is undoubtedly best to group together those spectra which can be attributed to chromium, those which are due to iron, to manganese, and so on; and this procedure was followed in earlier editions of this book. However, this method presupposes a considerable degree of knowledge on the part of the reader, and was abandoned in favour of an arrangement based on colour.

For practical purposes this has much to commend it. After all, absorption is definitely linked with colour, and the absorption bands seen in ruby, for instance, are quite different from those found in sapphire, though both belong to the same mineral species. Moreover, the whole descriptive portion of the book is arranged on a colour basis, in recognition of the fact that colour is the most obvious characteristic of gemstones, whether to the expert or to the tyro. All the characteristic spectra to be seen under each colour heading will therefore be found below, beginning with colourless stones and carrying through in the spectrum order red, orange, yellow, green, and blue. This naturally leads to a good deal of repetition in the case of a stone such as zircon. But this is no bad thing, since the spectrum shown by a white zircon is not to be described in quite the same terms as, shall we say, the spectrum of green zircon.

Wavelength figures may not mean anything to the beginner, but before long those who use the spectroscope come to know where to look for a line of wavelength 520 nm as compared with another line of wavelength 560 nm (though both must be described as 'green'), even though no wavelength apparatus be used.

Absorption spectra of the colouring elements

Completely colourless stones should, of course, show no absorption bands. But the presence of narrow bands has very little influence on colour, especially where, as in the case or zircon, there are several narrow bands evenly distributed across the spectrum. Stones which are nearly colourless are also included in this first group. Before describing individual spectra, a general account of the types of spectrum which we have learned to associate with the metallic ions which are known to give rise to colour in minerals may be given. The metals involved all belong to a group known as 'transition elements' and occupy consecutive places in the periodic classificaton of the elements, based on their atomic number. Beginning with titanium, atomic number 22, they continue in order, vanadium, chromium, manganese, iron, cobalt, nickel, and copper. Uranium in certain cases and some of the 'rare-earth' elements may give rise to absorption bands, but have little influence on colour. In the unique case of diamond, colour is due to defects in the crystal structure of the stone, often connected with the intrusion of nitrogen atoms into the carbon lattice.

Minerals can be classified into the idiochromatic ('self-coloured') type which owes its colour to an element that is an essential part of its composition – e.g. the iron in almandine garnet or peridot, the copper in malachite – and the allochromatic type, in which the colouring element is present in quite small quantity as an 'accidental' impurity. The majority of gem minerals are allochromatic: that is, the mineral itself has no distinctive colour, and is in fact colourless when pure, but exhibits a range of coloured varieties according to the presence of traces of different colouring elements. Quartz, beryl, corundum, tourmaline, topaz, spinel, zircon, and many others are in this category.

Returning now to the type of spectrum associated with some of the main transition elements, we will deal with **chromium** first, on account of its great importance as a colouring agent. To chromium (or chromium oxide) may be ascribed the finest reds and the finest greens seen in the mineral kingdom. The red of ruby and spinel, the green of emerald, jadeite, and demantoid garnet are all due to chromium, while in the alexandrite variety of chrysoberyl chromium produces a half-way colour which appears green in daylight and red by artificial light. Whether in red stones or in green, the absorption bands due to chromium show the same general characteristics and are very distinctive. There are narrow lines in the red end of the spectrum, the strongest of which form a doublet (two lines very close together) in the deep red, accompanied by two or more lines on the orange side of this; there is a broad absorption band in the centre of the spectrum, that is, in the yellow or green. The position, width, and intensity of this band largely determines the hue of the stone. There is strong absorption in the violet, and often narrow lines in the blue.

Chromium usually enters a gemstone by replacing to a small degree the aluminium which is an essential part of its composition. Examples of this are ruby, emerald, and alexandrite. In certain cases, however, magnesium may be the element replaced – e.g. in enstatite and Hawaii peridot. Colours produced by chromium are rich and clear because the absorption bands and lines are clear-cut and well defined, the unabsorbed colours being left at full strength. Colours due to iron, though often quite attractive, are not so brilliant, because, in addition to absorption in the main bands, there is some general absorption in almost all regions of the spectrum. To use a rough analogy, chromium bands are like cities within walls, having no suburbs, the countryside in-between being left pure and undefiled, whereas iron bands are like towns spreading out into suburbs, with scattered houses spoiling the country between the main urban centres.

There are two categories of **iron** spectra: those due to ferrous (divalent) iron and those due to ferric (trivalent) iron. As with chromium, the colours produced are mainly reds and greens, but the inky tint of blue spinel is also due to iron. The bands are seldom very sharp and are mainly in the green and blue parts of the spectrum. Minerals coloured by **manganese** are typically rhododendron-pink or orange in hue. There are bands in the blue, but the strongest bands (and these are often very intense) are found in the violet, and may extend beyond the visible spectrum into the ultra-violet region.

Cobalt minerals in nature are pink in colour, but none of these are used as gems. The familiar cobalt blue is found in cobalt glass and in blue synthetic spinel, both of which may be used to imitate sapphire, and also in synthetic blue quartz. The absorption spectra of these cobalt blues are very distinctive, with their three strong, broad bands in the yellow, green, and blue-green.

The other transition elements do not give rise to such distinct absorption bands, with the exception of **copper** in turquoise, which causes two narrow absorption bands in the violet, only one of which is usually visible, and **vanadium**, used to colour synthetic corundum of the 'alexandrite' type, where a single line is seen in the blue.

Bands due to **uranium** are found only in zircon, and being well defined and numerous can present a very striking absorption spectrum. The lines are so narrow and well distributed throughout the spectrum that they have little influence on the colour of the stone. Finally may be mentioned the narrow lines produced by the presence of two of the series of **rare-earth** elements – neodymium and praseodymium, conveniently known by the collective name **didymium** since they are inseparable in nature, though they have been isolated after laborious processes by the chemist. Didymium in glass, when present in quantity, causes strong absorption in the yellow, and produces a rather characteristic pink colour; in natural minerals these elements are present only in traces and the absorption bands or lines are so faint that they have no influence on the colour. With rare-earth elements the electrons involved in the absorption process are not those in the outer shell, and are thus less influenced by the surrounding electric field in the host crystal than are most of the 'colouring' elements. Thus the use of the term 'lines' rather than 'bands' is justified. Traces of didymium are most frequently present in calcium minerals, and cause a characteristic group of fine lines in the yellow region, and perhaps also in the green. Yellow apatite (and some green apatite) is the only natural gemstone to show didymium lines at all strongly, though the strongest of the lines have been detected in danburite, sphene, idocrase, fluorite, scheelite, and calcite. Didymium lines in strength, and even lines produced by other rare-earth elements, are sure signs of a synthetic crystal or glass. More details of such spectra will be found at the end of this chapter.

Colourless stones

Naturally enough, most colourless or nearly colourless stones do not show absorption bands, since in such stones the 'colouring' oxides which produce such bands are absent. Colourless zircons, however, do show characteristic lines, too faint and narrow to affect the colour; and off-white diamonds and synthetic rutile have absorption bands in the violet.

White zircon (uranium)

There is invariably a narrow line in the red at 653.5 nm, accompanied by a fainter line at 659 nm, very close to it. Others of the prominent zircon lines seen strongly in most coloured zircons (589.5 nm, 562.5 nm, etc.) are often visible. The spectrum is best seen by reflected light, as this enhances the strength of the faint bands.

Diamond (structure)

Most diamonds show a narrow absorption band at 415.5 nm in the deep violet which, when seen, is diagnostic for diamond. In 'Cape' stones this is very intense, and may be accompanied by five weaker bands at 478, 465, 451, 435, and 423 nm in the blue and violet. Of these weaker bands the 478 band is the most prominent, and due to its position in the spectrum is responsible for the yellow colour reaching the eye. Diamonds of the 'brown' series may show a fine line in the blue-green at 503 nm with fainter lines at 537 and 497 nm, but these are very difficult to see. On the whole, stones showing a 'Cape' spectrum have a blue fluorescence under ultra-violet light, while diamonds of the brown series have a green fluorescence. A blue (copper sulphate) filter will be found to make lines in the deep violet more easy to observe. It is worth noting that light is most easily transmitted through a diamond in the direction of the girdle. The stone should be held in this position by tweezers or on a glass plate with the aid of a small ball of modelling clay or 'Blu-Tack'. Alterations in the absorption spectrum of diamond induced by bombardment are described in Chapter 12.

Synthetic rutile

This spectacular synthetic gem has not yet been made free from colour. The yellowish tint is probably due to the presence of a powerful absorption band in the violet (about 425 nm) which terminates the spectrum.

Red stones

Ruby (chromium) Figure 10.9(1)

Natural and synthetic stones have the same spectrum, though since synthetics contain more chromium for a given depth of colour the spectrum lines in these tend to be more intense.

The ruby spectrum is a rich and complex one, but the distinctive features are clear enough. These are: a strong doublet in the deep red, at 694.2 and 692.8 nm, seen under normal conditions as a single *bright* fluorescence line; two fainter lines on the orange side of this (668 and 659.5 nm), which also may appear as fluorescence lines; a strong, broad absorption band, centred near 550 nm, covering most of the yellow and green, and three narrow bands in the blue, two of which are close together, the other distinctly separated, at wavelengths 476.5, 475, and 468.5 nm. There is strong general absorption of the violet. The fluorescence doublet is a sensitive test for traces of chromium in corundum. As little as 1 part of Cr_2O_3 in 10 000 Al_2O_3 in the corundum lattice will give rise to a red fluorescence in strong white light, ultra-violet light, or X-rays. One may thus see this bright line in the spectrum of many nearly colourless sapphires, and in blue sapphires from Sri Lanka in which there is a trace of chromium and not enough iron to damp the effect. In chrome-rich rubies or synthetic rubies the doublet may be seen as an absorption line in direct transmitted light. In Thai rubies, in which enough iron is present to 'damp' the fluorescence, the brightness of the fluorescence line is correspondingly diminished. This does not, however, provide a foolproof method of distinguishing between stones from the two main ruby localities. A copper sulphate filter assists in the observation of the fluorescence lines, as these are then seen against a dark background. It also helps one to see clearly the lines in the blue.

Ruby is strongly dichroic, and its absorption spectrum thus differs according to the direction in which the light travels through the stone. The main difference is seen in the broad central absorption band, which is much broader and more intense in the ordinary ray, thus giving it a purer red than the yellowish red of the extraordinary ray.

The only other red gemstone with which ruby may be confused on the basis of its absorption spectrum is spinel. Spinel also shows a red fluorescence exactly similar in colour to that of ruby, but the spectroscope analyses this into a series of bright lines, giving an organ-pipe effect in contrast to the appearance of one bright line accompanied by two much fainter lines on the orange side of this, which is seen in ruby. Another decisive distinction is the absence of lines in the blue in spinel.

This spectrum has been discussed at some length on account of its importance and interest. Even so, many of the subsidiary lines sometimes seen in the ruby spectrum have not been mentioned. To sum up, one can state that any red stone which, when illuminated, shows a bright fluorescence line in the red *and* narrow absorption lines in the blue is undoubtedly a ruby, either natural or synthetic.

A good idea of the nature and position of the absorption bands in ruby and the other spectra described hereafter will be gained by a study of the drawings specially made by Mr T. H. Smith, which are reproduced in Figures 10.9–10.12. Unlike most drawings which have been given in books, these are shown as through a prism spectroscope, since the author is convinced that for general identification purposes this form of spectroscope is the best to use. The

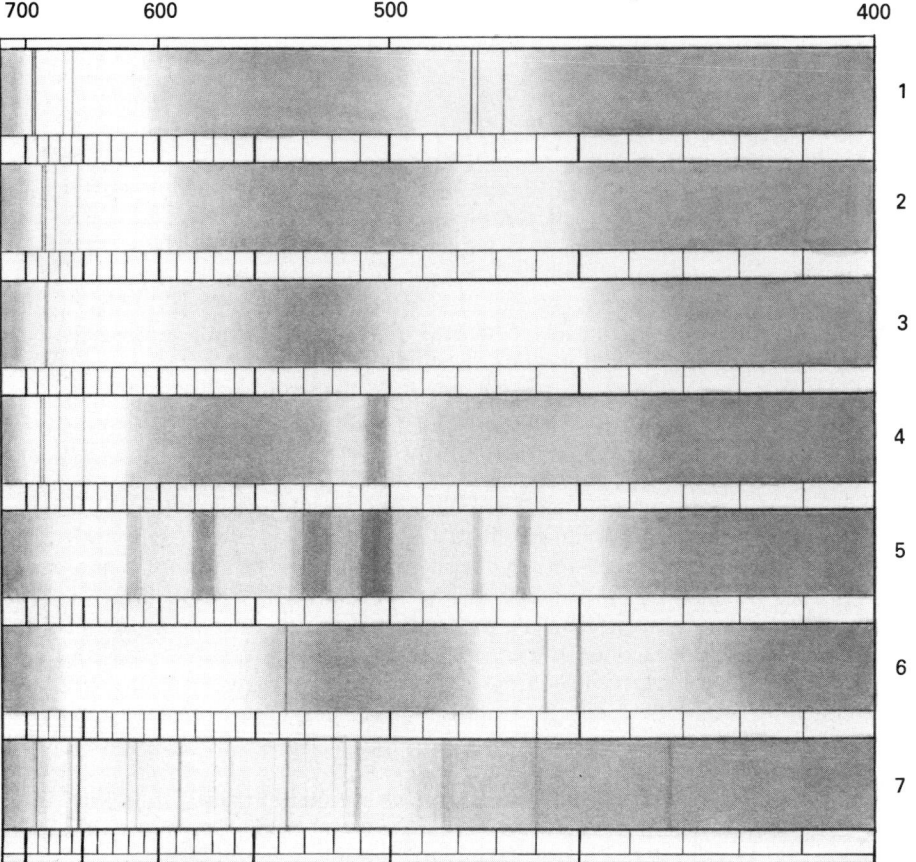

Figure 10.9(a) Absorption spectra of red stones as seen through a prism spectroscope. 1 Ruby; 2 spinel; 3 topaz; 4 pyrope; 5 almandine; 6 tourmaline; 7 zircon

wavelength scale above the spectra reveals how the red end of the spectrum is cramped and the violet extended when using a prism instrument.

For convenience in drawing, the doublet in the deep red is shown here as an *absorption* doublet, which will only be the case when a deeply coloured ruby or synthetic ruby is viewed by light which has been transmitted through the stone in a strictly direct beam. In ordinary practice the observer sees this as a *bright* line in exactly the same position. By juggling with the direction of the light passing through the stone under test it is often possible to see the change from dark line to bright line taking place.

Red spinel (chromium) Figure 10.9(2)

Apart from a broad absorption band in the green, centred at 540 nm, bands or lines in spinel are rarely noticeable, though in chrome-rich types from Burma a number of fine lines may be seen in the red. Red or pink spinels, however, can usually be determined with certainty on account of their very distinctive red fluorescence lines in the deep red. Unlike ruby, these form a group of five or

400   500   600   700

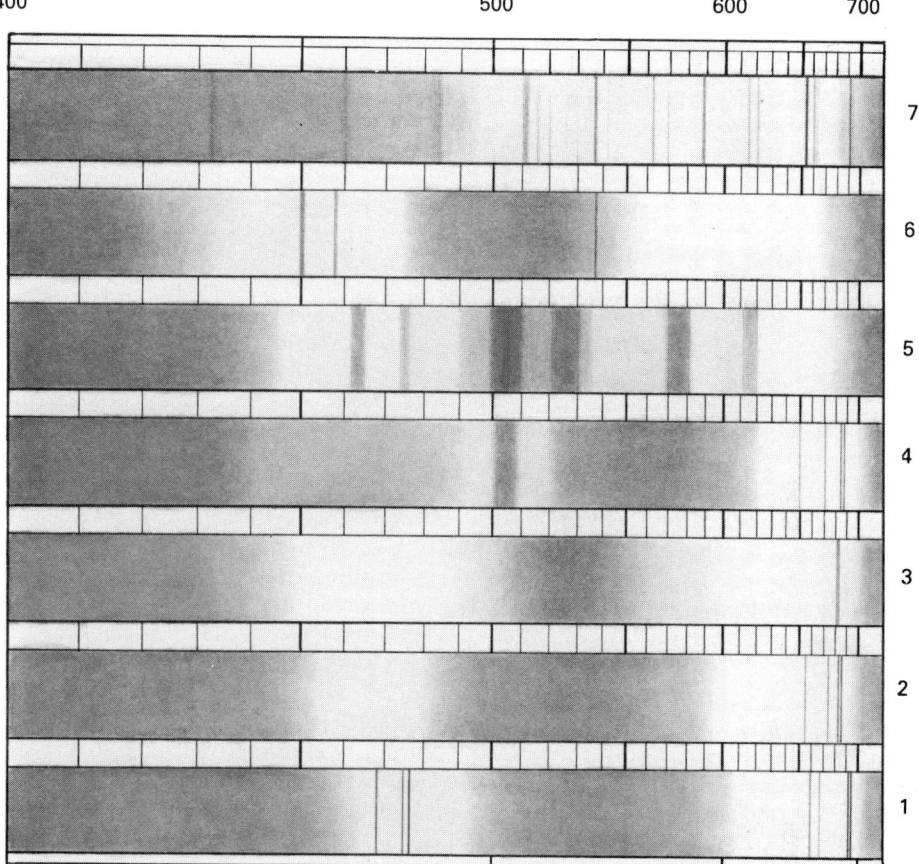

7

6

5

4

3

2

1

Figure 10.9(b) Reversed version of Figure 10.9(a). 1 Ruby; 2 spinel; 3 topaz; 4 pyrope; 5 almandine; 6 tourmaline; 7 zircon

more, reminiscent of a series of organ-pipes. Two central lines are brighter than the others: strongest of all is that at wavelength 686 nm, which is separated by a dark gap from the next most prominent line at 675 nm. To see this distinctive series of lines clearly it is essential to use a really powerful light source. A copper sulphate filter is also very helpful, as it shows the lines against a dark background. It is interesting to note that the seldom seen synthetic red spinels, whether made by the Verneuil process or by some other process of crystallization (see Chapter 9), do not show the 'organ-pipe' structure, but have their red fluorescence chiefly centred in a single line at 686 nm. When viewed through the spectroscope there is then a resemblance to ruby with these freak synthetics – but absence of absorption bands in the blue is a sure sign that they are not ruby.

Almandine garnet (iron) Figure 10.9(5)

The spectrum of almandine is one of the easiest to see and to recognize. With zircon, it shares the distinction of being the first gemstone in which dark bands were observed and recorded. This was in 1866, when Sir Arthur Church wrote an account of his discovery in a letter to the *Intellectual Observer*.

The main feature of the almandine spectrum is the presence of three broad, strong bands in the yellow (576 nm), green (527), and blue-green (505). There are several weaker bands, notably one in the orange (617) and one in the blue (462). Of the three main bands, the one at 505 nm is the most persistent, and can be faintly seen in almost all pyropes, as described hereunder, also in many spessartine garnets, especially those from Sri Lanka. Almandine has no fluorescence.

Pyrope garnet (chromium and iron) Figure 10.9(4)

The pure magnesium–aluminium garnet would presumably be colourless but such a mineral is never found in nature. All red garnets, in fact, are mainly mixtures of the almandine and pyrope molecules. For convenience we term those with low refractive index and SG 'pyrope' and those richer in iron, having higher refractive index and SG, 'almandine'. But it so happens that the pyropes most used in jewellery (from Kimberley, Bohemia, or Arizona) owe their rich colour more to chromium than to iron. This gives rise to a broad band in the yellow-green centred near 575 nm, which swallows two of the three main almandine bands which would otherwise be visible. The third almandine band at 505 nm can be seen rather faintly where the green merges into the blue. As with spinel, narrow chromium lines are seldom seen in the red. Distinction between pyrope and red spinel should be easy. With rare exceptions, pyrope has no fluorescence even under crossed filters. Moreover, the broad central absorption band is at the yellow end of the green, whereas in spinel it is 35 nm further towards the blue. It must be conceded that some red garnets which are clearly pyropes on the basis of their low SG and refractive index may contain little chromium and show merely a weak almandine spectrum.

Pink topaz (chromium) Figure 10.9(3)

Pink topaz, or the sherry-coloured topaz from Ouro Preto which turns pink after heat treatment, contains enough chromium to cause a red fluorescence between crossed filters, and this resolves itself through the spectroscope into a rather weak fluorescence line (probably a doublet) at 682 nm. In large and deep-coloured stones this may be seen as an absorption line, though never at any great strength. The topaz spectrum may not be very reliable as a diagnostic aid, but it is rather important for the gemmologists who use a spectroscope to know that pink topaz, under strong illumination, may show a narrow bright line in the red – otherwise they might mistakenly think such an effect must indicate pink sapphire. When the two are compared, there should be no confusion since, apart from a wavelength difference of 10 nm, the brilliance and sharpness of the fluorescence doublet in pink sapphire are incomparably greater, and two other, weaker fluorescence lines can be seen on the orange side of the strong doublet.

Red tourmaline (manganese) Figure 10.9(6)

Red and pink tourmalines usually show two rather narrow bands in the blue at 458 and 450 nm. These are not so narrow as the rather similar bands in ruby, they are much further towards the violet, and, of course, there is no fluorescence line in the red – so there is no real excuse for confusion between the two. In certain brownish-red tourmalines there is, in addition to the broad absorption band in the green which appears in all red tourmalines (indeed in all red stones), a narrow band at 537 nm appearing within the broad band towards its long-wave margin. This narrow band is not strong, but helps to make this spectrum completely diagnostic for this type of tourmaline. The absorption curves of pink and red tourmalines seem to show that their colour is largely due to manganese.

Red zircon (uranium) Figure 10.9(7)

Though some red zircons show the distinctive narrow band in the red at 653.5 nm, its fainter companion at 659 nm, and some ten other narrow bands distributed throughout the spectrum, due to uranium, there are some (notably those from Expailly) which show no bands at all. The drawing given shows the full spectrum. Wavelengths of these bands will be given in the description of the yellow zircon spectrum, which is more constant and reliable.

Of the other red or pink stones found in jewellery which show absorption bands one may briefly mention **red pastes** owing their colour to selenium, which have only a broad band in the green that varies in position according to the type of glass used, and the two translucent ornamental pink stones, **rhodonite** and **rhodochrosite**. These both owe their colour to manganese and have very similar spectra. Normally only a broad band in the green centred near 550 nm can be seen. Transparent specimens of rhodochrosite, which are very rare, show a band in the violet at 449 nm and an intense band at 410 nm in the deep violet. Transparent rhodonite shows a narrow band at 503 nm, a vague band at 455 nm, and very strong narrow bands at 412 nm and 408 nm. These spectra have little practical value, but are interesting in their relation to the spectrum of spessartine, the manganese garnet, the spectrum of which will be found under 'yellow stones' below.

Orange stones

Fire opal is more fiercely flame-red or orange than any other gemstone, and there are thus few coloured stones with which it can comfortably mix. It has no absorption bands beyond a general absorption of the whole spectrum except the red and orange. Orange glass shows a very similar effect, except that the cut-off to the spectrum is rather more sudden.

Orange synthetic sapphire, sometimes known as 'padparadsha', always contains some chromium, and thus shows the fluorescence line described under ruby. Spessartine often has a peculiar reddish yellow colour that might be described as orange – but this spectrum has been included under 'yellow stones'. Topaz, if orange, may show the fluorescence line described under pink topaz.

Yellow stones

Yellow diamond

A yellowish tinge is the commonest fault in the colour of gem diamonds and is typical of the 'Cape' series, in which stones are characterized by strong

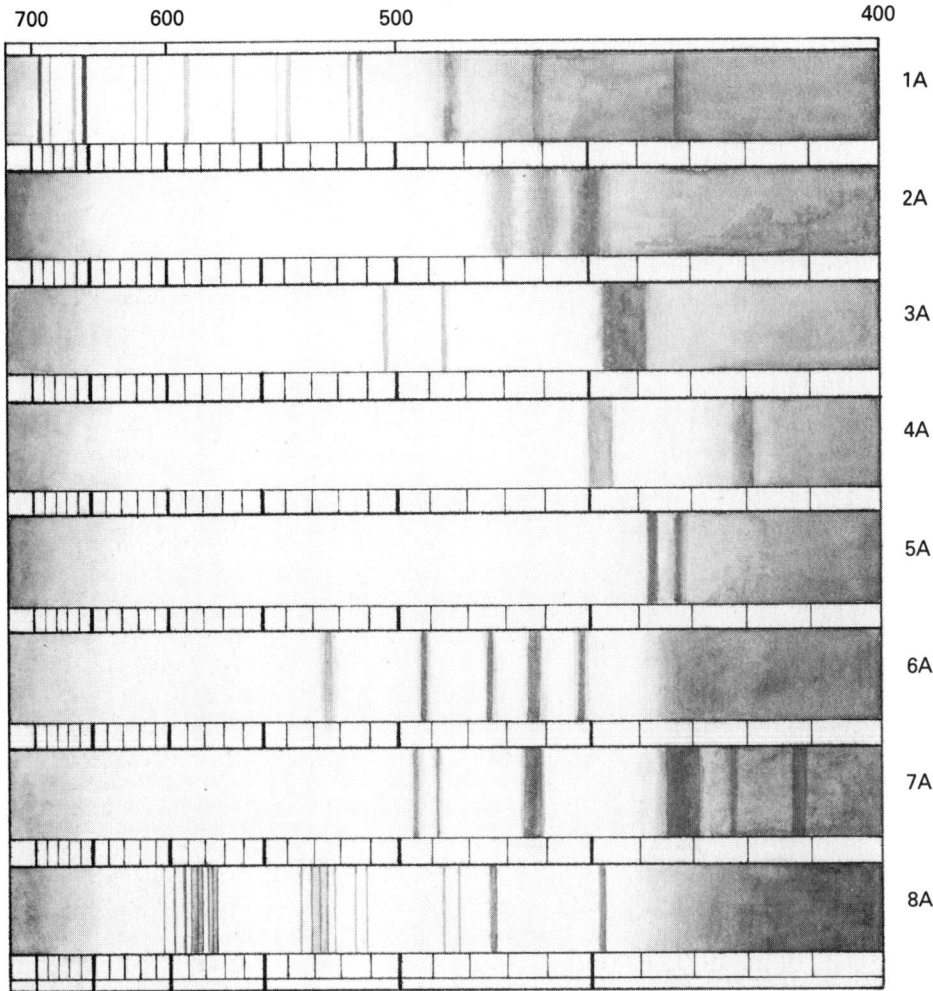

Figure 10.10(a) Absorption spectra of yellow stones. 1A zircon; 2A sapphire; 3A chrysoberyl; 4A orthoclase; 5A spodumene; 6A sinhalite; 7A spessartine; 8A apatite

absorption bands in the blue and violet region. The strongest and 'key' band to the series lies deep in the violet at 415.3 nm at 77 K, which can be seen, narrow and sharply defined, in most diamonds when a strong beam of blue light is passed through or internally reflected by the stone. Next in strength and easier to observe is a band centred at 478 nm, and there are weaker bands at 465, 451 and 423 nm. The stronger these bands, the deeper the colour of the stone – a fact that has been utilized by the firm of Eickhorst in an apparatus for grading the colours of gem diamonds. The deeper yellow stones may be acceptable as 'fancy' diamonds, but in modern times the colour has often been enhanced by neutron bombardment and subsequent annealing (the initial colour of bombarded stones being green). Though such treated stones are stable in colour and can legitimately be sold and used as such, laboratories are frequently asked to distinguish between natural fancy diamonds and those that have been

400   500   600   700

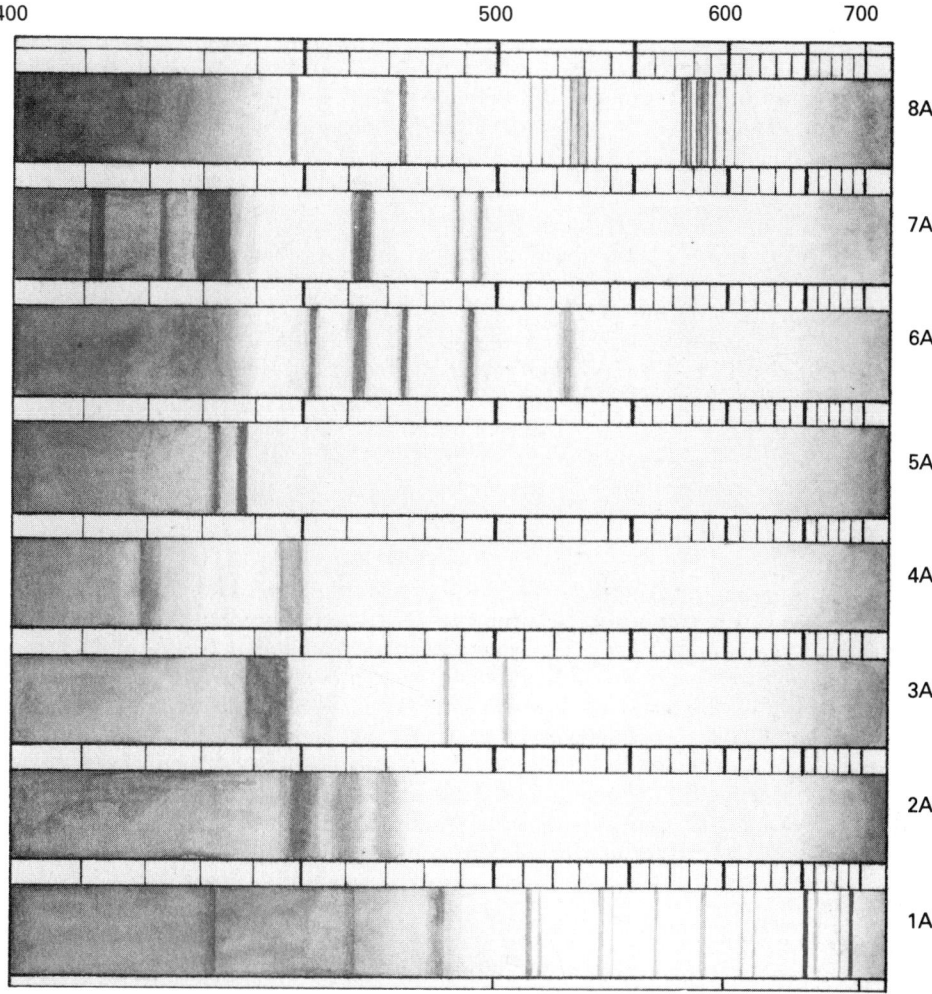

Figure 10.10(b) Reversed version of Figure 10.10(a). 1A zircon; 2A sapphire; 3A chrysoberyl; 4A orthoclase; 5A spodumene; 6A sinhalite; 7A spessartine; 8A apatite

bombarded, to protect the high rarity value of the former. This subject is discussed in Chapter 12.

There is also a much rarer class of yellow diamonds originally described by the author in 1943 and designated as the true 'canary' diamonds. These are distinguished by showing no visible absorption bands and by having a yellow fluorescence, where in 'Cape' stones the fluorescence is blue. Dr Raal has suggested that in these stones the colour may be caused by individual nitrogen atoms in the lattice, whereas in the far commoner 'Cape' series the nitrogen impurities are grouped in aggregates. This may be so in some cases, but a recent examination of the infra-red spectrum of a 'canary' diamond determined both the A and B aggregates of nitrogen to be present with no indication of dispersed nitrogen.

The yellow gem quality and commerically produced synthetic diamond made by Sumitomo has an infra-red spectrum typical of a pure Ib diamond and, as with natural Ib stones, its absorption in the visible spectrum is one of gradually increasing absorbance with shorter wavelengths from about 560 nm. The deeper the colour, the longer the wavelengths where total absorption takes place.

Yellow zircon (uranium) Figure 10.10(1A)

Yellow to brown zircons from Sri Lanka show a strong and very distinctive spectrum of narrow bands distributed remarkably evenly throughout the spectrum. The strongest bands are at wavelengths 691, 662.5, 659, 653.5, 589.5, 562.5, 537.5, 515, 484, and 432.5 nm. These and fainter bands can be seen in Figure 10.10. Golden yellow zircons produced by heat treatment of brown crystals found in Indo-China do not show so strong or full a spectrum. But the 653.5 band and its fainter 659 companion can always be seen, as narrow 'pencil lines' in the red. It should be remembered that the 'reflected light' technique should be used where bands are too faint to be easily seen in transmitted light.

Yellow sapphire (iron) Figure 10.10(2A)

Yellow sapphires from Australia, Thailand, or Montana contain enough iron to show quite strongly the group of three bands in the blue which are at their full strength in green sapphire. The strongest of these is at 450 nm, and this may be seen faintly even in some Sri Lankan yellow sapphires, which contain very little iron. The other bands are at 460 and 471 nm. The last-named band can always be seen as distinct from the other two, which tend to coalesce when at full strength. It is important to be able to distinguish this threefold complex band from the solid block of absorption which is seen (at shorter wavelength) in chrysoberyl. In addition to proving the stone to be a yellow sapphire, these bands, when seen, are also a guarantee that the stone is not synthetic. Synthetic yellow sapphires are coloured with nickel, not iron, and show no absorption bands.

Yellow chrysoberyl (iron) Figure 10.10(3A)

Pale yellow, golden yellow, greenish-yellow or brown chrysoberyl all show a broad band at the beginning of the violet region, centred at about 444 nm. The strength of this band increases with the depth of colour of the stone: it is, in fact, the main producer of the colour seen. In strongly coloured stones, two weaker and narrower bands may be detected in the green-blue region at 505 and 485 nm. The band at 444 nm can often be seen in the highly prized chrysoberyl cat's-eyes, thus distinguishing them beyond doubt from their cheaper rivals, cut from chatoyant quartz.

Yellow orthoclase (iron) Figure 10.10(4A)

Yellow orthoclase from Madagascar makes quite an attractive gem. Its colour is due to ferric iron replacing some of the alumina in the feldspar, and this gives rise to two rather vague bands in the blue and violet, wavelengths 448 and 420 nm, the latter being the stronger.

Yellow spodumene (iron) Figure 10.10(5A)

Yellow spodumene also owes its colour to ferric iron. It shows two quite narrow bands in the violet at 438 and 432.5 nm, which are almost exactly similar to those seen in jadeite, which has a closely analogous formula. The 438 band is considerably stronger than its companion.

Spessartine garnet (manganese) Figure 10.10(7A)

The rare manganese garnet, spessartine, is seldom pure yellow, having usually an orange tinge, or even a hint of red, due to admixture with other garnet molecules. Almandine bands are often faintly visible, especially the persistent 505 band. The specifically spessartine bands include two rather feeble bands at 495 and 485 nm, a stronger one at 462 nm in the blue, and a powerful band at 432 nm. Where visibility extends as far, two further narrow bands may be seen at 424 and 412 nm, the latter being very intense. This may be a puzzling spectrum to the beginner, but when seen in a stone which from its refractive index one might expect to be an almandine, it does serve to prove the presence of important amounts of manganese, and enable one to place the garnet with certainty into its correct category. Some of the garnets from the Umba river area on the Kenyan/Tanzanian border may contain almandine, spessartine, and grossular molecules. These stones may show rich and complicated spectra, and care is needed in interpretation.

Sinhalite (iron) Figure 10.10(6A)

Sinhalite has been fairly extensively cut as a gemstone in the past, and specimens, formerly mistaken for zircon, chrysoberyl, or peridot, keep cropping up. The most attractive specimens are a golden yellow with greenish tinge. The spectrum is very similar to that of peridot, but there is an 'extra' band in the blue at 463 nm, which is missing in the peridot spectrum. There is a weak band at 527 nm, the main bands being at 493, 475, 463 (mentioned above) and 450 nm. The spectrum ends with what may be a further band at 436 nm.

Yellow apatite (didymium) Figure 10.10(8A)

Though brittle and rather soft, yellow apatite can make quite an attractive collector's gemstone. It is one of several yellow stones containing essential calcium in which traces of the rare-earth elements neodymium and praseodymium known as didymium are present, giving a characteristic narrow-line spectrum. Didymium bands are very similar in type and position in whatever medium they occur; thus they do not form so positive a basis for identification as do bands due to the presence of transition elements. But the spectrum in apatite is so much more intense than in danburite or in sphene, for instance, that it does virtually serve as diagnostic for the mineral. The strongest didymium lines form a group in the yellow, with prominent members at 584 and 578 nm. Another group of lines in the green has its centre near 538 nm. The drawing in Figure 10.10 gives a good idea of this complex and beautiful spectrum.

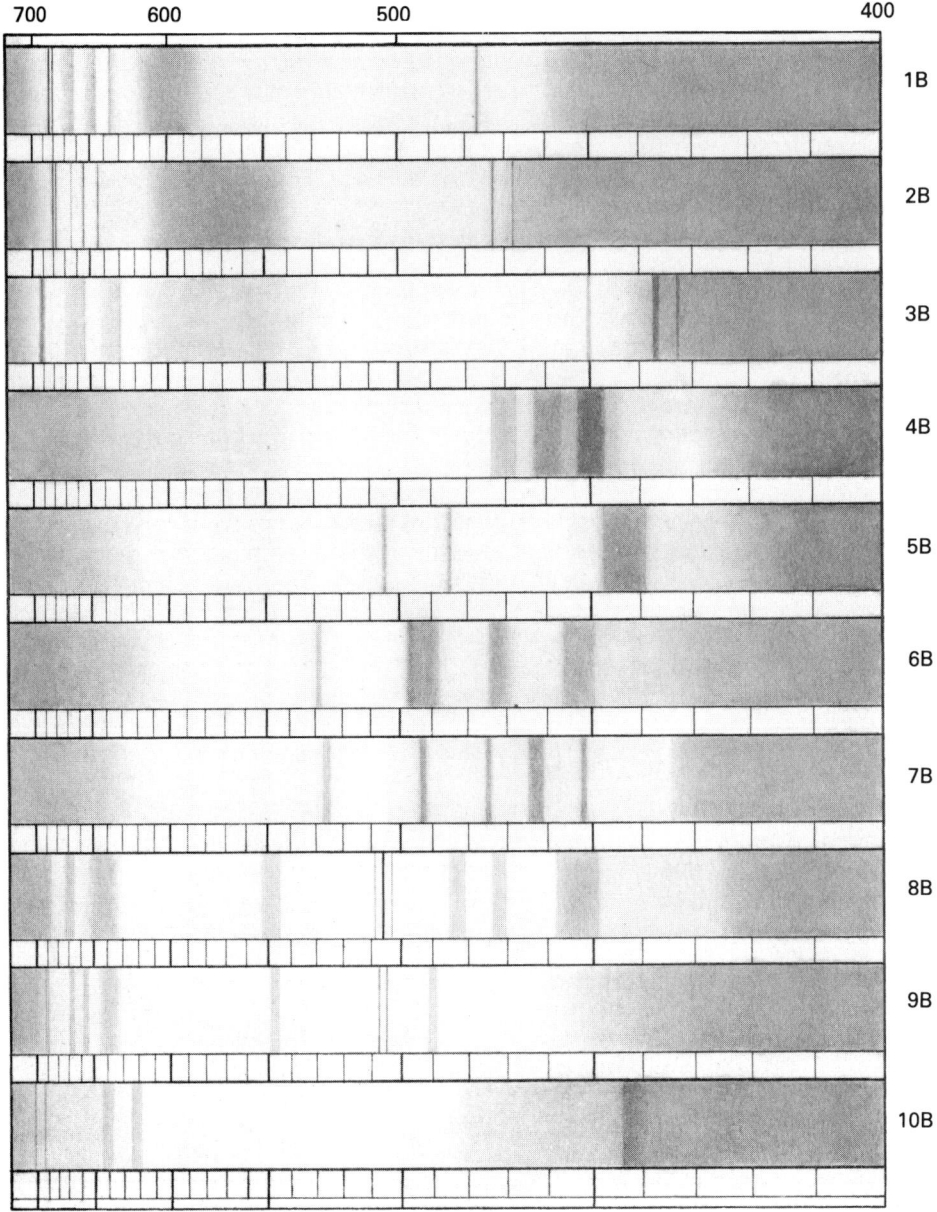

Figure 10.11(a) Absorption spectra of green stones. 1B emerald; 2B alexandrite; 3B jadeite; 4B sapphire; 5B chrysoberyl; 6B peridot; 7B sinhalite; 8B enstatite; 9B diopside; 10B demantoid

Green stones

Emerald (chromium) Figure 10.11(1B)

Emerald shows a typical chromium absorption spectrum, with a strong doublet in the deep red at 683 and 680 nm. The two narrow lines forming this doublet are

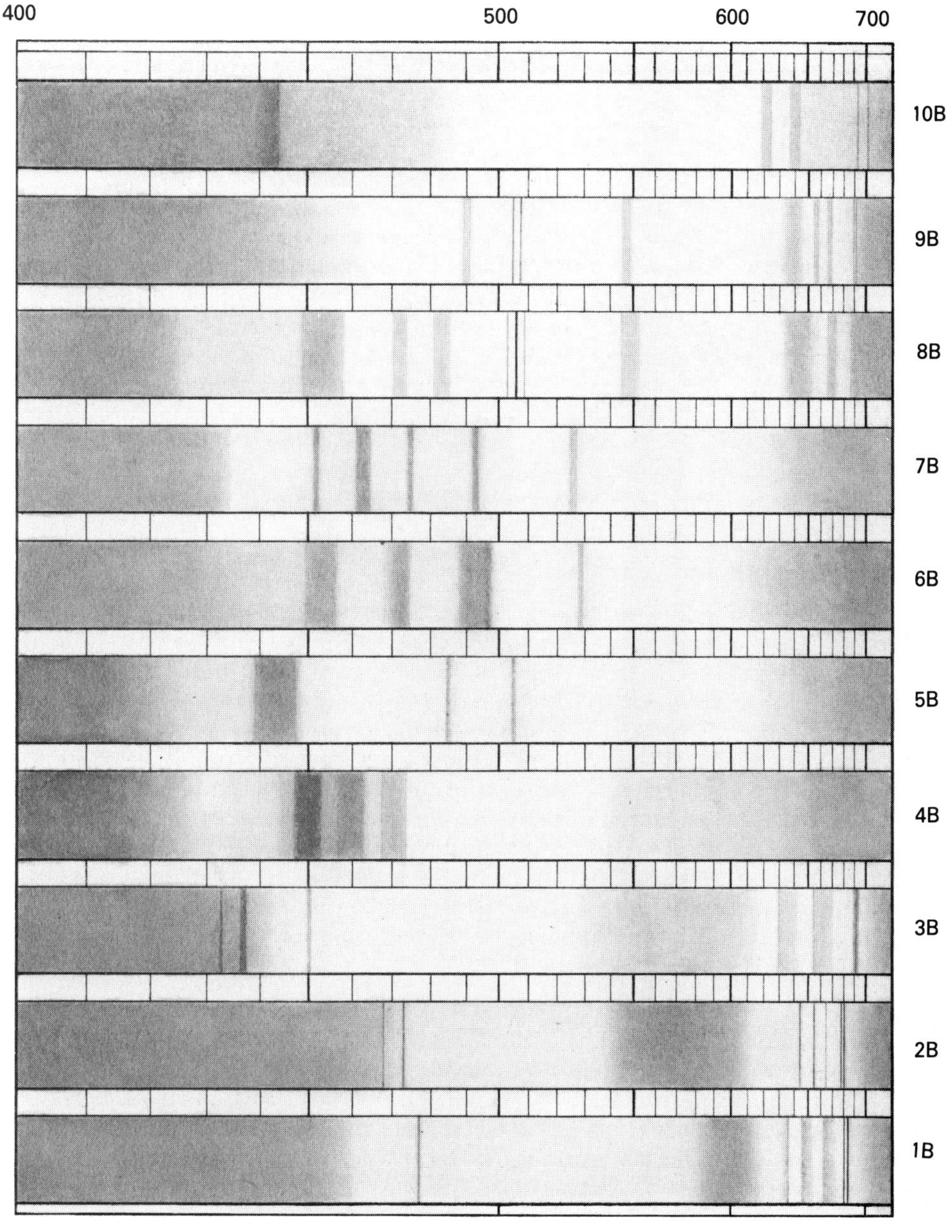

400   500   600   700

Figure 10.11(b) Reversed version of Figure 10.11(a). 1B emerald; 2B alexandrite; 3B jadeite; 4B
sapphire; 5B chrysoberyl; 6B peridot; 7B sinhalite; 8B enstatite; 9B diopside; 10B demantoid

twice as far apart as those in ruby, and can thus just be separated by a small
prism spectroscope. Two weaker and more diffuse lines appear at 662 nm and
646 nm, and beside each of these is a narrow region of high transparency, giving
a curious and characteristic appearance to the spectrum. Another line at 637 nm
is at its strongest in the ordinary ray, being then almost as strong as the doublet.

There is a broad but rather weak absorption band in the yellow – leaving the green unabsorbed: hence the colour of the stone. In chrome-rich specimens a narrow line in the blue at 477.4 nm may be seen in the spectrum of the ordinary ray.

Alexandrite (chromium) Figure 10.11(2B)

This is another typical chromium spectrum, with the lines more clearly defined than in emerald – sharper, perhaps, even than in ruby. In alexandrite the central broad absorption band which determines the colour is in a half-way position, centred at 580 nm, compared with the red ruby and the green emerald – hence its balance of colour between green (daylight) and red (tungsten light). The strong doublet in the deep red is at 680.3 and 678.5 nm, and is 'reversible' as in ruby, though not seen as a fluorescence line in ordinary circumstances. There are weaker lines at 665, 655, and 645 nm. Narrow lines can be seen in the blue at 473 and 468 nm. There is sufficient resemblance with the ruby spectrum to cause possible error if the spectrum is only carelessly examined, and the stone itself accepted as ruby on its appearance in artificial light only.

Jadeite (chromium and iron) Figure 10.11(3B)

Fine green jadeite owes its colour almost entirely to chromium, and typical chromium lines are seen in the red. There might be confusion here with translucent emerald, but in jadeite the lines are noticeably more diffuse. The doublet (not properly resolved) is centred at 691.5 nm, and two weaker lines at 655 and 630 nm. There is strong general absorption of the violet end of the spectrum, but if the slit of the spectroscope be opened to allow more light through, a powerful band at 437 nm can be detected lurking in the gloom. This is an iron band, and is seen at its best in the paler varieties of jadeite (white, lavender, pale green). It is narrow and well defined, and can be seen in light reflected from the specimen: it thus forms a useful test for jade beads and ornaments. A similar but far weaker band may be detected at 432 nm.

If the green colour is due to staining, a broadish absorption band in the red due to the dyestuff will reveal the fraud.

Chrysoprase (nickel)

Chrysoprase is rare, and highly prized compared with other forms of chalcedony; it owes its attractive apple- to jade-green colour to traces of nickel. A faint and ill-defined line in the red at 632 nm can be discerned and is due to this element. This would hardly warrant inclusion here were it not useful to be able to distinguish between true chrysoprase and the other green chalcedonies which are described below, by means of the spectroscope. There is no other easy means.

Chrome chalcedony (chromium)

Chalcedony of a rich green colour has recently been discovered and exploited in Zimbabwe but was also used by the Romans. It undoubtedly owes its colour to chromium. The absorption spectrum of this natural green chalcedony shows a rather strong and clear doublet in the deep red and little else, distinguishing it from **stained green chalcedony**, which shows a woolly version of the chromium doublet and two weaker bands on the orange side of this, each with a curious transparency patch alongside.

Green grossular garnet

Transparent green grossular garnet has been found in Pakistan, Tanzania, and in Kenya – the medium- to dark-green material being known as 'tsavorite'. It forms an attractive gemstone, though not approaching its illustrious cousin demantoid either in colour or dispersive power. The absorption spectrum shows little distinction in the specimens seen. The massive translucent green grossular found in the Transvaal and recently in Pakistan is more correctly called **hydrogrossular** as some of the silica is replaced by water molecules. This can grade into **californite**, the massive green form of idocrase, which also contains hydroxyl ions. When hydrogrossular is coloured green by chromium a broad band is seen in the orange, centred at 630 nm and with a noticeably sharp cut-off on the long-wave side. Californite, on the other hand, shows a strong band in the blue at 461 nm (due to iron) and a weaker band in the green at 530 nm.

Green sapphire (iron) Figure 10.11(4B)

The spectrum already described under yellow sapphire belongs in its full strength to green sapphire. Here the two strongest iron bands practically coalesce, but the third band (471 nm) is sufficiently detached to reveal that the band is a complex one, and not a single block, as in chrysoberyl.

Green chrysoberyl (iron) Figure 10.11(5B)

Green chrysoberyl gives merely a stronger version of the broad absorption band in the blue-violet already described under yellow chrysoberyl. This is centred at 444 nm. Weaker bands may be seen at 505 and 485 nm. The 444 band may often be seen in cat's-eyes, forming a useful means of confirming that they are not the less highly valued quartz cat's-eyes.

Peridot (iron) Figure 10.11(6B)

Peridot shows a very characteristic iron spectrum, the main feature of which is the series of three evenly spaced bands in the blue region. The first of these appears where the green changes to blue at 493 nm. The second is at 473, and the third at 453 nm. The edges of the bands are diffuse, but there is a fairly well defined 'core' to each – especially the first two. The appearance of these bands varies noticeably according to the vibration-direction of the light. The 'beta' spectrum is the strongest, and in this a weak band at 529 nm can be detected.

Sinhalite (iron) Figure 10.11(7B)

This gemstone has been already dealt with under 'yellow stones', but it is often greenish in cast. It is well to be clear on the difference between the spectrum of sinhalite and that of peridot, with which it was so long confused. The drawing reproduced in Figure 10.11 shows the two spectra side by side. The 'extra' band in sinhalite at 463 nm can be clearly seen.

Enstatite (iron) Figure 10.11(8B)

An outstanding feature in the absorption spectrum of enstatite is a clear-cut, narrow band at 506 nm. This is seen in tremendous strength in brownish stones from India, when weaker bands at 548, 483, 450 nm, etc. can also be seen. In the attractive green enstatites which, with fine red pyrope pebbles, were recovered from the diamond concentrates in Kimberley, the colour is enriched by the presence of chromium, and a doublet can be seen at 687 nm and weaker chromium bands on the short-wave side of this in the usual chromium manner. In these green enstatites the 506 band is still sharp and clear. There can be no doubt it is due to ferrous iron. This region of the spectrum, where green turns to blue, is often the position for the strongest iron band (cf. almandine, peridot, green tourmaline, diopside). In the 1980s colourless enstatites were discovered in southern Sri Lanka. In many stones described as colourless the 506 nm band is still present (albeit faintly).

Chrome diopside (iron, chromium) Figure 10.11(9B)

Attractive green diopside (often chatoyant) from Burma contains enough chromium to enrich the colour and show chromium lines in the red, with the doublet centred at 690 nm. Narrow bands in the greenish-blue are reminiscent of enstatite, but here there are two lines in place of one, about equal in strength, the wavelengths being 508 and 505 nm.

Demantoid garnet (iron, chromium) Figure 10.11(10B)

The most valuable and most attractive of the garnets, the green garnet demantoid, is a variety of the calcium–iron garnet andradite. Here again we have an iron spectrum (ferric) with enough chromium to enrich the colour and show chromium lines in the red. The iron band is an intensely strong one, centred at 443 nm. This is far enough into the blue-violet to be difficult to see as anything more than a sharp cut-off to the end of the spectrum, especially where chromium is present and causing general absorption of the violet. In paler stones, and with good lighting conditions, the band can be seen complete: though very strong, it is not very wide. The chromium lines are easily seen in good specimens: the doublet is very deep in the red (701 nm), and there are weaker and rather diffuse lines at 640 and 621 nm.

Zircon (uranium)

Green zircons can always be relied upon to show absorption bands, but the type of spectrum seen varies enormously with the degree of internal breakdown (metamictization) of the zircon crystal concerned. Most Sri Lankan green or greenish zircons show a strong ten- or twelve-band spectrum, with the strongest band in the red at 653.5 nm and the other bands distributed throughout the spectrum as described under 'yellow zircon'. But even in these it will be noticed that the edges of the bands are 'woolly', and not sharply defined as in the fully crystalline 'high type' zircons. Such stones have usually a rather cloudy or sleepy appearance. In the really 'low type' zircon with SG 4 or under and no appreciable birefringence there may remain only a diffuse and rather broad band near 653 nm. In some low types, however, a narrow band at 520 nm in the green is very noticeable. After heating at 800°C these and other metamict zircons having a SG below 4.02 develop a strong anomalous series of absorption bands at 687, 668.5, 652.5, 589, 574, 560.5, 473.5, and 451 nm. This spectrum is rarely met with in zircons as they occur in nature. The anomalous spectrum is shown in Figure 17.1, together with the normal zircon spectrum for comparison.

Zircons of rather dark brownish-green colour are found in Burma and these have suffered no metamictization and contain more uranium than those from Sri Lanka. The result is a spectrum showing not only the dozen or so of main zircon bands at great intensity but other fainter lines between, bring up the number to some forty bands in all, as seen through a small spectroscope. To the practised eye all the above variations on the main zircon theme spell 'zircon', especially when the appearance of the stone is taken into account. The key position to look for is the 653.5 nm region in the red: where the other bands in the green and blue are seen also there can be no mistake. If these are missing, the beginner may confuse the narrow uranium bands for a chromium spectrum. But the strong chromium doublets are always far deeper into the red, and the accompanying features of each chromium spectrum are sufficiently distinctive to provide the necessary supporting evidence in case of doubt.

Green tourmaline (iron)

Tourmalines of full-bodied green or greenish-blue absorb all the red part of the spectrum down to about 640 nm. In that critical position where the green gives way to the blue a narrow absorption band can be seen (497), which can be ascribed to iron.

Aquamarine (iron)

Sea-green aquamarine is seldom used in jewellery now, the stones being usually heat treated to produce a pale blue colour which is more popular. In green aquamarine and sometimes in other pale beryls a distinctive narrow line can be seen at 537 nm in the green. This is only visible in the extraordinary ray (see also blue aquamarine), and is usually more easily detected by scattered or reflected light.

Epidote, clinozoisite (iron)

Epidote commonly absorbs so much light that it can seldom be cut into an attractive gem, beautiful though its appearance may be as a well-crystallized mineral. The colour is typically a deep brownish-green. The blue and violet are both strongly absorbed, but if enough light can be transmitted to enable them to be seen, there is an intense band at 455 nm, and one less intense at 475 nm. The closely related mineral clinozoisite contains less iron and is more transparent. This shows the 455 band very clearly.

Andalusite (manganese)

The brownish-green andalusites showing a hint of red, which are rare enough to be sought after by collectors, may show an absorption band in the blue at 455 nm, but this is not very reliable as a test. Quite a different spectrum is seen in a rare green type of the mineral from Brazil. Because of the extreme fineness and delicacy of the bands, this was formerly assumed to owe its absorption features to some unidentified rare earth. But this beautiful spectrum can now be ascribed to manganese, and was first reported as a feature of viridine, a manganese-rich form of andalusite. In this spectrum there is a sharp edge of shadow at 552.5 nm in the green, shading away on the yellow side, while on the other side is a narrow line at 549.5 nm. There is another line nearer the blue at 517.5 nm, and strong general absorption of the blue and violet.

Fluorite

Weak bands may be seen in green fluorite at 634 and 610 nm in the orange, at 582 nm in the yellow, and 446 and 427 nm in the violet. These bands may help to identify a large translucent fluorspar ornament or figure, which may be confused with emerald or jadeite. In cut specimens (which, in any case, are only collector's items, as the stone is far too soft for wear) these bands may be too faint to be noticed.

Kornerupine

Only in fine specimens is this mineral attractive as a gem, and its rarity precludes its use in commercial jewellery. It is pleochroic, and the absorption bands vary considerably according to direction.

There is a band in the blue-violet at 446 nm which is fairly strong in the 'beta' ray, and weaker bands at 540 and 503 nm in the green and 463 nm in the blue. There is another weak band at 430 nm in the violet. The bands are probably due to iron. The spectrum is not distinctive enough to form more than confirmative evidence of the identity of the mineral.

The list of absorption spectra shown by green stones could be extended still further, but all the most important and reliable instances have been given above. Those translucent stones which occasionally show bands (e.g. nephrite, Connemara marble, etc.) and are not mentioned above will be noted in the descriptive text later in the book where this is thought to be helpful.

Blue stones

Sapphire (iron) Figure 10.12(1C)

One of the most important spectra in the whole series for the practising gemmologist is that of sapphire, for it not only proclaims the species of the stone, but it can also reveal whether it be natural or synthetic. The spectrum, which is due to iron, is seen at full strength in green sapphire and in varying strength in yellow sapphire, as has been described above. In blue sapphires the bands are again very variable in strength. Australian sapphires show all three bands quite strongly, while in Sri Lankan sapphires, which contain very little iron, only the 450 band can be seen, and even this is often exceedingly faint. The band is an 'ordinary ray' band, and therefore shows at maximum strength in the direction of the optic axis. In a cut stone this direction may usually not be known, but it is worth while, if the band is only very faintly seen, to turn the stone on edge and on end in an attempt to see the band more strongly. A 'Polaroid' disc turned to the correct angle over the eyepiece of the spectroscope may also help to increase the apparent strength of the band. Passing light through a copper sulphate filter flask is here very helpful, enabling the observer to see any bands in the blue region with greater clarity. With Verneuil type synthetic sapphires, although iron oxide as well as titanium is used to produce the blue colour, virtually all the iron evaporates from the boule or is concentrated in the surface layers which are removed when cutting is carried out. Three very weak and vague bands in the blue have often been detected by the author in Verneuil type synthetic sapphires, the central one of these being very nearly in the same position as the 450 band of natural sapphire. But there should be no confusion between the two, as it must be emphasized that the 450 band is quite a narrow one when it is only faintly developed. Indications of the 460 band to one side of the main band can also be seen in sapphires from any other locality than Sri Lanka, and this adds to the distinctiveness of the spectrum. It is well worth the gemmologist's while to practise the observation of this band, as it will save much time and trouble in testing sapphire jewellery. While the synthetic blue sapphire produced by Chatham is only rarely seen as a cut stone, it is worth noting here that the spectrum of this synthetic could be confused with that of a stone from Sri Lanka.

Blue spinel (iron) Figure 10.12(2C)

The absorption spectrum of blue spinel is another which is well worth study. It is somewhat complex and difficult to describe, and may even vary a little in its nature from specimen to specimen. But when one has one's eye in, so to speak, the nature and distribution of the bands are very distinctive. Indications of the 'blue spinel' bands can be seen in many purplish or brownish-red spinels also. The key bands to note are a broad one in the blue centred at 459 nm and a narrow band of about equal strength at 480 nm, on the green side of this. Rather elusive bands in the green (555 nm), yellow (592), and orange (632) complete the pattern, and give the impression of quite a rich spectrum. Nuances of absorption which can hardly be counted as bands add to this impression of richness. Distinction between the blue spinel and sapphire spectra can be made with complete certainty by taking the trouble to apprehend the different nature and pattern of the bands in the blue, quite apart from their different wavelengths. Similarly, the vague bands in the green should not be confused with the cobalt bands in synthetic blue spinel which may also occur in the rare cobalt-bearing natural blue spinel. But possible mistakes of this kind should be borne in mind, and accessory tests carried out where the observer does not feel quite sure.

Aquamarine (iron) Figure 10.12(3C)

Mention has already been made of the spectrum of green aquamarine. The blue aquamarine now so popular shows two rather ill-defined and weak bands in the blue (456 nm) and violet (427). The 427 band is fairly strong in large specimens (and aquamarines used in jewellery are usually large) while the other band, weak though it is, helps to make the spectrum recognizable. Aquamarine from the Maxixe mine has a most peculiar spectrum, with strong bands in the red at 697 and 657 nm, and a weaker band in the orange at 628 nm. In the mid-1970s beryls of an exceptionally fine dark blue colour appeared on the market. These showed the strong absorption bands in the red just described as typical of beryls

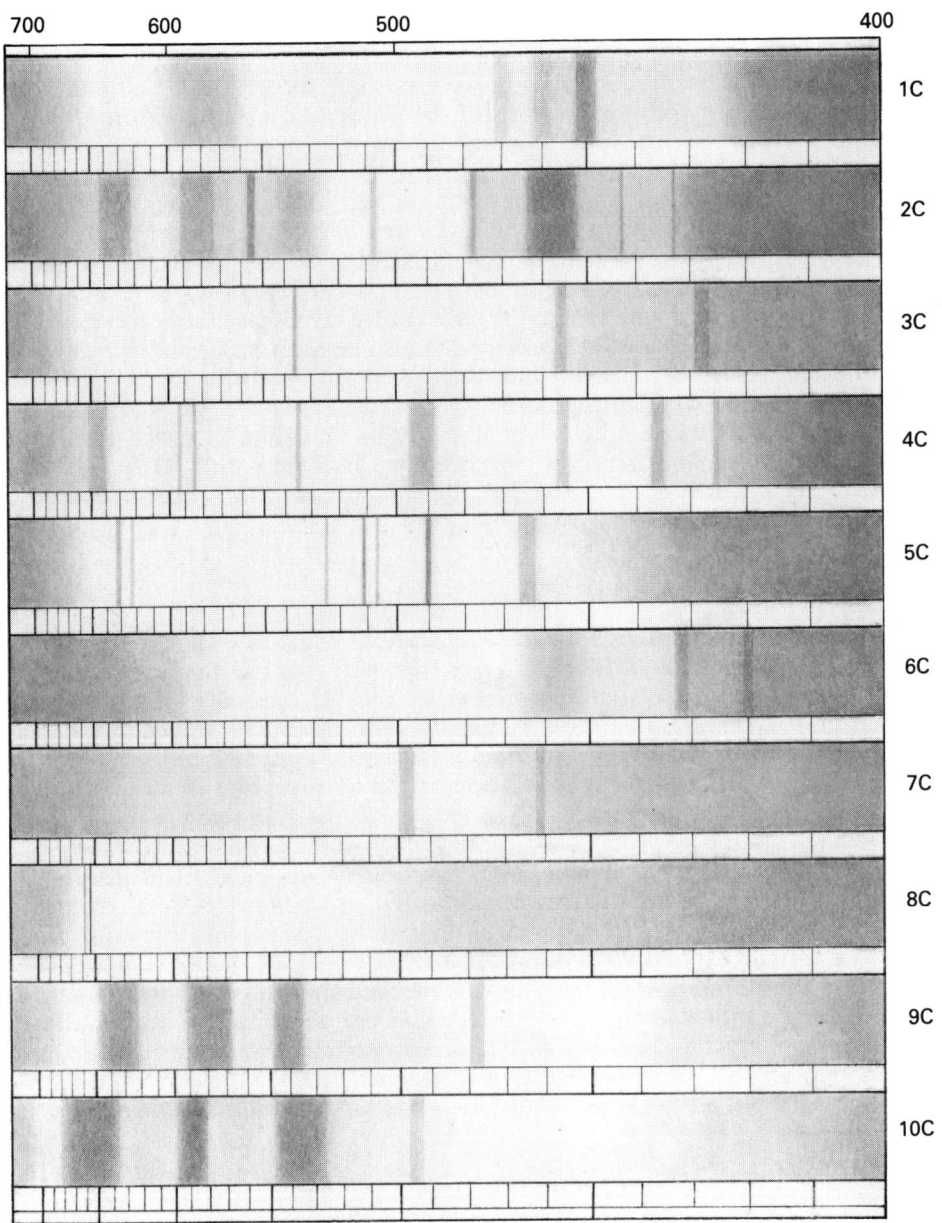

Figure 10.12(a) Absorption spectra of blue stones. 1C sapphire; 2C spinel; 3C aquamarine; 4C iolite; 5C apatite; 6C turquoise; 7C tourmaline; 8C zircon; 9C synthetic spinel; 10C cobalt glass

from the old Maxixe mine, and also the optical peculiarity by which the strongest blue colour belonged to the ordinary ray rather than to the extraordinary ray, as in normal aquamarine. These blue beryls, called for convenience 'Maxixe type' by gemmologists, unfortunately lose their lovely colour when exposed for some hours to bright sunlight, or for a longer period to tungsten light, and this

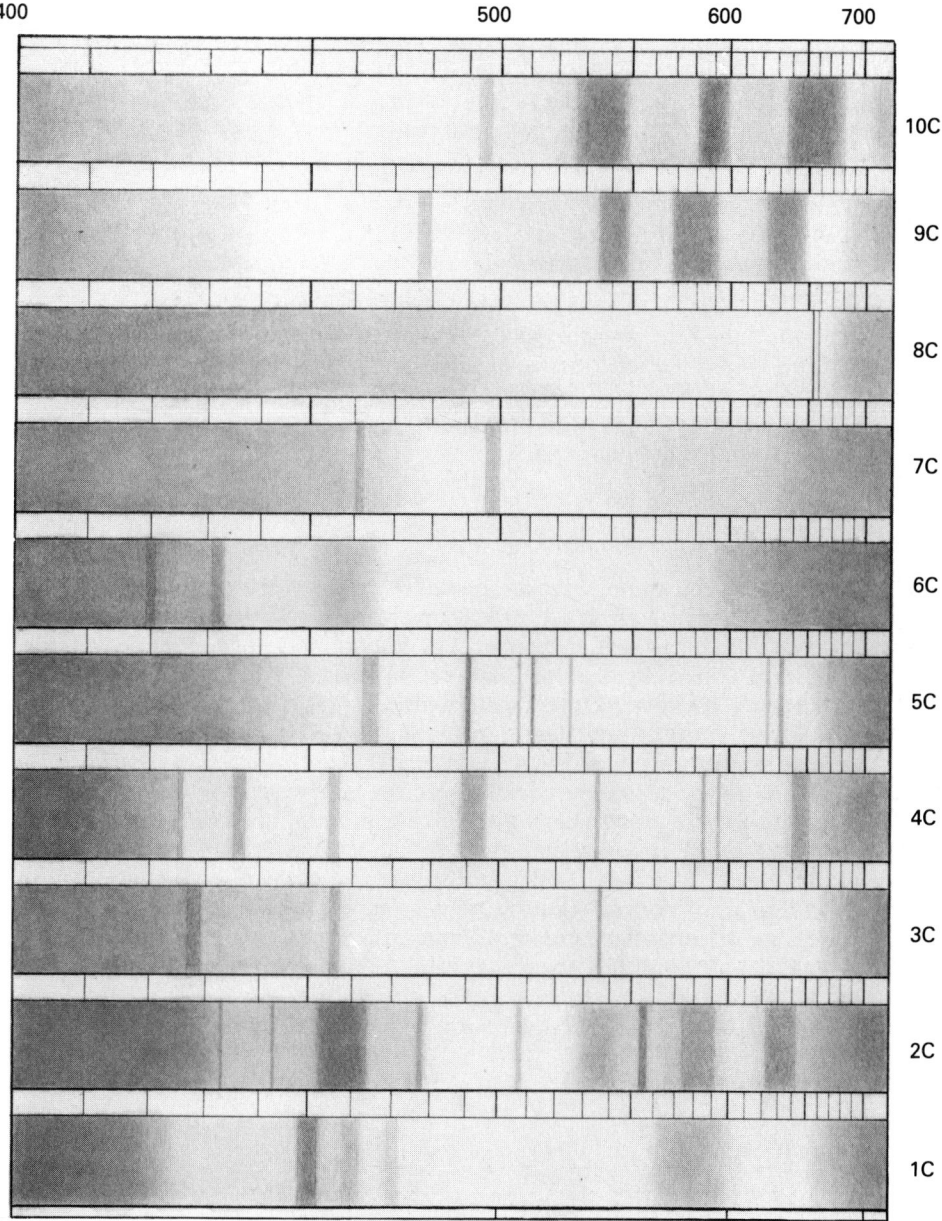

Figure 10.12(b) Reversed version of Figure 10.12(a). 1C sapphire; 2C spinel; 3C aquamarine; 4C iolite; 5C apatite; 6C turquoise; 7C tourmaline; 8C zircon; 9C synthetic spinel; 10C cobalt glass

considerably reduces their desirability as a gem variety. It seems probable that the deep colour has been induced by some form of radiation treatment. The narrow line at 537 nm, mentioned under 'green aquamarine', does not seem to appear in the heat-treated blue types. In the drawing, this line has been included, however, to give an idea of its position and nature.

Iolite (iron) Figure 10.12(4C)

As one might expect in so pleochroic a gemstone, the absorption bands seen in iolite vary a good deal with direction. In the blue ray there are vague bands in the blue and violet regions at 492, 456, and 437 nm. In the yellow ray there is a distinctive double band, each member being quite narrow, in the yellow region (593, 585 nm). There is also a narrow band or line at 535 nm, reminiscent in position and nature of the green aquamarine line mentioned previously.

Apatite Figure 10.12(5C)

Curiously enough, blue apatite does not often show the didymium bands so typical of the yellow variety. Instead, there are bands at 631 and 622 nm in the orange, at 511, strong and fairly narrow in the green, a strong band at 490, and a broad, weak band at 464. These bands belong to the 'ordinary' ray which is yellowish in colour when isolated, as it is in the direction of the optic axis.

Turquoise (copper) Figure 10.12(6C)

The absorption spectrum of turquoise, discovered by the author during the 1940s, has proved to be a most useful diagnostic feature in a gem material which is notoriously difficult to test. There are two bands of almost equal appearance and strength (as shown by a photograph) at 432 and 420 nm, of which only the 432 can normally be seen. Only through thin edges is turquoise translucent enough to allow enough light through to render this violet band visible, but fortunately it shows very well by reflected light. There is a vague band in the blue at 460 nm, which helps to provide a distinctive pattern for the mineral. No bands similar to these have been seen in any minerals or counterfeits which resemble turquoise in appearance.

Turquoise which has been impregnated with plastics material has a more intensive colour than normal (though no colouring matter has been added). In such stones the absorption band at 432 nm is seen much more strongly than usual. Visual acuity is low in the violet region, and many gemmologists have difficulty in seeing this useful turquoise spectrum. The use of blue light, concentrated through a flask of copper sulphate solution, is of particular value in this case.

Tourmaline (iron) Figure 10.12(7C)

Blue tourmaline shows the same narrow band at 497 nm mentioned under 'green tourmaline'.

Zircon (uranium) Figure 10.12(8C)

Blue zircon owes its colour to heat treatment of reddish-brown rough. At least two of the strongest of the zircon lines – those at 653.5 and 659 nm – can be seen, as narrow 'pencil lines', especially when reflected light is used. Other of the main zircon bands may also show faintly in the yellow and green.

Zoisite

The deep violet-blue form of zoisite discovered in Tanzania, for which the name 'tanzanite' has been suggested, has a rather distinctive spectrum as found in nature, with a moderately strong, broad absorption band in the orange at 595 nm, and weaker broad bands centred at 528 nm in the green and in the blue at 455 nm. This last band might be thought dangerously near the well-known band at 450 nm in sapphire, which is so useful an aid to its recognition, but in the heat-treated blue zoisites used in commerce the band is hardly visible, and is, in any case, far less narrow and well defined.

Synthetic blue spinel (cobalt) Figure 10.12(9C)

Cobalt-blue is very rarely found in natural minerals, and when seen usually heralds the presence of some artificial product – notably blue synthetic spinel or cobalt glass. The cobalt absorption bands seen are very similar in each case, consisting of three broad main bands in the orange, yellow, and green, but in synthetic spinel the bands are more closely clumped together (see drawing) and the central band is the widest of the three, whereas in glass the centre band is the narrowest. Wavelengths of the synthetic spinel bands are 635, 580, and 540 nm. The width and intensity of these bands vary proportionately to the depth of the blue colour. In deep blue specimens the bands almost amalgamate into a single block. Absorption of the yellowish-green, together with free transmission in the deep red make these cobalt-coloured artefacts appear strongly red under the Chelsea filter.

Cobalt glass (cobalt) Figure 10.12(10C)

The position of cobalt bands in glass varies a little with the composition of the glass concerned. Average measurements are 655, 590, and 535 nm. Cobalt glass is often used as the base of blue garnet-topped doublets: the superposition of a faint almandine spectrum on that of the cobalt glass may give a rather confusing effect unless the nature of the stone is apprehended.

Blue synthetic quartz (cobalt)

Synthetic quartz, coloured blue with cobalt, has been encountered as a mounted gemstone. The typical three broad absorption bands of cobalt were observed in this artefact, in position very much as in cobalt glass. **Lithium niobate**, another synthetic, has also been produced in a cobalt-blue variety. The dictum still remains true, that a cobalt spectrum seen in a blue 'gemstone' is a strong sign that it is a glass, glass-backed doublet, or some form of synthetic stone.

Rare-earth spectra

The beautiful absorption spectrum consisting of groups of closely spaced, narrow lines, due to the two rare-earth elements collectively known as didymium, has already been mentioned in previous pages, and wavelengths given for the main bands as seen in yellow apatite. A closer study of the didymium spectrum, and its slight but possibly significant variations in different media is now becoming of some practical value for those engaged in serious gem testing because some of the new synthetics such as scheelite, YAG (yttrium aluminate), and cubic zirconia are frequently 'doped' with one or other of the rare-earth oxides.

In nature, the two elements neodymium and praseodymium are inseparable, though the relative proportions of each in a mineral are found to vary slightly, with neodymium always the dominant partner. On its own, neodymium induces a lilac colour in the host crystal or glass and its strongest lines are grouped in the yellow and in the green, while praseodymium produces a green tint in its host material, with its strongest lines fairly well spaced in the blue and violet. Measurements for bands due to each of these elements in YAG can be given as follows: neodymium – 732 nm in the very deep red, 594, 592, 589, 585, 580, 578, 573.5 and 569 nm in the yellow, and 531 and 528 in the green, with many fainter lines; praseodymium – two lines in the yellow at 589 and 583 nm, and a much stronger series of three lines in the blue and violet at 485, 477, and 449 nm. The sharpness of these lines is very noticeably greater in the case of the artificial garnets such as YAG than it is in apatite: lines in scheelite are also sharp, suggesting that a high refractive index may be a related factor.

It is worth noting that the presence of these rare-earth ions often gives rise to a 'line' fluorescence under suitable stimulus, and this may help to differentiate natural from synthetic material on occasion: for instance, synthetic scheelite doped with neodymium showed no fluorescence under crossed filters, while natural scheelite containing traces of didymium showed a line fluorescence with wavelengths 650, 620, and 558 nm, probably due to the praseodymium content.

Several other rare-earth elements give rise to line spectra, though these do not appear in any natural gems. When seen they will surely give warning of a man-made crystal. The spectra of erbium and of dysprosium as developed in artificially produced garnets can be seen in the drawing reproduced as Figure 10.13.

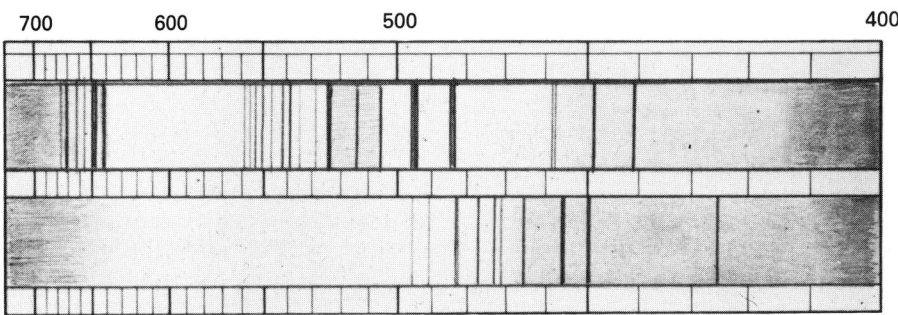

Figure 10.13 Erbium bands (*top*) and dysprosium bands in synthetic rare-earth garnets

All absorption spectra of any importance have been described above. In many of the species faint bands additional to those listed have been seen and measured, but these add nothing at all to the recognition of the stones in question. Those interested in complete descriptions will find them in a series of forty articles in the *Gemmologist* on 'The spectroscope and its applications to gemmology', which appeared from September 1953 to December 1956. An index to the series was given in the issue for January 1957.

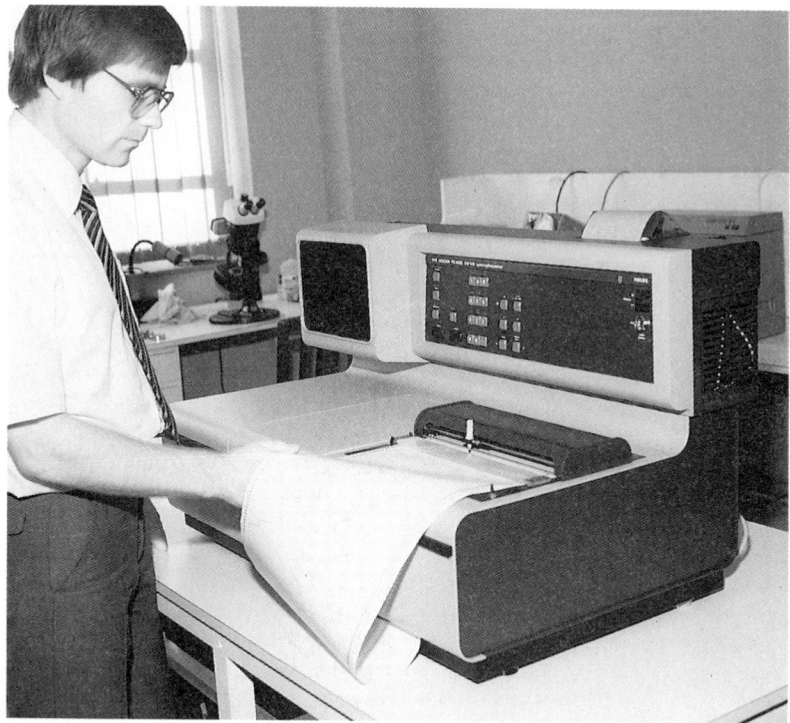

Figure 10.14 The Pye Unicam PU 8800 ultra-violet/visible spectrophotometer

Figure 10.15 The Nicolet model 510 Fourier Transform infra-red spectrometer

Beginners are urged to confine their attention to simple spectra such as those of almandine and ruby and some zircons before attempting less well-defined absorption effects. A really strong source of light and a properly adjusted and suitable spectroscope are the most important keys to eventual success in this beautiful and important method of gem testing.

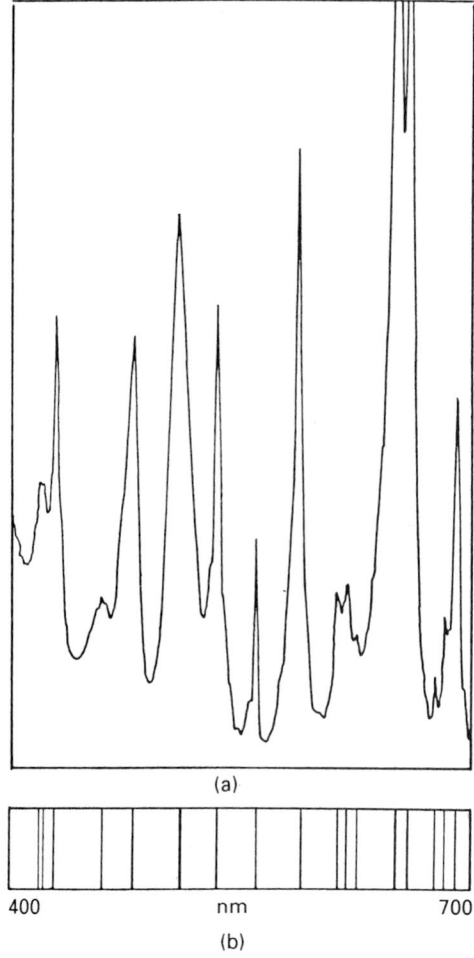

(a)

400   nm   700

(b)

Figure 10.16 Absorption curves of a zircon (a) as plotted by a spectrophotometer and (b) as seen in a hand-held diffraction grating-type spectroscope

At the close of this chapter it should be mentioned that sophisticated and expensive infra-red spectrometers and ultra-violet-visible spectrophotometers are being used increasingly in professional gemmological laboratories. Their use is especially desirable when it is necessary to distinguish between clean natural rubies, emeralds, and sapphires and their ingenious synthetic counterparts, and to detect the presence of impregnating plastics in apparently fine opals. They are also extremely helpful in categorizing natural and irradiated coloured diamonds. There are many other uses and more will be found. To describe these operations would be far beyond the scope of this book, but illustrations are provided in Figures 10.14 and 10.15 of the instruments currently (1989) in use in the Gem Testing Laboratory of Great Britain in London.

A diagram is also provided (Figure 10.16) which shows the relationship between the traditional representations of the absorption spectrum of a zircon using a hand spectroscope and the chart (graph) produced by a spectrophotometer.

11

Fluorescence as an aid to identification

Though fluorescence in gemstones can seldom in itself serve as a definitive test, it can act either as a most useful indicator or to provide confirmatory evidence in certain cases, and laboratory workers would certainly feel handicapped if deprived of their ultra-violet lamps. In what follows, a short description of fluorescence as a testing method will be given, though in accordance with the principles of this book only those luminescent effects which the author has himself observed and found useful or interesting will be included.

First, what exactly is fluorescence? It is the emission of visible light by a substance under the stimulus of visible or invisible radiations of shorter wavelength. If the fluorescent glow persists for an appreciable time after the stimulating rays have been cut off, this afterglow is termed 'phosphorescence'. Sometimes the phosphorescence may differ in colour from the original fluorescence. This happens when some of the wavelengths composing the fluorescent glow are extinguished as soon as the stimulus is removed, while others decay more slowly, and persist. 'Luminescence' is a convenient term to cover both fluorescence and phosphorescence. 'Thermophosphorescence' is seen when certain stones, at some time after irradiation, are seen to emit visible light on warming. The phosphorescence in such cases is considered to be 'frozen in', and needs heat energy to release it.

In the London Gem Testing Laboratory four types of radiation are in common use for producing fluorescence, and their effects are often distinctly different on any one specimen. Each has its own special value in certain cases.

The first to consider, on account of its simplicity, is merely visible light of the blue and violet regions, produced by passing a powerful beam of light from a 500-watt projection lamp through a flask containing a strong solution of copper sulphate. The flask should conveniently be one of 600- or 750 ml capacity, and this serves as a filter and condensing lens combined. It is well worthwhile making the solution with distilled water to avoid any trace of milkiness – and, even so, filtering may be necessary to obtain a crystal-clear liquid. The copper sulphate absorbs all red, orange, and yellow light, so that the filtered rays are invisible when viewed through a good red or orange filter. If any substance, therefore, which is placed in a beam of the blue light and is not otherwise illuminated is seen to glow when viewed through one of these filters, it must be exhibiting fluorescence – i.e. be absorbing energy from the blue-violet light and re-emitting part of the energy as light of longer wavelength.

This technique of using 'crossed filters' to observe fluorescence was first devised by the great British physicist George Gabriel Stokes in 1852, when he established the true nature of fluorescence for the first time. The author first adapted the method as a lecture demonstration, but found it so sensitive and useful for testing gems which have a red or orange fluorescence that it has been constantly used in the London Laboratory ever since.

The second and most popular method for stimulating fluorescence is by means of a high-pressure mercury discharge lamp housed in a fused quartz tube, the light from which is filtered through 'Wood's glass'. Wood's glass is a dark glass containing cobalt and some 4 per cent nickel, and is named after a great American pioneer in physical optics, R. W. Wood, who first proposed its use. This filter cuts out most of the four strong mercury lines in the visible spectrum, and transmits those in the near ultra-violet, notably a powerful line at 365 nm, which does most of the work in stimulating fluorescence. Hanovia Ltd and a few other manufacturers supply several forms of such 'analytical' lamps, which have considerable use in industry for detecting impurities etc. in manufactured products. Medical 'sun lamps' are of this type, but lack the Wood's glass filter. Without the filter any fluorescent effects would, of course, be lost in the intense glare of the unshielded lamp.

Rather less effective lamps can be home-made by using an arc lamp with special charged carbons emitting a rich ultra-violet spectrum – or even a 'Photoflood' lamp – and filtering the light through a piece of Wood's glass mounted in an opaque screen.

The third form of radiation is short-wave ultra-violet light emitted by a low-pressure quartz-mercury lamp in which the 253.7 nm mercury line is preponderant. Wood's glass does not transmit this line, and a special filter such as Chance's OX7 glass has to be used. Unfortunately, a good deal of visible light also passes through this filter. Convenient short-wave lamps such as the 'Mineralight' are made in the USA; but a small 'germicidal' lamp made by Phillips, used with an OX7 filter, has proved very effective, despite its low power. Effects are in most cases less spectacular than under long-wave ultra-violet, but in some important instances are more diagnostic.

Another and very useful form of ultra-violet lamp consists of a low-pressure mercury vapour tube or tubes in which short-wave ultra-violet light is emitted and induces in the special internal coating of the glass filter tubes the emission of long-wave ultra-violet rays. The principle is similar to that used in ordinary fluorescent lighting. Lamps of this kind give results which are in most cases comparable with those obtained under high-pressure mercury (long-wave) lamps combined with a Wood's glass filter, though they may be rather more successful in showing differences in fluorescent response of natural rubies and emeralds and their synthetic counterparts. A number of dual lamps incorporating both long- and short-wave sources, enabling fluorescent effects under each type of radiation to be compared at the touch of a switch, are available from laboratory suppliers (Figure 11.1).

X-rays, the fourth source of radiation, are not usually available to the amateur gemmologist, and moreover, they are dangerous in use unless very carefully

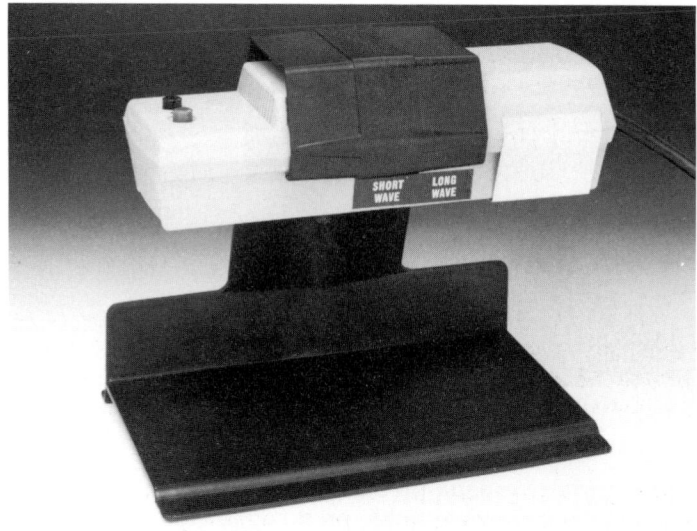

Figure 11.1 Combined long- and short-wave ultra-violet lamp (courtesy: Gemmological Association)

shielded. Fluorescent effects under these rays must be viewed through a thick lead-glass window. Several gemstones are known to alter in colour when exposed to X-rays, and even ultra-violet light may cause discoloration in white and blue zircons, so that prolonged exposure must be carefully avoided.

Broadly speaking, for gems showing a fluorescence at the red end of the spectrum the 'crossed filter' technique is the most convenient, and the most sensitive. For stones showing green or blue fluorescence, long-wave ultra-violet rays are generally the best to use, while only in special cases need short ultra-violet or X-rays be resorted to because of some specific reaction not found under the other radiations.

Crossed filters

A red fluorescence is one of the peculiarities of minerals owing their colour to a trace of chromium, provided that iron is not present to inhibit the glow. Thus it is that ruby, red spinel, alexandrite, emerald, and pink topaz all show up well between crossed filters (see Figure 11.2). Synthetic sapphires of the 'alexandrite' type, most synthetic blue spinels and synthetic green sapphires also glow red under the same conditions. Of these, Burma and synthetic rubies and clear red or pink spinels glow the most brightly: synthetic red spinels are also quite strongly fluorescent. The spectroscope is valuable here in making the test more specific, and enables one to distinguish between ruby and spinel. Ruby and synthetic ruby have an identical fluorescence spectrum, in which the doublet at 692.8 and 694.2 nm appears as a single intense line. Only faint indications can be seen of fluorescence lines on either side of this. Natural spinel, on the other hand, displays a group of several narrow lines, like a set of organ pipes, of which two of almost equal strength are in the middle. Most curiously, synthetic

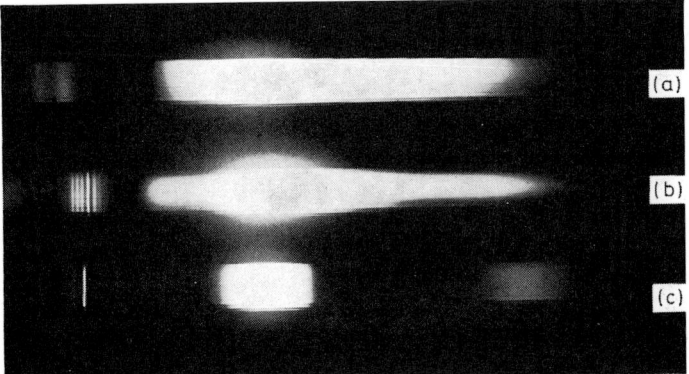

Figure 11.2 Fluorescence spectra of (a) emerald, (b) red spinel, and (c) ruby, all excited by blue light

spinel, whether it is made by the Verneuil process or by crystallization, has not got quite the same spectrum, but shows a single predominant line near 685 nm against a background of fainter lines. Thus in appearance these synthetic spinels show a spectrum rather closely resembling that of ruby. There should be enough of the 'organ pipe' effect to enable a distinction to be made, but a glance in the blue part of the spectrum (transmitted by the copper sulphate flask filter) will settle the issue, as ruby will show its three strong narrow absorption lines in this region, whereas there are no lines in the blue in spinel of this colour.

Alexandrites and emeralds vary in their response according to their chromium content and the degree to which iron is present. In each case the red fluorescence will serve to distinguish between the chrome-bearing varieties of emerald and alexandrite and ordinary beryl and chrysoberyl, respectively. Synthetic emerald shows a stronger glow than most natural emeralds, while those from South Africa or India, which contain iron, display practically no fluorescence.

Pink topaz shows quite a distinct glow, emanating chiefly from an unresolved doublet at 682 nm, and yellow Brazilian topaz shows a similar but weaker effect. One must guard against confusing pink topaz with pink sapphire on this account. Pink sapphire has a far brighter fluorescence line, but it is better to check any suppositions by means of the refractometer or some other definite test. Some kyanites, whether blue or green, contain enough chromium to show a dim red fluorescence.

So much for chromium fluorescence under crossed filters. Such fluorescence is truly red, but the red seen through the viewing filter in certain other cases may be only part of an extensive fluorescence system which, if seen *in toto* under ultra-violet light, would give a different colour effect. For instance, between crossed filters strontium titanate shows a faint red glow – but so do many diamonds which in ultra-violet light are strongly blue-, green-, or yellow-fluorescent.

A test between crossed filters is of some practical value in the case of black pearls. If these are of natural colour they show a dim red glow through the viewing filter: if stained with silver nitrate this effect is not apparent.

By using saturated copper sulphate solution and a carefully chosen yellow filter one can still maintain the 'crossed' effect, and this enables orange- and yellow-fluorescent subjects to show good results – Sri Lankan yellow sapphire, for instance. Ammoniacal copper sulphate, which has a far darker colour, enables one to cover the green also. However, the strength of the transmitted light is much diminished, and the particular advantages of the crossed filter technique compared with ultra-violet light are largely lost.

Long-wave ultra-violet light

In theory, all those materials which show a red fluorescence due to chromium between crossed filters should also show the effect under the conventional 365 nm light from a mercury lamp filtered through Wood's glass. With ruby and red spinel, indeed, the red glow is quite spectacular, but the dimmer effects with such stones as emerald and alexandrite are undeniably more easily observed by the crossed filter technique.

Fluorescence displays in museums or gem exhibitions will usually be found to contain some half-dozen mineral varieties which can be guaranteed to yield a fine show of colours under ultra-violet light. Though these are hardly gem materials it may be useful to list some of them here, as anyone investing in an ultra-violet lamp will enjoy assembling an attractive group of specimens to show it off.

Fluorite is an essential constituent in any such display. The very name 'fluorescence' is derived from the magnificent violet glow that most specimens of this attractive mineral emit under these invisible rays. The fluorescence is also excited by ordinary daylight; and often for this reason green crystals of fluorite have a purple appearance unless light is transmitted directly through them, in much the same way that reddish-brown lubricating oils appear green by reflected light because of their fluorescence. Curiously enough, the massive banded fluor mined in Derbyshire which is known as 'Blue-John' and has been used for centuries for vases, boxes, and ornamental objects, is not at all fluorescent under ultra-violet rays. An interesting fact about some purple fluorite is that after stimulation with ultra-violet light it continues for a long time to emit an invisible phosphorescence of ultra-violet rays – a fact easily demonstrated by placing an irradiated specimen on a photographic film in a light-tight box for some hours. After development an 'auto-image' will be clearly seen where the crystal made contact with the film. The effect is not invariable: recent experiments by the author on some necklaces of green gem fluorite failed to give this result, though the violet fluorescence was quite intense.

After fluorite, the crystals most worthy of a place in the display are perhaps the beautiful pseudo-hexagonal crystals of aragonite which are found together with native sulphur at Girgenti in Sicily. These crystals have a lovely rose-pink fluorescence, partially masking a green component, which, when the crystals are removed from the rays, persists as a green phosphorescence for a few seconds – an astonishing phenomenon when first beheld.

Massive scapolite from Ontario provides another fine patch of colour under an ultra-violet lamp – in this case a bright orange-yellow. When examined with a spectroscope this light is seen to have a banded spectrum, reminiscent of the fluorescent light emitted by uranium compounds in so-called 'canary' glass, which was rather popular for ornamental purposes in Victorian times. However, it has been proved by W. A. Runciman that a more likely cause of the yellow fluorescence in scapolite is the presence of a small percentage of sulphur ions. Yet another bright fluorescent colour is given by willemite from Franklin Furnace, where a trace of manganese in this zinc silicate mineral causes it to glow a brilliant green. Calcite from several favoured localities provides a beautiful rose-red glow.

Turning now to gem materials, synthetic yellow-green spinels which contain traces of manganese have a bright green fluorescence. Kunzite and yellow topaz from Brazil show a rather weak orange glow, while white zircon and yellow apatite show a yellow fluorescence. The light emitted by zircon under the rays shows some exciting bright lines through the spectroscope, but it is unwise to expose the stone for long, as the action of the rays tends to cause it to revert to the brownish tint of the mineral as originally found. Heating to dull red heat will usually restore any spoiled stones.

Yellow and off-white sapphires from Sri Lanka yield a characteristic apricot-yellow fluorescence. This is of practical value in distinguishing them from synthetic yellow corundums which, in general, show no curved growth lines and, if free from bubbles, are difficult to prove. Synthetic orange sapphires will give the red fluorescence of chromium under the rays.

Silicon carbide (carborundum), particularly after roasting, has an orange-yellow fluorescence, which is a useful means for detecting its presence as an impurity in diamond powders.

The fluorescence of diamond itself is mentioned last, though it is undoubtedly the most interesting and important of all. Luminescence in diamond has been known and studied for a century or more, and still provides a major puzzle for the physicist. The majority of gem diamonds show a sky-blue to violet fluorescence under long-wave ultra-violet light, but the strength of this fluorescence varies enormously (and quite unpredictably) in intensity: some stones, in fact, are virtually inert under the rays. In the past, the numerous attempts to connect fluorescent effects in diamond with traces of impurities or to correlate it with stones from particular localities were conspicuously unsuccessful. However, recent work by Lightowlers, Collins, and others at King's College, London, has shown that diamond fluorescence can reasonably be ascribed to the presence of nitrogen atoms acting as 'donors' and boron atoms as 'acceptor' centres. In type I diamonds nitrogen is now known to be a major impurity. The blue fluorescence is connected with certain definite wavelengths where discrete emission bands can be seen. The key position of the series is at 415.5 nm, which is also the wavelength of the most important absorption band in diamond. There are also broad regions of continuous emission extending as far as the red end of the spectrum. A photograph of fluorescent diamonds is shown in Plate 1.

Spectrophotometer curves of the absorption and emission of light by diamonds have shown that there are two main 'systems' hinging on the two key positions at 415.5 nm in the violet and 503 nm in the blue-green. At liquid nitrogen temperatures ($-190°C$) the effects are sharper and stronger than under ordinary conditions, and the curves then show a series of sharp 'peaks' which represent lines or narrow bands. These peaks reveal a striking mirror-image symmetry between the absorption bands on the short-wave side of the two key positions, and fluorescence bands or lines on the long-wave side. The 415.5 and the 503 lines can themselves appear either in absorption or emission, according to conditions, rather in the manner of the deep red doublet in ruby.

Under room conditions the sharp peaks observable at low temperatures show only as ripples on the smooth curve of the blue and the green emission systems, and this is borne out by the effects seen with a hand spectroscope when light from a strongly blue- or green-fluorescing diamond is studied. A 'striped' effect shows the existence of ill-defined maxima, but these are hardly measurable as lines.

However, in certain types of diamond there are emission lines in the yellow and red which are not so temperature-sensitive, and can be seen and measured at room temperatures. Some of these bright lines were noted and measured by Sir William Crookes in the strong glow of some diamonds under cathode rays – admittedly a much more vigorous stimulus than ultra-violet rays. Crookes recorded lines at 503, 513 and 537 nm in a stone fluorescing with a pale green light which was strong enough to read by, and in another diamond, lines in the yellow which must have been at 574 and 589 nm, since he ascribed them (wrongly) to yttrium and to sodium. These data are given in his remarkable little book *Diamonds*, published in 1909, but the work was done much earlier.

The most striking series of lines seen in the London Laboratory was emitted by a certain type of pink diamond, several examples of which have been seen. These type IIa diamonds show a strong orange or apricot-yellow fluorescence under all radiations. The fluorescence spectrum of these showed a strong and fairly narrow line at 575 nm in the yellow, with several other lines extending towards the red in diminishing strength, the approximate measurements being 587, 598, 617 nm. These pink stones belonged to the type IIa category so far as their ultra-violet transmission is concerned.

Search amongst diamonds of good industrial quality from Sierra Leone, Ghana, etc. led to the discovery of several stones which showed the 575 line weakly but very sharply – either under the stimulus of strong blue-violet light or long-wave ultra-violet. In these diamonds, which seemed to be 'intermediate' in that they transmitted light to about 240 nm, the 575 line was not at the head of its series, but was accompanied by an equally sharp line at 537 nm, which is known as an absorption line in diamond and has also been given by Crookes (see above) and by more recent workers as an emission line.

The only simple way in which gemmologists can study diamond fluorescence at low temperatures is by making use of solid carbon dioxide ('dry ice') – a fuming white solid which is much used in refrigeration. This provides one with a local temperature of −70°C. A pound or so of the solid can be stored in a large vacuum flask, or even in a cardboard box heavily wrapped in newspaper, and should provide enough material for a day's experiments, lumps of suitable size being removed as required. Provided it is handled with rubber gloves or with non-conducting utensils it is not at all dangerous in use. Fluorescent diamonds for study should have been picked out beforehand and these can be placed directly on a flat lump of the ice as this has no fluorescence of its own, and placed under the various radiations. Before long, a layer of hoar frost will have coated the stone, which will have to be wiped clean again. Laboratory workers with access to cylinders of liquid nitrogen or other non-dangerous gases can locally reduce the temperatures of gemstones by directing a jet of cold gas on them. Care must be taken that sudden changes in temperature do not induce many unwelcome cleavages, and immersion in liquid gases is definitely *not* recommended.

Under these conditions the fluorescence effects are notably stronger and the lines sharper. In particular the author was delighted to see a structure of fine lines between 600 and 700 nm in a true canary-type yellow-fluorescing brilliant, a kind in which no absorption or fluorescence lines had been previously detected. At low temperatures a line of 427 nm has now been observed in 'true canary diamonds', and this puts them in the type Ib category.

The above discussion and description of diamond fluorescence as observable with simple instruments is very partial and incomplete. Even so, it may seem to have occupied more space and ventured into deeper waters than is justified on the practical value of such a study purely for identification purposes. But a thorough knowledge of all the known behaviours of diamond as they can be observed by the gemmologist is potentially important in an age when colour in diamond can be artificially changed and synthetic gem diamond is now available.

It may also have served to indicate the complexity and rich interest of the subject. Diamond, so deceptively pure and simple in its composition and structure, when suitably stimulated proves to be full of glimmering lights and shadows – full of interest and mystery.

When a strongly blue-fluorescent diamond is removed from the rays and observed within the cupped hands, a fairly persistent *yellow* phosphorescence can be observed in all cases – a valuable factor in the recognition of diamonds in jewellery, since no other blue-fluorescent mineral shows this effect.

Broadly speaking, blue-fluorescent diamonds belong to the 'Cape' series. Diamonds belonging to the 'brown diamond' category tend to show a green fluorescence, which is linked to an emission and absorption band at 504 nm in the same way that blue-fluorescing stones are linked to the 415.5 nm band. In many stones the two systems are existent together. A third important fluorescent colour in diamond is a warm yellow – seen more often in industrial stones than in gem diamonds.

Robert Webster proposed making use of the highly variable fluorescent phenomena in diamond as a means of identifying important pieces of diamond-set jewellery. Each piece under ultra-violet light will show a pattern of bright, medium, and dimly fluorescent diamonds which will be unique for that particular ornament, so that photographs of the piece in ordinary and in ultra-violet light will provide a foolproof identity document should the piece be lost or stolen and subsequently recovered.

Short-wave ultra-violet light

Effects seen under short-wave mercury light, suitably filtered, with the 253.7 nm line as the main activating radiation, are in a few cases more diagnostic than those given by long-wave ultra-violet light. The mineral scheelite, for instance, a calcium tungstate only occasionally cut as a collector's item, is inert under 365 nm rays, but glows a bright whitish blue under the 253.7 nm radiation.

Of more practical value to the gemmologist is the reaction of most white synthetic spinels, which fluoresce with a strong bluish white glow under the short waves. Under these conditions synthetic white sapphires show a very weak deep blue. A short-wave lamp is thus of some assistance in testing jewellery set with small colourless baguettes purporting to be diamond, which may be difficult to identify by conventional methods. The rare blue gemstone benitoite shows a bright blue fluorescence under short-wave ultra-violet light.

Under short-wave light, synthetic rubies show a markedly brighter red glow than the majority of natural rubies; this acts as a useful guide and check in difficult cases. Most synthetic blue sapphires show a curious green surface fluorescence under the rays, or a whitish blue glow. This glow seems to be due to titanium, and where Sri Lankan sapphires are exceptionally free from iron they may also show a green fluorescence, as do the majority of heat-treated stones from this locality. However, when synthetic sapphires are examined carefully with a lens while fluorescing, the curved structure bands can often be clearly seen, and this, of course, provides positive proof of their origin in the Verneuil furnace.

Quite apart from fluorescence effects, the value of a short-wave lamp in differentiating between synthetic and natural rubies and emeralds has already been mentioned earlier in the book. In each case the synthetic stones are much more transparent to short-wave light than the natural specimens, and the degree of transparency of any sample can be quickly determined by the method suggested by Mr Norman Day. The stone or stones are placed (in a darkened room) on a piece of contact printing paper immersed in a dish of water, and exposed for a few seconds to the rays from a short-wave lamp placed some 38 cm (15 in) above the dish. The developed paper will show quite clearly the relative transparency of the stones to the rays. Since the exposure time is rather critical for this experiment, it is wise always to include known synthetic and natural stones to act as standards for comparison.

X-rays

In general, fluorescence under X-rays is very similar to the effects seen under ultra-violet light. Rubies and synthetic rubies, for instance, show a red glow due to their chromium content. When the rays are switched off, however, Verneuil type synthetic rubies show a distinct afterglow when observed with dark-adapted eyes, and this phosphorescence may persist for ten seconds or in some cases for very much longer. With natural rubies, or some of the more modern synthetics, such as some varieties of the Ramaura, on the other hand, the glow is usually extinguished instantaneously. Verneuil type synthetic rubies can be relied upon to show curved structure lines when viewed at the correct angle, but when mounted in, say, an eternity ring, the viewing angle is very limited and no structure lines are seen. In such cases the test for phosphorescence under X-rays provides a much-needed confirmatory method. Orange, yellow, or even colourless synthetic sapphires often contain traces of chromium and will then also show a red fluorescence and afterglow under X-rays.

We have noted how variable in intensity and, to some extent, in colour is the fluorescence of diamond. Under X-rays the effects seen are far more uniform: almost all stones glowing with a chalky blue tint of moderate intensity. There is seldom any appreciable phosphorescence. Some diamonds which show a green fluorescence under ultra-violet light give a similar glow of colour under X-rays. Synthetic white spinels usually display a bright greenish fluorescence under X-rays, followed by a prolonged phosphorescence. It is curious and worth noting that diamond, when it is blue-fluorescent under ultra-violet light, shows a prolonged afterglow, but the blue glow stimulated by X-rays is seldom followed by any phosphorescence.

An interesting and unexpected orange fluorescence is seen when massive green hydrogrossular garnet (the so-called 'Transvaal jade') is placed under X-rays. This effect is distinctive enough to form a useful test for this rather attractive jade substitute, in particular, to separate it from the californite variety of idocrase which has very similar properties.

It has long been known that cultured pearls tend to show a yellowish fluorescence under X-rays, while most natural salt-water pearls show a glow so dim that it can only be seen with dark-adapted eyes. The glow in cultured pearls is due to the presence of traces of manganese in the nucleus which is usually fashioned from mother-of-pearl derived from freshwater mussels. It is hardly surprising that natural freshwater pearls should also show a yellowish fluorescence for the same reason. Careful observation will show that the fluorescence is chiefly at the surface of the pearl, while in cultured pearls the glow comes from within.

A still brighter fluorescence (with phosphorescence) is shown by freshwater cultured pearls from Lake Biwa in Japan and by those from China. Since these contain no nucleus and their internal features are not always easy to distinguish on a radiograph, this characteristically strong X-ray fluorescence provides a valuable indicator of their origin.

Though useful as a confirmatory test or as a warning sign, X-ray fluorescence should never be considered as a substitute for the more stringent methods of testing pearls outlined in Chapter 28. With drilled or partly drilled pearls the fluorescence test under X-rays can be made more specific by observing the relative brightness of light from the drill-hole and the surrounding pearl, as explained in that chapter.

Only the more outstanding fluorescent effects have been mentioned above. Those seeking more complete data should consult the series of eleven articles written by Mr Robert Webster, which appeared in the *Gemmologist* from June 1953 to April 1954, or, more conveniently, the relevant sections in his book *Gems* (3rd edition, 1975; 4th edition, revised by B. W. Anderson, 1983).

12

The identification of diamond

This book has now reached its half-way stage, in terms of both pagination and subject matter. The description of apparatus and of the testing methods recommended has been completed, and it now remains to show in some detail how these methods can be used in the identification of individual gemstones and how each can be distinguished not only from any other natural stones that may resemble them but also from man-made products intended as substitutes for the real thing.

Since this is not an academic treatise but a practical guide for the jeweller–gemmologist, the order in which the individual stones are dealt with is roughly based on their commercial importance. Diamond quite clearly takes precedence over any other, followed by ruby, sapphire and emerald, these four being the classical 'precious stones' in the days when all others (except pearl) were relegated to the 'semi-precious' ranks. Thereafter the order of the chapters becomes less predictable, but if the reader quickly fails to find the stone in which he or she is interested the index can be consulted.

Diamond accounts for over 90 per cent by value of the world's trade in gems, and is one of the few stones that every jeweller handles and hopes to be able to recognize at sight with tolerable certainty. There have been many cases, however, when even experienced dealers have made costly (and humiliating) mistakes – very often because their suspicions have not been aroused. In recent years things have certainly been made more difficult by the production of synthetically made substitutes showing fire, brilliance, and lustre very close to those seen in diamond itself.

The distinctive appearance of diamond itself is due to the combined effect of its 'adamantine' surface lustre, the perfection of its polished surface, its brilliance, and its 'fire'. These are bound up with its supreme hardness, high refractive index, and dispersion – also the skill with which it is cut.

The proportion of light reflected from the surface of diamond is higher than that for any other natural colourless stone (Figure 12.1), and this, combined with its hardness and the superb flatness and high polish of the surfaces prepared by the skilled diamond cutter, gives rise to the peculiar 'adamantine' lustre which

Figure 12.1 A selection of rough diamonds showing crystal form

belongs to diamond alone. Synthetic rutile may reflect more light, and strontium titanate as much light as diamond, but their softness does not permit so flat and mirror-like a surface to be prepared, nor such sharp angles between the facets. If one tilts a diamond until the reflection of window frames or of an electric light bulb be seen in the table facet, it will be noticed how free from distortion the reflected image is.

Diamonds are so fashioned that practically all light entering the front of the stone is totally reflected from the back facets as from a series of mirrors (the rays striking them at angles greater than the 'critical angle', which for diamond is only 24½°). Thus a well-cut brilliant, viewed from the back and held up to the light, will show only a pinpoint of light from the culet, and nothing more. Also, looking down on a brilliant-cut diamond mounted in a ring, one cannot (because of this total reflection effect) see the wearer's finger below the stone, as one usually could in the case of stones of lower refracting power.

Further, white light entering a diamond is dispersed into its spectrum colours to a considerable degree, giving flashes of pure colour from the smaller crown facets. This 'fire' in diamond, combined with its superb optical purity, constitutes one of its main claims to beauty.

The high refractive index of diamond makes the stone appear much shallower than it really is when viewed through the table facet; and this, together with the fact that the refraction is *single*, forms another distinguishing sign.

Other revealing features which can often be detected with a lens are small portions of the original crystal surface, remaining on the girdle. The 'naturals', on which a series of small triangular markings known as 'trigons' may be present, hardly affect the beauty of the stone, and are often purposely retained by diamond cutters to enable them to know at a glance the direction of the 'grain' while the diamond is being cut and polished. Tiny 'nicks' in the girdle can also often be detected, and these may show flat, angled surfaces which follow the planes of perfect octahedral cleavage of the diamond. The inclusions, too, are frequently distinctive, shining flakes of graphite or other forms of 'carbon' being perhaps the commonest of these. Diamond has a great affinity for grease, and the surface of a cut diamond will reveal a greasy film after being handled. This is another characteristic worth noting.

So much for examination by eye or lens: now for some specific tests.

Hardness

Hardness is a quality difficult to define and to measure in gemstones, and hardness tests should normally be avoided where more accurate tests, less liable to damage the stone, can be applied. But the case of diamond is unique. More than 150 years ago, when Mohs devised his 'scale' of scratch hardness (which has proved so useful that it is still in constant use by mineralogists and gemmologists), he designated diamond as hardness 10. Next on the scale came corundum (sapphire) with hardness 9. These were followed by topaz 8, quartz 7, feldspar 6, apatite 5, fluorite 4, calcite 3, gypsum 2, and talc 1 – figures that should be engraved on the memory of anyone who handles precious stones. Mohs was careful to choose minerals that are easily obtainable. He made no claim that his 'numbers' had any quantitative significance, merely that any mineral on the scale can scratch another having a lower number and can itself be scratched by a mineral higher on the scale. Many attempts have been made to obtain accurate relative figures for hardness in minerals by controlled scratching under a given load, by measuring the loss of weight after controlled grinding and, most conveniently and with the least destructive effect, by indentation measurements. Results from one method are not strictly comparable with those from another since the type of hardness measured is not the same, but the general conclusion is clear on two points: (1) that hardness in a crystal is a *directional* property, varying with each crystal face and in different directions on the same face, and (2) that the difference in hardness between diamond (10) and corundum (9) is far greater than that between the other minerals chosen by Mohs.

Although some artificially prepared abrasives such as silicon carbide (carborundum) and boron carbide exceed corundum in hardness, and the high-pressure form of boron nitride ('borazon') may even equal diamond in hardness, diamond still remains the only gem material which can scratch ruby or sapphire.

Polished pieces of synthetic corundum are not difficult to obtain, and, failing these, any lapidary can provide a polished piece of natural poor-quality sapphire for a consideration. If a facet edge or girdle of a diamond be applied carefully but firmly to the surface of a polished corundum 'test piece' it will be found to 'bite' and to leave a decided scratch mark, which will not disappear when rubbed with a moistened finger. By this simple procedure the stone will have been proved beyond doubt as a veritable diamond, and, with reasonable care, no harm can come to the stone tested, whether it be diamond or not. Strontium titanate, a synthetic material which closely resembles diamond and has virtually the same refractive index, is very soft (not much more than 5 on Mohs' scale) and a gentle touch with the point of a steel needle will leave a mark on the surface of one of these stones.

Refractive index

Diamond is also remarkable for its high single refraction (2.42), and though this is not unique it still forms a valuable test. The highest reading obtainable on a standard refractometer is limited to 1.81 by the contact liquid used. Diamond will thus give a 'negative' reading, though it would be unwise to place a stone thought to be a diamond on a refractometer, as damage to the soft glass of the instrument is difficult to avoid when so hard a stone is applied.

Only three other natural gemstones giving negative readings are likely to be encountered. These are zircon (1.926 to 1.985), demantoid (1.89), and sphene (1.90 to 2.03). Of these, only zircon is colourless, and only demantoid is singly refracting. With a lens, zircon and sphene show strong doubling of the back facets, while demantoid may show typical 'horsetail' inclusions or a distinctive spectrum.

In recent years, two substances have been artificially prepared which match or exceed diamond in refractive index. These are synthetic rutile with indices 2.61 and 2.90, and strontium titanate, which is singly refracting and has an index (2.41) very near to that of diamond. Rutile can at once be recognized by its huge double refraction and its stupendous 'fire', which causes it to flash with colour almost like an opal. Also, at best, rutile has a distinct yellowish tint, even when a thin film of sapphire has been deposited on the stone by a special sputtering or evaporation process, to improve its hardness and appearance. Strontium titanate resembles diamond much more closely: it is nearly colourless and is singly refracting. But this, too, has an excess of 'fire' by diamond standards, and a comparison between the two reveals this difference at once. In a mounted stone, the high SG of this synthetic (5.13) cannot aid one as a test. A careful hardness trial, using a steel needle, as indicated above, is very practical. Inclusions when visible, may be distinctive, and, in common with all the limitations of diamond, strontium titanate lacks the transparency to X-rays that is so notable a feature of diamond (transparency of diamond to X-rays is illustrated in Figure 12.4).

Several colourless man-made crystals have been used recently to simulate diamond, the most important in the mid-to late 1970s being yttrium aluminate commonly called 'YAG' and today cubic zirconia, both of which have already been described in Chapter 9. These are singly refracting and beyond the range of the normal refractometer; one of the reflectivity meters or a thermal probe will give useful information in such cases.

Refractive index forms the basis for the distinction of diamond from the more common and long-established synthetics, colourless sapphire, and spinel. These are often used for 'diamond' clusters or surrounds in cheap jewellery. The stones may well be too small to place easily on a refractometer, but immersion of the whole piece of jewellery in di-iodomethane will immediately show that the suspected stones are not diamond. Both sapphire and spinel virtually 'disappear' in this liquid, which approaches them so closely in refractive index, while diamond will still show its facet edges in clear relief.

Fluorescence

The fluorescent glow emitted by diamond under short- or long-wave ultra-violet light is very variable, but often distinctive, and forms a useful test on many occasions. This is particularly true in the case of an ornament containing many diamonds. Placed under a quartz mercury lamp with a Wood's glass filter, or under a 'black bulb' lamp which incorporates such a filter, some of the diamonds will be seen to emit a bright sky-blue fluorescent light, others a feebler effect,

while some may seem almost inert. One or two may show a yellow or
yellowish-green fluorescence. If none of the stones in such an ornament shows a
fluorescent glow, this is a suspicious sign; but if they *all* show a uniform glow of
any kind they certainly cannot be diamonds. A photograph of rough diamonds
fluorescing under ultra-violet light is shown in Plate 1.

Though useful as a guide, fluorescence tests seldom provide the gemmologist
with complete proof of the nature of the stone tested. However, if a diamond
shows an appreciable blue fluorescence under long-wave ultra-violet light, the
test becomes specific if a yellow afterglow can be detected when the stone is
quickly removed from the rays or the lamp switched off. This is one of the few
invariable properties of blue-fluorescent diamonds, and the stronger the
fluorescence, the stronger is the phosphorescent afterglow. Diamonds which
display a greenish-yellow fluorescence also show an afterglow, but the strength
of this is less predictable. Under X-rays diamonds show a more uniform
behaviour – almost invariably showing a strong blue glow, but, curiously,
without an afterglow.

The afterglow after exposure to long-wave ultra-violet light is so useful a test
that it is worth taking some trouble to ensure its success. Though rather
persistent, the intensity of the phosphorescence quickly falls and can easily
escape notice unless the room be well darkened and the eyes dark-adapted.
However, if the specimen is held in cupped hands and, immediately after it is
removed from the rays, the eye is applied thereto (the hands forming a small
completely dark chamber), the glow of the diamond can quite easily be detected.

Specific gravity

Though the 'constants' of diamond are seldom used for testing purposes, even
by laboratory workers, they should be stated here. The SG of gem diamond is
usually given as 3.52 and this is a good working value. Careful measurements on
fine diamonds have established 3.515 as the most accurate figure. Nearest to this
among possible 'rivals' is the SG of sphene (3.53) but the double refraction,
pleochroism, and low hardness of this mineral do not allow for any real
confusion with diamond. More worth noting is that white topaz pebbles, which
are often assumed to be diamond by hopeful prospectors and amateur
collectors, have a SG near 3.56 though one low value of 3.545 was recently
obtained by the author. The fact that topaz pebbles are water-worn, while
diamond rarely is, and the lack in topaz of the distinctive diamond lustre, should
preclude any false hopes when topaz is encountered in the field.

The more recent imitations of diamond – strontium titanate, yttrium
aluminate, and cubic zirconia – all have densities far above that of diamond, as
noted in Chapter 9 and tabulated later in this chapter, and without recourse to
any density determination an unmounted specimen would reveal itself as a fake
to experienced diamond dealers when they came to weigh it in the ordinary
way, since the weight would be so much more than expected. A brilliant-cut
stone of each of the above, if cut to the size of a 1 carat diamond, would weigh
1.45 carats if it were strontium titanate, 1.30 if it were YAG, and as much as 1.60
if it were cubic zirconia.

Dispersion

Though diamond is justly famed for its 'fire', its dispersion of 0.044 for the B–G
interval is in fact unusually low for a stone of such high refractive index (see

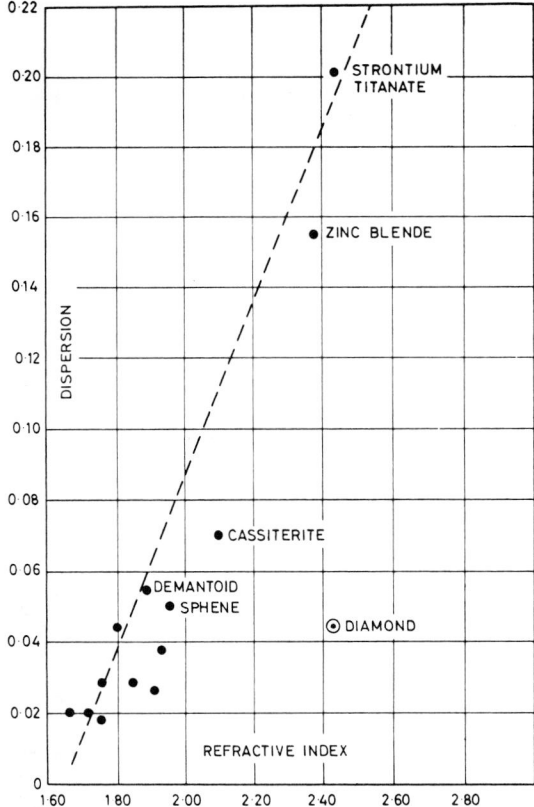

Figure 12.2 Diagram showing the relatively low dispersion of diamond compared with other gemstones of high refractive index

Figure 12.2). Sphene (0.051), demantoid (0.057), and cassiterite (0.071), with much lower indices, all exceed diamond in dispersion, while for those stones of comparable index zinc blende with its dispersion of 0.156 and strontium titanate with 0.200 must be considered normal. To most people's taste, however, the degree of fire shown by diamond is amply sufficient to give it beauty, while strontium titanate and synthetic rutile look opalescent and gaudy by comparison (see Chapter 3).

The diamond substitute 'yttrium aluminium garnet' (commonly referred to as YAG) errs in the other direction, though its dispersion of 0.028 is in keeping with its refractive index of 1.834. The lack of fire is less noticeable in step-cut specimens than when the stone is brilliant cut, and one must admit that in fire and lustre it exceeds synthetic white spinel, which in itself caused quite a stir when promoted in 1935. YAG shares with synthetic spinel the virtues of clarity, complete absence of colour, considerable hardness, and single refraction.

Another synthetic gem substitute, lithium niobate ('linobate'), launched some years ago, does not seem to have been used to any great extent. Brown, yellow, or orange scheelite, natural or synthetic, can resemble coloured diamonds when finely cut, though its dispersion of 0.026 is not really high. In the matter of dispersion the newest diamond substitute, cubic zirconia, with a figure of 0.060, comes nearest to diamond (0.044).

Birefringence

All diamonds show some degree of strain birefringence when examined between crossed polars, with bands of shadow alternating with brilliant patches of colour. This effect can be seen readily in a brilliant-cut diamond when held in tongs between crossed polars and viewed in the direction of the girdle, and forms a useful accessory test. It must be remembered that two at least of the materials used to imitate diamond, synthetic spinel and paste, also show strain birefringence. But when carefully observed the effects are not very like those seen in diamond.

The reader may think it strange that a crystal of the cubic system should not be truly isotropic. However, it must be remembered that, under ordinary conditions, graphite is the stable form of crystallized carbon – diamond only being truly stable under the enormous pressures at which it was formed at great depths below the earth's surface. Also, it was under these extreme conditions that the growing diamond engulfed tiny crystals of other minerals which were forming contemporaneously. None of these can have the low coefficient of expansion of diamond, and thus each is the centre of local strain, as can often be seen under low magnification between crossed polars (see Figure 12.3).

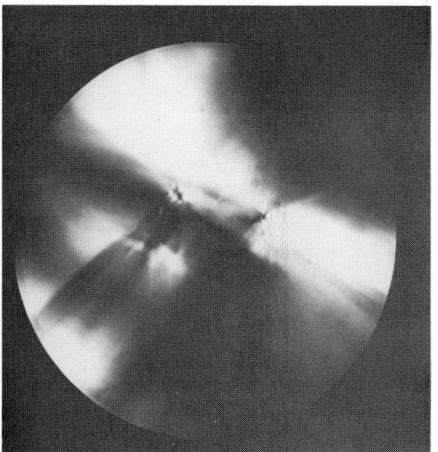 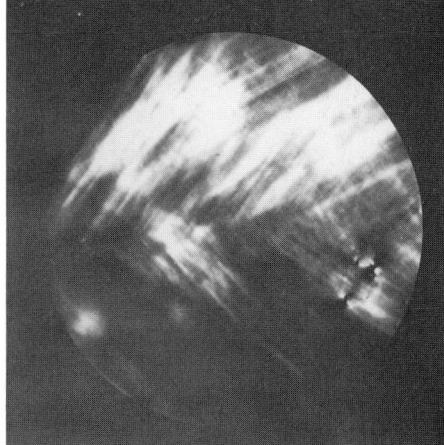

Figure 12.3 Diamonds between crossed polars, showing strain birefringence: (*left*) Type I; (*right*) type II

Absorption

As stated in Chapter 10, most diamonds show a narrow absorption band in the deep violet at 415.5 nm when light which has been transmitted through them is examined through a suitable spectroscope. This is most easily seen by transmitting blue light (filtered through a flask of copper sulphate) across the pavilion facets parallel to the girdle plane. When seen, this band is diagnostic for diamond. In stones of the 'Cape' series this main band is accompanied by others in the blue and violet. The strongest and most easily visible of these is centred at 478 nm. The feeble bands seen in green-fluorescing diamonds of the 'brown' series are discussed later in the section on treated diamonds.

Transparency to X-rays

For the laboratory gemmologist with even the simplest of X-ray equipment a conclusive test for diamond is available, thanks to the extreme transparency of the mineral to X-radiation (see Figure 12.4). The degree of transparency of a

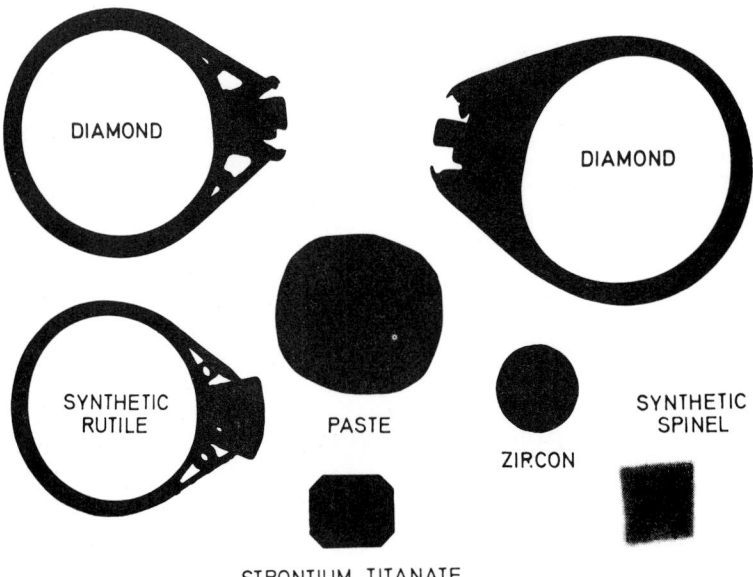

Figure 12.4 Diamonds are transparent to X-rays: their imitations are not (positive print)

substance to these short waves depends (for a given thickness of material) upon the relative weight of the atoms through which they have to pass. Though dependent to some extent upon the metal used for the anti-cathode of the X-ray tube, in general the opacity increases with something like the fourth power of the atomic number (which is roughly half the atomic weight) of the element or elements concerned. Thus carbon (atomic number 6) is far more permeable by the rays than other gem minerals which contain elements such as silicon (atomic number 14) or aluminium (13) let alone such metals as calcium (20) and iron (26). Thus a ten-second exposure of a film on which any stones in question are resting will reveal in very clear fashion a difference between diamond and any of the stones which resemble it. This test has the advantage of leaving a visible record calculated to convince bench or jury in cases of legal or criminal dispute, and also of being applicable to rough or even opaque industrial diamonds such as the coke-like, impure multi-crystalline carbonado, which was formerly greatly prized for setting in rock drills on account of its great toughness.

Inclusions

All but the highest qualities of gem diamonds contain minute inclusions of other minerals which, though counted as blemishes when visible with a ×10 lens or (still worse) with the unaided eye, are often characteristic enough to aid in satisfying the trained observer that the stone is indeed a diamond.

Identification of the minerals included in diamond has been the subject of intensive study during the past few decades. Absolute proof of identity by study under the microscope alone was not always possible, and some false assumptions were made which it will take some years to eradicate from the textbooks. The more stringent recent work has employed such techniques as crushing diamond samples to isolate the included particles, and also the use of X-ray diffraction methods in order to obtain cast-iron proof of identity. More than twenty mineral species have now been positively named as inclusions in diamond. Among the more important are black inclusions of magnetite and graphite, dark red grains of chrome spinel and pyrope garnet, green diopside and enstatite, and colourless octahedra of diamond itself.

An ingenious recent practice, which has proved successful in improving the appearance of diamonds containing unsightly dark inclusions, has been to penetrate the stone by means of a carefully aimed laser beam from the table to the point where the inclusion lies. Curiously, this in itself improves matters by making the inclusion paler and less obvious, but subsequent treatment with acid may clean out the tiny cavity more effectively. The channel burnt by the laser may be only some 0.025 mm (one-thousandth of an inch) in diameter, so it may be surprisingly difficult to see, even under lens examination. Viewed from the side with a low-power microscope the trace of the beam should be obvious; often there may be several beams in the one stone, each directed to a different inclusion. Needless to say, if such laser-beam traces are observed in gemstone it can only be diamond, since this is the one crystalline gemstone that is combustible.

Synthetic diamond

Following their earlier success (1955) in manufacturing **synthetic diamonds** in very small sizes for industrial purposes, the General Electric Company of America announced in 1970 the experimental production of synthetic diamonds of gem quality in various colours and sizes (for most crystals) of up to 1 carat.

In 1985 Sumitomo of Japan announced the production of gem-quality synthetic diamond and cut stones up to 0.2 carat have appeared in New York. In 1987 De Beers revealed that they had grown diamond crystals up to 11 carats in size during the past few years in Johannesburg. They have now released for examination gem-quality crystals in excess of 5 carats and cut stones up to 9 carats. Both Sumitomo and De Beers material are distinctly yellow or brownish-yellow in colour, and the De Beers stones also come in very deceptive greenish-yellow colours.

The identification of these synthetic diamonds may pose problems for the gemmologist in the future. The commercially produced Sumitomo and De Beers synthetic diamonds so far examined are of Type Ib and contain nitrogen dispersed as single atoms substituting for carbon in the crystal structure. Most natural diamonds are of Type Ia and contain nitrogen atoms in small clusters. This has its effect upon the absorption spectra and provides a useful clue in identification. The synthetics do not show sharp bands, but display a general and increasing absorption towards the violet. Most natural yellow diamonds display a sharp band at 415 nm in the violet and particularly if the colour approaches the depth seen in the synthetic stones may show further bands in the blue-violet. One should mention, though, that there are many natural Type Ib yellow and brown stones to be found on the fancy diamond market and, therefore a lack of sharp absorption features does not indicate a synthetic stone, but it does mean that further tests, such as infra-red spectroscopy and ultra-violet fluorescence or microscope observations need to be made. However, if a 415 nm line is observed the stone is, at the present state of knowledge, highly unlikely to be a synthetic.

Many synthetics show a zoning structure related to their octahedral and cubic internal symmetry, a combination that is rare in nature, and this may appear as a 'bow-tie' or 'hourglass' structure.

Synthetic diamonds so far examined are inert under long-wave ultra-violet radiation, but natural stones may or may not fluoresce. Under short-wave radiation Sumitomo stones fluoresce moderate to intense yellow or greenish-yellow. Under this same radiation De Beers stones fluoresce shades of yellow (from brownish-yellow and greenish-yellow stones) but yellow stones may be inert. Under short-wave radiation many synthetic diamonds show distinctly zoned yellow fluorescing geometric areas, with intervening inert areas. This zoning and short-wave ultra-violet illumination coupled with the absence of sharp absorption lines may be useful features in indicating a synthetic diamond.

Since they are made by a high-pressure flux method, inclusions of metallic flux are often seen in these synthetics and they may display distinct magnetism resulting from the traces of included flux. Many synthetics show areas of fine pinpoint inclusions, sometimes as 'clouds'.

Diamond substitutes

During the above account of the properties of diamond, most of the possible substitutes have been mentioned and their relevant properties indicated, but it may be useful at this point to run through the whole list, indicating the discriminative features of each.

The other materials to consider are **white zircon, synthetic rutile, strontium titanate, synthetic white spinel, synthetic white sapphire, diamond doublet**, and **paste** (lead glass) imitations. Recent competitors include the synthetically made 'YAG' (yttrium aluminate), 'GGG' and **cubic zirconia**. The natural minerals **sphene, scheelite, blende**, and **demantoid garnet**, when finely and freshly cut might, on occasion, be mistaken for brown or green diamonds.

YAG (yttrium aluminate) is marketed under the name 'Diamonair'. Being hard, transparent, colourless, singly refracting, and with fair dispersion, this material is similar to diamond in appearance and, since it has a refractive index (1.834) beyond the range of a standard refractometer, the identification of a mounted specimen may prove rather difficult. Immersed in 1.81 index refractometer fluid it would practically disappear: under X-rays it has a mauve fluorescence and under ultra-violet rays it shows a yellow fluorescence (compared with the usual blue of diamond), and between crossed polars it is liable to show a pseudo-uniaxial interference figure, like some pastes. It would, of course, fail to scratch corundum and, as a laboratory distinction, it would be virtually opaque to X-rays whereas diamond is more transparent than any other to these radiations. The SG of YAG is 4.57.

GGG is a convenient acronym for 'gadolinium gallium garnet', another of the 'synthetic garnet' range in which the name garnet merely implies that these man-made crystals have the same crystal structure as the true silicate garnets found in nature. This gadolinium compound has achieved a certain reputation among the many attempts to simulate diamond, and having a higher refractive index (2.03) and dispersion (0.038) than YAG it would be slightly more convincing to the eye were it not for the brownish tinge that is frequently noticeable. YAG, unless deliberately 'doped', is completely free from colour. GGG has the exceptionally high density of 7.05. Both of these have now been overshadowed by cubic zirconia.

Cubic zirconia, the latest and perhaps the most effective to the eye of all diamond substitutes, is a cubic form of zirconium oxide. The stable form of this compound has a monoclinic structure and is found in nature as the mineral baddeleyite. To render the cubic form stable a proportion of calcium oxide or yttrium oxide has been added to the compound by its manufacturers. Its properties vary somewhat according to the 'stabilizer' used: the CaO version, for instance, has a density of 5.65, as against 5.95 for the version containing the heavier oxide of yttrium. The refractive index is 2.17 and dispersion 0.060. The new material inevitably brings with it yet more fancy names devised by the manufacturers. 'Phianite' is the selling name used by the Moscow producers; the Swiss firm Djevahirdjian (already famous for its vast production of Verneuil synthetics) has coined the term 'Djevalite', while a New York firm supplies it as 'Diamonesque'. The last name at least has the merit of avoiding the '-ite' termination, which should properly be reserved for natural minerals.

All the synthetically made substitutes for diamond can easily be distinguished from diamond itself, when free from a setting, by their much higher density. Mounted stones should be examined carefully through a lens to ascertain whether they have the perfectly flat faces and the sharp interfacial angles that are virtually unique to diamond. If the stone is large enough, a reading from the table facet with a reflectivity meter or from any facet with a thermal probe should be decisive; failing such an instrument, a comparison of the appearance of the stone when immersed in di-iodomethane with that of a known diamond should reveal a much lower contrast of the facets than with diamond. The natural inclusions typical of all but the purest diamonds are missing, and sometimes microscopic bubbles may be discerned.

In addition to the slightly rounded facet junctions that seem to be a common feature of cut 'CZs', it is worth noticing that the girdle usually shows vertical striations due to rapid grinding, which are quite different from the surface left by the diamond polisher. Some specimens of zirconia have shown narrow absorption bands, either in the yellow-green or in the violet part of the spectrum. Where seen, such signs are an additional warning that the stone is not a diamond, but the majority of new synthetics show no signs of absorption in the visible region.

White zircon is the only common natural colourless stone with sufficient 'fire' to pass for a diamond among the unwary or the inexperienced. In the past, most fraud cases involving so-called diamond rings sold in public turned on the fact that zircon has a certain resemblance to diamond. To the gemmologist the distinction could hardly be more simple. The strong double refraction of zircon makes doubling of the back facet edges obvious under a good × 8 or × 10 lens, and a test with the spectroscope can be used in confirmation. The calcium tungstate mineral **scheelite**, well known as an important ore of tungsten, is sometimes found in transparent pieces which may be colourless, yellow, orange, or brown. When well cut these stones can present a magnificent appearance, and bear a considerable resemblance to diamond. Though too soft (about 5 on Mohs' scale) and too scarce to be used in commercial jewellery, scheelite should not be ignored (as it usually is) in books on gems. It seems appropriate to mention it here, as a well-cut brilliant scheelite, when suitably mounted, might easily pass for diamond unless critically examined – and even if it were realized that the stone could not be diamond, it is so little known that it might prove puzzling to identify.

The refractive indices of scheelite are 1.920 and 1.937, and thus beyond the range of a standard refractometer. The double refraction of 0.017 is similar to that of tourmaline, and doubling of the back facet edges should be visible with a ×10 lens unless the stone is small. The dispersion of 0.026 is sufficiently high to give the stone considerable life. The SG is high, being near 6, and should certainly help in identifying a loose stone.

A well-known peculiarity of scheelite lies in its repsonse to ultra-violet light, since it is inert to the longer waves but shows a bright whitish-blue fluorescence under short-wave ultra-violet. Being a calcium mineral, scheelite frequently contains traces of the rare-earth metals, giving rise to the well-known 'didymium' absorption spectrum of groups of narrow closely spaced lines, of which the strongest group is in the yellow. Unfortunately, scheelite has been manufactured by the 'pulling' method for use as laser material, and clear faceted specimens have been sold fraudulently to collectors as natural stones. Curved lines and bubbles (almost as in Verneuil synthetics) can be seen in some of these: in others, an exceptionally strong rare-earth spectrum or the complete absence of rare-earth absorption lines can provide warning features.

A quite common ore of zinc, **blende** or **sphalerite**, which is usually black and opaque, can also occur in transparent yellow, orange, or brown specimens which superficially resemble fancy-coloured diamonds when they are freshly and skilfully cut. Blende has the same cubic structure as diamond, and its high refractive index (2.37) provides it with a splendent lustre, though its easy cleavage in six directions makes it a difficult stone to polish, and its softness (3½) gives it a slightly resinous appearance and debars it from any length of life as a jewel. The SG of blende is constant at 4.09; it is isotropic, and its dispersion of 0.156 is very much higher than that for diamond, though not very noticeable on account of its colour. In some pieces of blende narrow absorption bands can be seen in the deep red – possibly due to cadmium – which are mentioned here on account of the danger that gemmologists mistaking a cut blende for zircon and going to the spectroscope to confirm their opinion might be confirmed in their error by a hasty glance at these bands, though the clearest and narrowest one is at 667 nm and the two others still deeper in the red.

Synthetic rutile has even greater double refraction, and its tremendous fire makes it full of opalescent colours; moreover, its body colour is never free from a yellowish tinge, and its surface has a greasy lustre compared with the hard clean-looking surfaces presented by diamond. **Strontium titanate** has much more serious claims to be an effective substitute, being nearly free from colour, singly refractive, and with virtually the same index of refraction as diamond. Its softness, however, imparts a certain greasiness to the surface polish, and forms an effective test if a needle point be carefully applied – while the fact that the dispersion, though less than that of rutile, is still four times that of diamond and gives it a false and over-coloured appearance to the informed eye. Strontium titanate has no fluorescence under ultra-violet light, and is opaque to X-rays. Tiny 'centipede' surface markings are often visible (see Chapter 9). It is only possible seriously to confuse **synthetic white sapphire** and **synthetic white spinel** with diamond when they are cut as small baguettes or very small brilliants and mounted in jewellery, though it should be remembered that synthetic spinel formed the basis for an absurd 'synthetic diamond' scare in 1935. Where a refractometer reading is possible, a sure means of identification is available in each case. Where small mounted stones are in doubt the whole piece can be immersed in di-iodomethane. In this liquid both sapphire and spinel will show very low relief, the facet edges practically disappearing. Diamonds, on the other hand, will stand out almost as boldly as in air, with their facet edges plainly visible. For distinguishing between the two categories of synthetics a test between crossed polarizers is helpful. Synthetic white sapphires will transmit light except in the four extinction positions, while synthetic spinels will show the peculiar and characteristic form of strain birefringence to which the writer has given the descriptive term 'tabby extinction'. Diamond, it should be noted, also shows strain birefringence between crossed polarizers, as mentioned earlier. **Natural white sapphires** are seldom free from some tinge of colour, but are sometimes used as a substitute for diamond in native jewellery.

Diamond doublets are occasionally seen, and can be puzzling, though they do not look quite 'right'. Usually they consist of a crown of diamond cemented to a pavilion of white synthetic sapphire, quartz, or even glass. Such composite stones are usually met with in a mount, and as the crown of the stones shows the typical surface lustre, inclusions, and general appearance of diamond, it is easy to accept them as poor-quality genuine stones unless one's suspicion is aroused. At a certain angle the edges of the table facet can be seen reflected in the surface formed by the junction layer (Figure 12.5). Under the microscope,

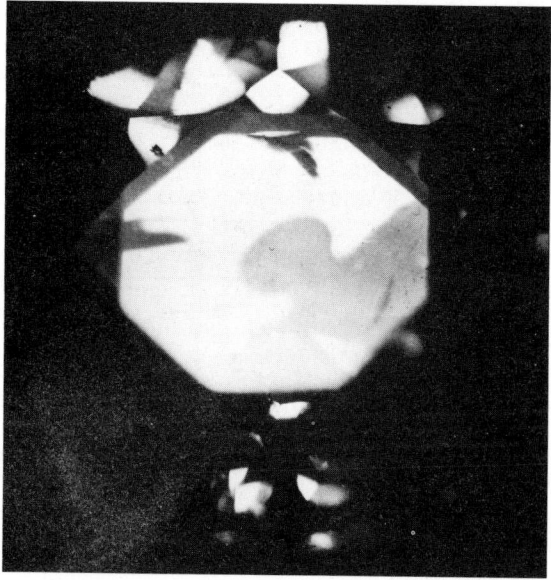

Figure 12.5 Diamond doublet, showing reflection effect when looking down on table facet

bubbles may also be seen in this layer. Immersion in di-iodomethane will make the refractive index difference between the crown and pavilion of the stone very obvious.

Lead-glass imitations of diamond (commonly known as '**pastes**') can make a very effective display when they are new and well cut, and these are very extensively used in 'diamanté' costume jewellery. They are seldom intended to deceive, and the poor quality of their surface polish, their glassy lustre, and their softness all reveal their nature. The refractive index of these glasses varies, but is usually between 1.62 and 1.68. The refraction is, of course, single, and under the microscope glassy bubbles and often 'swirl marks' are visible.

On account of their high refractive index and appreciable fire, the yellow or brown **sphene** and green **demantoid garnet** might be mistaken for coloured diamonds. Sphene has distinctively strong double refraction and pleochroism, while the singly refracting demantoid has a powerful absorption band in the violet and almost invariably shows fibrous 'horse-tail' inclusions which are an effective aid to its identification.

The properties of diamond and its possible substitutes are shown in Table 12.1. Topaz and quartz are added to the list, though their lack of brilliance and fire make it very unlikely that they would be confused with diamond. The

Table 12.1   Diamond and its possible substitutes

Stone	H	SG	RI	DR	Dispersion
Diamond	10	3.515	2.417	None	0.044
Zircon	7½	4.69	1.926–1.985	0.059	0.039
Cubic zirconia	8	5.65	2.15	None	0.060
GGG ($Gd_3Ga_5O_{12}$)	7	7.05	2.03	None	0.038
YAG ($Y_3Al_5O_{12}$)	8	4.57	1.834	None	0.028
Strontium titanate	5½	5.13	2.41	None	0.200
Synthetic rutile	6½	4.25	2.61–2.90	0.287	0.300
Lithium niobate	5½	4.64	2.21–2.30	0.090	0.120
Scheelite[a]	5	6.1	1.918–1.934	0.016	0.026
Blende	3½	4.09	2.37	None	0.156
Demantoid	6½	3.85	1.89	None	0.057
Synthetic spinel	8	3.63	1.727	None	0.020
Sapphire[a]	9	3.99	1.760–1.768	0.008	0.018
Topaz	8	3.56	1.61–1.62	0.010	0.014
Quartz	7	2.65	1.544–1.553	0.009	0.013
Paste (typical)	5	3.74	1.635	None	0.031

[a] Natural and synthetic.

properties of paste vary considerably in accordance with its composition, but the data for a representative specimen of good quality are included in the table. In the last column are listed figures for dispersion, which give a numerical basis for the effect of 'fire'. The figures represent the difference between the refractive index of the stone in question for violet and for red light (B–G in the solar spectrum). The value for the dispersion of rutile is approximately that for the ordinary ray; for the extraordinary ray it is even higher. The value for strontium titanate is estimated.

Before closing this description of those natural or man-made stones that may be mistaken for diamond, and the means by which they may be distinguished, the reader should be reminded of the sections on reflectivity meters and thermal probes in Chapter 2. In contrast to these sophisticated instruments there are several useful tests which can aid the gemmologist in assessing diamonds and their simulants.

First we have the 'tilt' or 'light-spill' test, which relies upon the optics of the correctly proportioned diamond. Light entering such a diamond will be totally internally reflected and virtually none will pass through the stone. If a diamond is viewed at right angles to its table facet and against a black background the stone will appear uniformly bright because of the rays being reflected from the pavilion facets. If now the stone is tilted away from the observer, the brilliant reflections will persist through quite a large angle before light begins to 'spill out' of the stone. Simulants with lower RIs will lose light at much lower angles and an increasing number of upper pavilion facets will appear black as the stone is tilted. The same effect can be obtained by placing a diamond and several simulants close together with table facets upwards and then moving the eye away from the vertical position so as to increase the oblique angle of view. Stones of lower indices will 'spill out' light as the angle of view is increased (see Figure 12.6) and comparisons may be made of the various simulants (and their RIs).

Figure 12.6 The 'tilt' test in action. (*Top row: left to right*) GGG, lithium niobate; (*centre*) YAG; (*bottom row: left to right*) cubic zirconia, diamond (photo: P. G. Read)

The second test, again relying upon the optics of the 'correctly' proportioned diamond, is the 'dot ring' method. The gem under test is placed table facet down on a black spot on a piece of clean white paper. With low RI simulants the 'dot' will appear as a ring around the pavilion facets, but a diamond will transmit no light and the dot will be invisible.

Coloured treated diamonds

The question of coloured diamonds has assumed new importance because of the modern practice of inducing colour in off-coloured stones by submitting them to some form of atomic bombardment. Perfection colour in diamond is a complete absence of colour. However, on account of their rarity, there has always been a market, among those who can afford such luxuries, for diamonds of decided hue (see Plate 3). Of these, yellow 'canary' diamonds and brown 'cinnamon' diamonds are the best known – though even of these perhaps only a few hundred are found in the millions of stones mined annually – while very few indeed are known which have a pronounced green, blue, or red colour. Today, either by bombardment with neutrons in an atomic pile or by charged particles from a cyclotron, diamonds of commercially unfavoured yellowish or brownish tint can be turned green, and stones of brown or golden yellow hue can be produced from these by subsequent careful heating. By electron bombardment, stones of distinctly blue or bluish-green colour can be obtained.

These changes in colour are ascribed to 'radiation damage', that is, to the displacement of atoms in the diamond crystal, leaving gaps in its structure which affect the absorption of light.

Since the beginning of the twentieth century, following experiments by Sir William Crookes, it has been known that the colour of diamond could be changed to green by exposure to radium or its salts. Quite a number of stones were so treated, and retain their colour to the present day. What is more surprising is that the treated stones also retain a marked radioactivity, emitting chiefly 'alpha particles' (helium nuclei) with a range in air which indicates that they are emitted by polonium, one of the later products of radium emanation. Crookes was unable to destroy this radioactivity by treatment with acids or even fused alkalis, and it would seem that a proportion of radioactive atoms were able to penetrate to a minute degree below the surface of the stone, where their disintegration continues. Lind and Bardwell found that the colour of radium-treated stones could be removed by heating for several hours at 450°C, or by recutting the stone.

The radioactivity of these radium-treated stones made them very easy to test, since in contact with a photographic film or plate they yield an 'autoradiographic' image in a few hours (see Figure 12.7), and produce scintillations on a

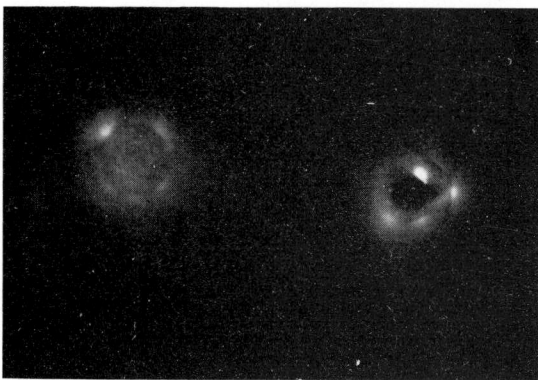

Figure 12.7 Autoradiograph of brilliant and drop-shaped radium-treated diamonds (60 hours' exposure on bare plate)

suitable zinc-sulphide screen. Incidentally, the radioactivity was sufficiently strong to cause skin damage to anyone who wore such a stone constantly in a ring.

Diamonds can also be turned green by bombardment by charged particles in a cyclotron. When the stone is irradiated from the base a curious 'umbrella' effect can be seen if the culet is viewed through the table facet (see Figure 12.8). When the stone is irradiated from the top, a dark rim of colour can be seen if the stone is viewed with the table facet downwards. This form of treatment has been largely superseded by neutron bombardment (usually with subsequent annealing) in an atomic pile.

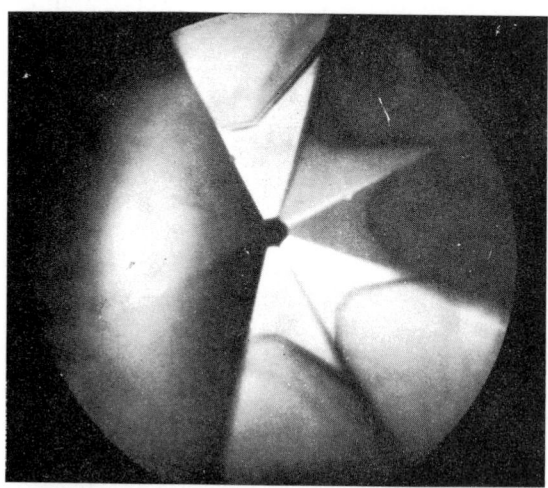

Figure 12.8 'Umbrella' colour margin seen near culet in cyclotron-treated diamond. When seen, this is proof of a treated stone (photo: R. Webster)

The modern irradiated stones are no longer perceptibly radioactive, except for a few hours after they have been treated, so that this simple test is no longer operative. In the case of green stones, the colour tends towards the dark bluish-green of tourmaline, and is not like any hue seen in natural diamonds. The yellows and browns are more convincing in appearance, but can usually be detected by abnormalities in their absorption and fluorescence characteristics, as described later. As for blue diamonds, Dr J. H. F. Custers, of the Diamond Research Laboratory in Johannesburg, found that all natural blue diamonds belong to an unusual structural type, which he designated 'Type IIb', which have certain characteristic properties, including transparency to ultra-violet light down to a wavelength of about 250 nm, have no absorptions due to nitrogen in the infra-red, phosphorescence under short-wave ultra-violet rays, and behaviour as semiconductors. Treated stones will not have these properties, and the 'make-up' of the colour is spectroscopically not the same as in the rare natural blue stones.

Much work will have to be done before these colour changes, or indeed the causes of colour in natural diamonds, are properly understood.

Robertson, Fox, and Martin were the first to discover (1934) that diamonds can be divided into two main categories according to their transparency to ultra-violet light, their infra-red absorption, and various other properties. Stones of the commoner type (designated Type I) are transparent to about 300 nm only and are usually blue-fluorescent to a greater or lesser degree. The rarer 'Type II' diamonds are transparent to about 225 nm and are for the most part non-fluorescent.

As indicated above, Dr Custers has found that Type II diamonds can be further subdivided into Type IIa stones, which will not conduct electricity, and Type IIb stones, which, while being virtually inert during exposure, are phosphorescent after exposure to short-wave ultra-violet light and can carry a strong electric current when under a potential difference of 250 volts.

Stones transparent to short-wave ultra-violet light are not so rare as some authorities imagine. For instance, of four flat 'portrait-stone' diamonds in the author's collection three transmit below 250 nm, and only one cuts off at the 'normal' wavelength of near 300 nm. These four stones also exemplify another fact – that there are intermediate degrees of transparency which do not neatly fit into either category. This is clearly demonstrated in Figure 12.9. The four stones

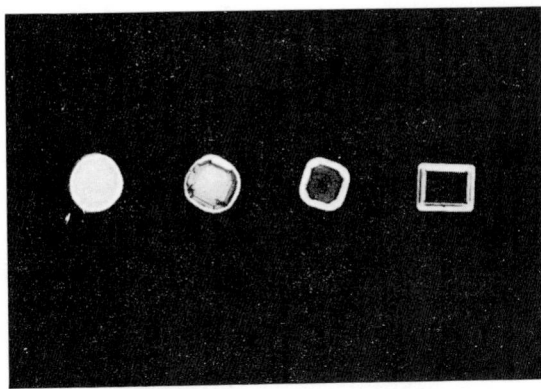

Figure 12.9 Variations in transparency to short-wave ultra-violet light shown by four portrait-stone diamonds

were placed on contact paper in a dish of water and exposed for a few seconds to rays from a short-wave ultra-violet lamp with Chance OX7 filter, and the paper developed. This gave a negative image, so that the rectangular stone on the right of the picture with its dark centre represents the stone most transparent to the rays.

The surprising discovery by Kaiser and Bond (1959) of free nitrogen as a constant impurity in Type I diamonds was shown to account for many of the absorption bands in the infra-red, visible, and ultra-violet of this common form of the mineral, which differentiate these diamonds from stones of Type II.

It might be thought that these strange differences in diamond (unaccompanied, it may be said, by any difference in appearance) are of purely scientific interest, and are out of place in a book devoted to simple methods of gem testing. However, they may have practical importance in the distinction between stones which owe their colour to atomic bombardment and those which are naturally coloured. If it is true, for instance, that all natural blue diamonds belong to Dr Custers' IIb category, then any blue diamond which is a non-conductor and is opaque to short-wave light and has nitrogen characteristics in its infra-red spectrum (thereby showing itself to be a Type I diamond) must owe its colour to electronic bombardment. So far, no exceptions have been found to this rule, though some diamonds which are *not* blue have been found to have Type IIb characteristics.

Turning to colours other than blue produced by bombardment: R. A. Dugdale working at the Atomic Energy Research Establishment at Harwell found that diamonds which had been 'greened' by bombardment and subsequently annealed at suitable temperatures to produce brown and golden yellow stones developed certain absorption lines not originally present in the stone. The most important of these from our point of view is a narrow and rather fugitive line in the yellow at 595 nm (previously reported as 592 nm (USA) and 594 nm (UK)), which is very rare in untreated diamonds. Other lines developed by the treatment are those at 503 nm and 496 nm (the former previously reported as 504 nm) in the blue-green part of the spectrum. It must be noted that some (notably brown and fluorescent green) untreated diamonds also show the line at 503 nm.

Mr G. R. Crowningshield, Vice-President of the GIA Gem Trade Laboratory in New York, was able to observe the 595 ('592') line in thousands of treated yellow diamonds, and the author's own experience confirms this. The presence of the absorption line in the yellow accompanied by lines at 503 and 496 nm in a yellow or brown diamond is in fact virtual proof that the stone has been treated. The lines are more prominent by reflected light, that is, by light reflected from within the stone. The 595 band in particular can easily be missed in the transmitted light spectrum, and it may disappear altogether when the stone becomes warm.

Another interesting and important point in assessing the cause of colour in diamond concerns absorption bands in the blue and violet which are invariably associated with diamonds of the 'Cape' series. These bands include the important one at 415 nm in the deep violet, which is actually at the head of a series extending into the near ultra-violet, and another in the blue at 478 nm, with its weaker associates 465, 451, 435, and 423. Though the band at 478 nm is neither so clear-cut nor so powerful as the 415 band, it is often more easily seen since it is in a region of greater visual acuity. These 'Cape' bands are clearly linked with very stable defects in the diamond lattice, since they are not affected by the intense bombardment and subsequent heating undergone by 'treated' stones. Their presence thus reveals to the skilled observer who has also detected the tell-tale lines at 595, 503, 496 nm that the stone which is now a fine golden yellow was originally a common off-coloured diamond of the 'Cape' series. Of natural yellow diamonds, some show bands of the 'Cape' series (but without the 'danger' lines) while others which are the true 'canaries', show no absorption bands at all and have a yellow or orangeish-yellow fluorescence. (See the upper spectrum in Figure 12.10.)

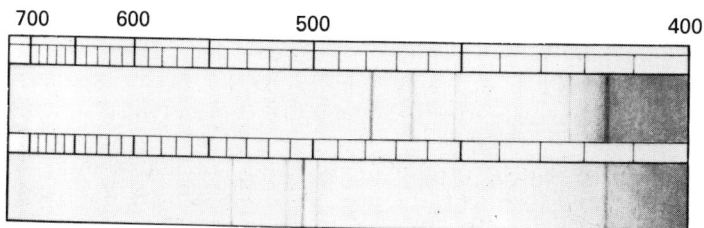

Figure 12.10 Absorption spectra of (*top*) 'Cape' diamond and (*bottom*) diamond of 'brown' series

In nature, diamonds showing the 'Cape' spectrum have a blue fluorescence, though this may vary greatly in strength. On the other hand, diamonds showing the 503 band (also sometimes a weaker line at 515 nm) belong to the 'brown' series and are properly green-fluorescent. It can be noted that treated stones in which the defects giving rise to this line have been artificially induced also tend to show a green fluorescence under long-wave ultra-violet light. The line at 496 nm (known also as the H4) is rare in naturally coloured diamonds, but a 'side band' of the 503 nm line (known also as the H3) located just to the long-wave side of the 496 nm, at approximately 497 nm, has often been confused with it. The difference between these two lines as seen in the hand spectroscope is quite clear; the 496 is a sharp line with almost no fuzziness at its edges whereas the 497, though it can be quite strong, if the 503 is also strong, has a certain fuzziness about its edges. The 496 is often quite strong in treated diamonds which also have a strong Cape spectrum, and since it is these off-coloured stones that are mostly irradiated, it is often found that in treated diamonds the 496 is stronger than the 503. In naturally coloured diamonds that have absorption lines present on the blue-green border it is usual that the 503 is the line that predominates and there is a number of 'side bands', including the 497 to its short-wave side.

The distinction between treated and natural 'fancy' diamonds is a matter for the specialist. No hard and fast rules can be laid down as, to a certain extent, each diamond is a law unto itself: but enough has been said to indicate the main lines on which one must work. Physicists, using spectrophotometers, and often working at very low temperatures to sharpen the effects, have been able to show characteristic differences in the absorption curves of treated compared with untreated diamonds – particularly in regions beyond the limits of the visible spectrum. Gemmologists have to do their best with quite simple apparatus, but are sometimes rewarded by observing aspects of diamond behaviour which have escaped the observation of the scientists.

Testing industrial diamonds and powders

It is well known that a large proportion of the diamonds recovered from mines or alluvial fields are not suitable for cutting as gemstones. Taking a 1989 estimate of annual world diamond production of some 95 million carats, about 50 per cent are classed as 'industrials' and 50 per cent are cuttable, but probably only about 10 per cent by weight end up in jewellery. In contrast, of the 22 million carats recovered annually from the vast alluvial diamonds fields of Zaire, 75 per cent are unfit for use as gemstones.

Although the price of these lower-grade diamonds is naturally less than for those of gem quality, their value is quite considerable. Dealers in industrial diamonds therefore have to exercise vigilance against fraudulent substitution, and may occasionally seek help from a gemmological laboratory. Literature on such matters is hard to come by, and a few notes based on the author's experience may be of value to those faced with some of the commonest problems in this field.

In the identification of diamond as found in the field, all that should be needed is a practised eye, aided with a head loupe, so distinctive are the forms and lustre of diamond crystals even when they are opaque and quite unlike the layman's idea of 'diamond'. Where any doubt exists, the unique hardness and unique transparency to X-rays are useful as check factors for a laboratory worker.

Diamond abrasive powders

In the experience of the London Gem Testing Laboratory, however, the most constant pleas for help made by dealers in industrial diamonds concern the purity and sometimes the grain size of diamond grits or powders, which are used in vast quantities in grinding, polishing, and finishing metals in industry. If these are adulterated to any extent their effectiveness is, of course, greatly reduced, and if the grain size is not as claimed the fine finish of polished work-pieces may be marred.

For effective tests to be carried out a sample of at least one carat should be made available; any not used in the tests can, of course, be returned. In a routine check the following stages have been found to give satisfactory results:

1. As a preliminary measure, a very small quantity of the powder should be transferred to a clean microscope slide, and the grains dispersed and made individually visible by smearing with a finger tip. The slide should then be viewed under the microscope at a magnification of about fifty diameters. Unless the particles are exceedingly small they should be clearly visible as individual grains under these conditions. In powders derived from crushed diamond boart these grains are angular and transparent and are bounded by cleavage planes, giving them a very distinctive appearance with which operators should make themselves familiar by a study of diamond abrasives of known purity. Synthetic diamond grit, on the other hand, usually reveals crystal forms (deliberately controlled for the purpose by the makers) darker in colour, less transparent, and usually showing magnetism when approached by a small powerful magnet.
2. If the appearance of the powder seems satisfactory under the microscope, a portion of about half a carat is scattered evenly along the bottom of a porcelain combustion boat (6 cm × 1.5 cm (2½ in × ½ in)) which has previously been carefully weighed. The boat is then re-weighed to determine the amount of powder present. The charged boat is then placed in a small electric furnace and heated to at least 900°C for several hours. The completion of the burning of the diamond particles can usually be judged by inspection on opening the furnace door while still at maximum temperature. The coarser the powder, the slower the combustion rate.
3. After allowing the furnace to cool to room temperature, the combustion boat is carefully removed with crucible tongs and weighed. In the case of a pure diamond powder the loss should, of course, be 100 per cent, but a slight residue amounting to 1 or 2 per cent of the original weight is to be expected, due to the impurities inherent in boart of crushing quality. Where there is a considerable residue it is wise to carry out a second run to confirm the amount of impurity found; in fact, where time and the number of porcelain boats can be spared, it is a good plan to test two samples of each powder in the initial firing.

Common impurities

When the dealers concerned learn that their samples of supposed diamond powder are seriously adulterated, they may well decide merely to return them to the suppliers without bothering to enquire the nature of the impurity or impurities. But while gemmologists can hardly be expected to function as analytical chemists, their natural curiosity may prompt them to investigate further if this can be done without too much time or trouble. Of the wide variety of adulterants encountered in the London Laboratory some had clearly been added with a purpose to deceive – barium sulphate, powdered glass, and quartz being examples – while in other cases the powders represented an attempt to reclaim waste diamond grit from diamond-cutting benches, and the removal of iron particles, etc. has been incomplete. Silicon carbide (carborundum) was a frequent impurity, easily recognized under the microscope by its blue-green colour, and under long-wave ultra-violet by its orange fluorescence.

Grain size

The most favourable grain size for diamond abrasive powders obviously depends upon the particular purposes for which they are intended, just as various grades of sandpaper or emery paper are selected for abrasive or finishing processes. By various methods – sieving, sedimentation, elutriation, or centrifuge – the principal producers of diamond powders are able to provide products having a remarkably even grain size. A glance under the microscope at a dispersed sample will confirm this. A badly graded powder will appear crude and uncontrolled by comparison.

Occasionally a large consignment of diamond powder may be offered to one of the major companies from an outside source at a temptingly low price, and before concluding the purchase laboratory help may be sought to ascertain not only the purity but also the grain size and consistency of grain size. In the London Laboratory it was found to be a fairly simple matter to check on the average and extreme grain size of such samples. This was done by dispersing a tiny sample of the powder in oil, as previously explained, and measuring the longer dimensions of extensive samples of grains by using a micrometer eyepiece in conjunction with an objective of moderate power. Any good standard microscope will enable one to do this work. In the case of extremely fine powders of sub-micron particle size (1 micron = 1 micrometre = one-thousandth of a millimetre) individual particles are difficult to discern except at very high magnification. It may be wise, before estimating the purity of such a sample by ignition, to ensure that it is indeed diamond by taking an X-ray powder photograph of the material, since it must be borne in mind that not only diamond but *any* organic substance will burn with no residue in a furnace at 900°C.

13

The identification of ruby

After diamond, the most important precious stones from a commercial standpoint are ruby and sapphire which, despite their very different appearance, are both varieties of the same mineral species, corundum. Crystals of ruby and sapphire are shown in Figure 13.1. Let us consider ruby first, and see by what simple means it can be recognized and also by what tests the other stones which to a certain degree resemble it may be distinguished.

Figure 13.1 Crystals of ruby (*top right, bottom left*) and sapphire (*top left*: polished to show hexagonal zoning)

Corundum is simply crystallized alumina (Al_2O_3) and, when quite pure, is entirely devoid of colour. The rich hue of ruby is due chiefly to small amounts of chromic oxide, which is able to replace part of the alumina without disturbing the trigonal crystal structure. The source for the world's finest rubies has always been the Mogok district in Burma, where so many other interesting gem minerals are found. Rubies from Thailand*, owing to the presence of traces of iron, tend to have a less attractive colour, resembling garnet, and are not popular in the trade; while rubies from Sri Lanka* are often more pink than red. Nevertheless, fine rubies from these less-regarded districts are occasionally found, and should be sold as such, on their merit. To vary the price of a stone, not on account of its quality but on account of its fancied origin, is one of those illogical habits which the trade would do well to eradicate.

A more recent prolific source of rubies is East Africa (Tanzania) where large transparent pieces showing little crystal form have been found in surface deposits accompanied by garnets of very similar appearance. The colour of these African rubies is very pleasing, though being due as much to iron as to chromium the hue is not so rich as in those from Oriental sources. At least the new locality has provided rubies of important size at a reasonable price. Large rubies of good quality from the better-known localities are so rare as to be far too costly for all but the very wealthy.

There is another and quite different deposit of ruby in Tanzania where the mineral occurs in well-formed hexagonal crystals (see Figure 13.1), often of considerable size, accompanied by a bright green zoisite and a dark amphibole mineral. These crystals are for the most part opaque and of a not very pleasing colour. Some have been cut as cabochons and can show a crude type of asterism. Sometimes transparent layers have been discovered within these opaque crystals from which small gem rubies resembling dark, very chrome-rich Burma material have been successfully cut.

Rubies of good quality from other sources – Kenya, Nepal, Afghanistan, and Pakistan – have been discovered and mined in recent years, but political difficulties have often prevented a regular supply of these stones from reaching the main markets. Dealers in the trade are mostly conservative, and prefer to handle stones from recognized sources that have an established reputation. At their best, the rubies from Kenya approach most closely the rich blood-red stones for which Burma has always been so famous, and the colour may even show something of the 'treacly' distribution that has always been recognized by gemmologists as a hallmark of Burma rubies. The Pakistan stones mostly have a multitude of inclusions and the colour is not so rich, but apparently these are very iron-free, with the result that the stones may show a distinct phosphorescence after exposure to X-rays, which can be disquieting for those who have always regarded this as a useful diagnostic feature for synthetic rubies made by the Verneuil process.

The properties of an average Burma ruby may be briefly stated as follows: hardness, 9 (next to diamond on Mohs' scale); refractive indices, 1.765 and 1.773; double refraction, 0.008; SG, 3.99 to 4.00. The dichroism is quite strong, the twin

---

* It was the widespread practice in the trade to retain such terms as Siam ruby or sapphire and Ceylon ruby or sapphire, although the names of the countries referred to have long been changed to Thailand and Sri Lanka. However, times change, and the terms Thai ruby and Sri Lanka sapphire are now widely used.

colours being pale yellowish red and deep carmine, and the absorption spectrum is highly diagnostic notably by reason of the narrow *bright* line (doublet) seen in the deep red and the narrow dark lines in the blue when the stone is powerfully illuminated. The former is due to the bright red fluorescence of ruby, but the distinction from synthetic ruby, which also has the properties so far mentioned, has still to be considered.

Apart from this, the materials most likely to be confused with genuine ruby are **spinel, garnet, tourmaline, pastes**, and **doublets**. A careful refractometer reading using either sodium light or a good red colour filter is in itself conclusive, since the only red stone in which the refraction is similar is garnet. Only pyrope garnet, with a *single* index between 1.74 and 1.75, approaches the best ruby in colour, though other garnets of the almandine–pyrope series, having higher indices, may be confused with Thai rubies where colour is the only guide. Perhaps the simplest test to apply, though not so conclusive, is the test for dichroism, remembering always that it is essential to tilt the stone and examine it from different angles to get the strongest effect. Spinel, garnet, paste, or the ordinary doublet will show no trace of dichroism, and the only dichroic red stone apart from ruby is the rubellite variety of tourmaline, which seldom, if ever, is seen in the same shade of red as corundum. The dichroic colours of red tourmaline, dark red and pink, are, however, rather similar to the colours seen in Burma (and synthetic) ruby, noted above.

Another simple aid to the recognition of ruby is its appearance under the Chelsea colour filter. Viewed through the filter, under a bright light, ruby shows a peculiar vivid fluorescent red, and this, combined with its dichroism, should leave little doubt in the mind of the observer, even if a refractometer is not available.

The red fluorescence of ruby is a spectacular sight under ultra-violet light (quartz-mercury lamp with Wood's glass filter) – but visible green, blue, or violet light will also stimulate this fluorescence in ruby; hence the appearance of the bright fluorescence doublet when examining the spectrum of the stone, and its fluorescent red appearance when viewed through the Chelsea colour filter. A splendid exhibition of the fluorescence of ruby and synthetic ruby is provided by viewing specimens between 'crossed filters' – the stones being illuminated by strong blue light through a flask containing strong copper sulphate solution, and viewed through a good red gelatine filter. None of the blue light can pass through the filter, but the red fluorescent glow from ruby most certainly can: stones are thus seen glowing like coals against a dead-black background – a sight so beautiful that it still delights the author after years of repetition. Red spinel shows a bright red fluorescence under the same conditions, which to the unaided eye is quite indistinguishable from that of ruby. The spectroscope, however, enables the two species to be separated with certainty. In ruby, the red glow is seen to consist almost entirely of light from the strong doublet at wavelengths 694.2 and 692.8 nm, which appears as a single line in a small prism spectroscope. In spinel, there appears a whole group of bright lines, rather like a set of organ pipes, with the two strongest in the centre of the group. This 'organ pipe' fluorescence is a very sensitive test for spinel. No other red stones, garnet, zircon, tourmaline, nor any red pastes or doublets, show any red fluorescence.

The presence of iron has an inhibiting effect on fluorescence, and for this reason the red glow from Thai rubies is, on the whole, less brilliant than in those from Burma. It is less generally realized that if a ruby is too chrome-rich this also lessens the strength of its fluorescent glow. This applies to some of the Burma stones, which are often mistakenly classed as Thai rubies on this basis.

In some of the large East African rubies, though not in all, the fluorescence is very dim and the absorption spectrum only shows the chromium lines very weakly. In such stones iron probably plays as large a part as chromium in producing the rather subdued red colour.

Even if we have assured ourselves that the stone examined is indeed a ruby, we have still the important question to answer – 'is it a natural or a synthetic stone?' – since in all the physical properties already mentioned the **synthetic ruby** does not differ appreciably from the natural. The whole question of the discrimination between natural rubies and sapphires and their synthetic counterparts manufactured by the Verneuil inverted blowpipe process has already been discussed in Chapter 9, but since the problem is one which so constantly worries the jeweller a few more words on the subject may be welcomed.

In the first place it is well to recollect that flawless rubies of rich colour and important size are, in nature, extremely rare. This fact alone should make one suspicious of a large clear specimen of unknown origin.

In most cases the mere appearance of the stone is enough to warn the trained observer that a stone is synthetic. This is partly due to a slight but real difference in the nature and proportions of the colouring oxides present in the stone, but is perhaps mainly due to the fact that a synthetic ruby is seldom cut in such a manner as to display the best colour. The most attractive colour in ruby is seen in the direction at right angles to the basal plane of the crystal, which is the direction of the 'optic axis' – that is, of single refraction and *no dichroism*. Thus in a properly cut ruby strong dichroism should not be observable directly through the table facet when the stone is viewed with the dichroscope. In a synthetic stone, due to the shape of the original boule, the direction of cutting is almost invariably wrong, so that the fine purple-red of the 'ordinary ray' is diluted with the unpleasant yellowish-red of the 'extraordinary ray', and the two tints can be seen side by side in the two images of the dichroscope window when looking directly through the table facet.

Another point to realize is that a natural ruby is hardly ever quite 'clean', which is to say that it almost always encloses small crystals of other minerals in the form of pale angular grains, cavities of irregular shape, often relatively large, and patches of fine crisscrossing canals, or of fine reddish rutile needles, which give a silky effect by reflected light and in consequence are known as 'silk' (see Figure 13.2). When the presence of either of these forms of inclusion can be detected with a lens there can be no doubt that the stone is genuine. Typical Burma ruby inclusions are illustrated in Figure 13.3. Care must be taken, however, not to confuse the crystalline inclusions mentioned with the spherical or elongated *gas bubbles* which are often a feature of synthetic stones (see Figure 9.10). These are seldom so large in size as the crystals enclosed in Burma ruby, and under the lens are more likely to appear as little clouds of dust-like particles. Sometimes a dense cloud of microscopic bubbles may cause a milky reflection not unlike the 'silk' seen in natural stones.

The distribution of colour is also revealing: a curious feature of the natural Burma rubies is the presence of patches of deeper colour in the form of wisps and swirls to which the name 'treacle' has been aptly applied. A hint of 'treacle' in a ruby used to be a sure sign of its genuine origin but nowadays similar

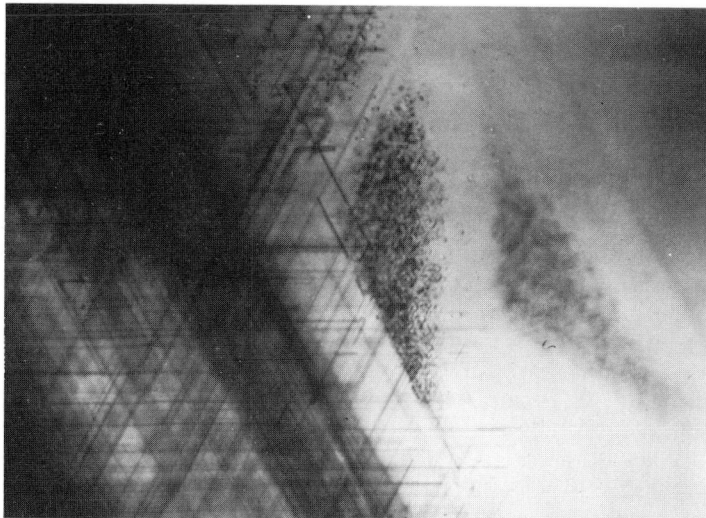

Figure 13.2 Typical patch of 'silk' (rutile needles) in Burma ruby

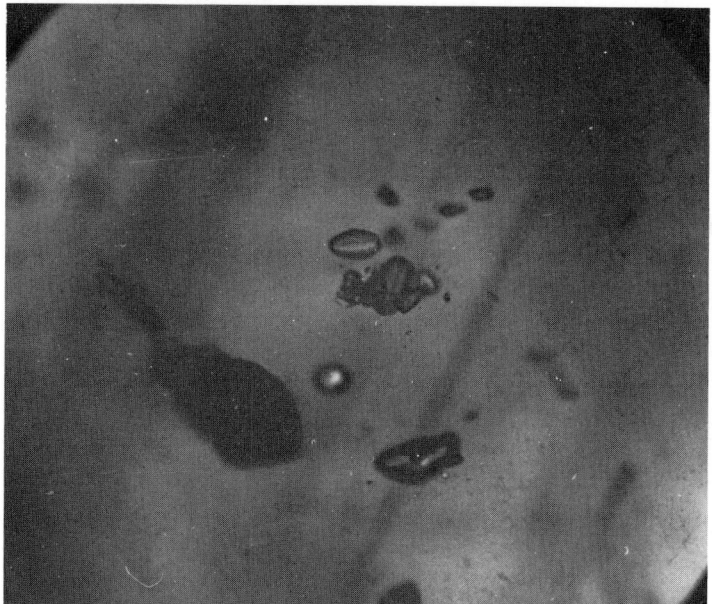

Figure 13.3 'Silk', zircon, and other inclusions in Burma ruby

structures may occur in some flux-grown rubies. Certain other stones owing their colour to chromium sometimes show a rather similar effect – e.g. red spinel, pyrope, and emerald. Verneuil grown synthetic ruby, when it is viewed with a lens or microscope in the right direction, always reveals the finely spaced *curved* structure lines which represent the successive layers of growth of the boule as it grew under the blowpipe flame. All these internal features can be recognized with far greater ease and certainty under even the simplest form of microscope, which in addition opens up a whole new world of interest and beauty, for indeed the inclusions in most gemstones, and particularly in ruby, are both extremely interesting and beautiful (Figure 13.4). It is of great assistance in studying internal features to immerse the specimen in a glass-bottomed cell containing a highly refractive liquid such as 1-bromonaphthalene. If the specimen is a synthetic free from bubbles one must rely on finding the curved lines, and this may entail examining the stone out of its setting at several angles until the lines become visible. The curves are usually more clearly seen when the lighting is not too brilliant, and the mirror of the microscope should be tilted to get the most marked effect, and the condenser lowered.

In pre-Verneuil synthetics, which are sometimes still met with in old jewellery, the curved lines are much more pronounced, and by no means rigidly parallel. The photomicrograph reproduced in Figure 13.5 shows the curved lines in a small synthetic ruby (0.10 carat) held obliquely by a strip of foil. For comparison, a photomicrograph of a small Burma ruby is shown (Figure 13.6) in

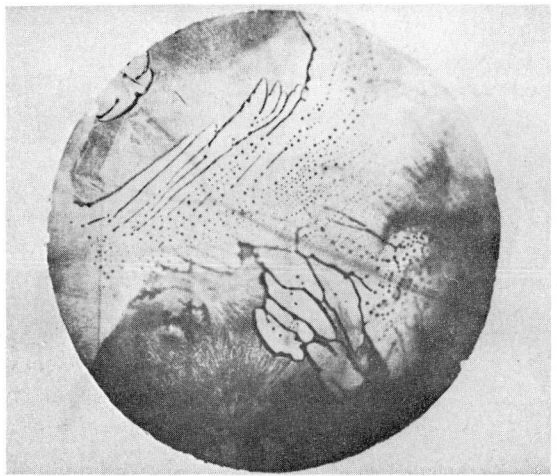

Figure 13.4 Unusual bubble formations and gas-filled tubes in synthetic ruby

Figure 13.5 Curved lines in small synthetic ruby

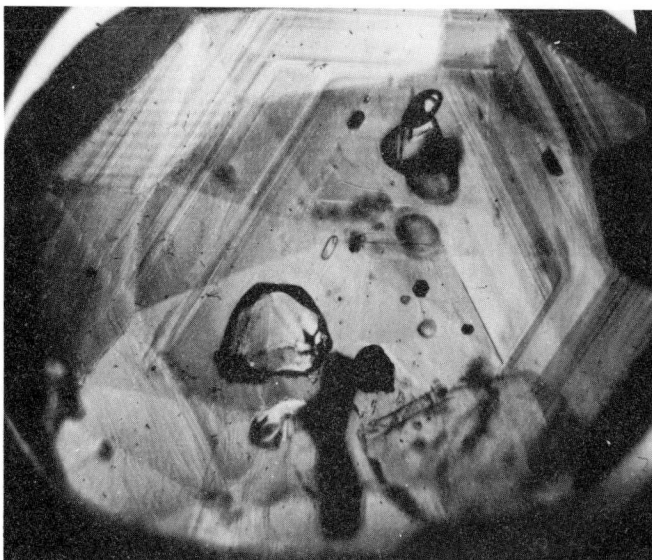

Figure 13.6 Crystal inclusions and zoning in Burma ruby

which not only the relatively large crystal inclusions can be seen but also *straight* lines of zoning parallel to the hexagonal outline of the original crystal. Sometimes only very small inclusions may be present, and these may easily be mistaken for bubbles by an inexperienced worker. In all difficult cases it is wiser to submit the stone for a laboratory test.

Rubies from Thailand can, on occasion, be remarkably 'clean', but their lack of fluorescence under X-rays and short-wave ultra-violet is then a revealing feature. The most typical inclusions in Thai rubies are round and opaque, always surrounded by a roughly circular 'feather' (see Figure 13.7); when these are small, careful examination is needed to differentiate them from the bubbles seen

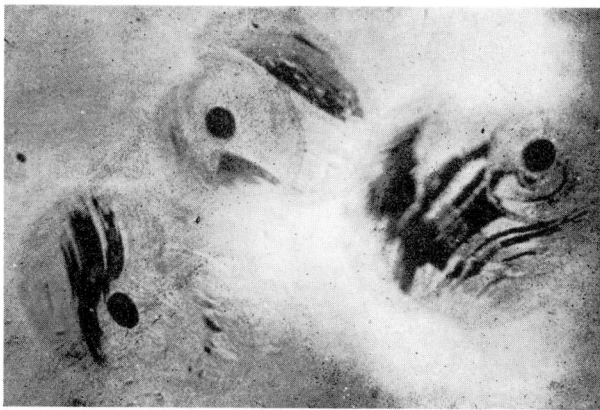

Figure 13.7 Opaque crystals surrounded by liquid feathers, which are the most typical inclusions in Thai ruby

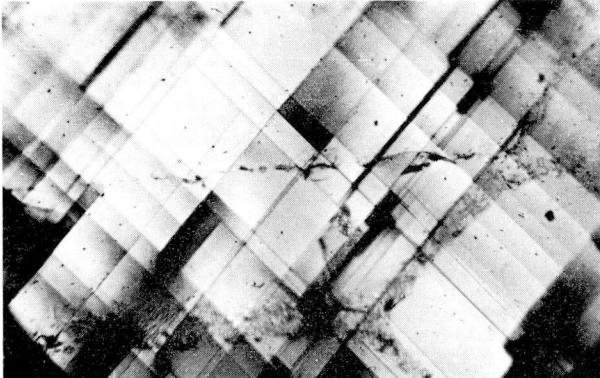

Figure 13.8 Twin planes in Thai ruby

in synthetic stones. Another feature of Thai rubies consists of twinning planes, sometimes intersecting, as in Figure 13.8.

The results of a brilliant investigation by Dr E. Gübelin as to the nature of the rounded dark inclusions so typical of Thai rubies may be briefly summarized here. Identification was made possible by means of microprobe analysis of the composition of the mineral grains which had been exposed at the surface of the stone by controlled polishing. Some, showing a hexagonal outline and a metallic lustre by reflected light, were proved to be the iron sulphide mineral pyrrhotine, sometimes known as 'magnetic iron pyrites'. Others were of almandine garnet, and some were yellow hexagonal crystals of apatite. These inclusions are thought to be primary in origin, formed by the incorporation of islets of impurities during the growth of the crystal.

Synthetic rubies show a brighter fluorescence than natural stones under short-wave ultra-violet rays, and though this test should not be too heavily relied upon it can serve a useful purpose in indicating which stones in a large parcel of calibré rubies or in an eternity ring should be examined with special care.

The greater transparency of synthetic ruby to short-wave ultra-violet light has already been described in Chapter 9 and provides another most valuable background test in those few cases where there is still some uncertainty after an examination under the microscope.

In the above discussion it has been assumed throughout that the 'synthetic rubies' referred to were stones made by the Verneuil method, which has been operating in the factories of several large producers for three-quarters of a century. But while 99 per cent of the synthetic stones used in jewellery today are still manufactured by this inexpensive process, the growing band of crystal growers in the world today has succeeded in producing laboratory-made rubies that entirely lack the well-known Verneuil characteristics and tax the skill even of laboratory gemmologists to determine. Most of these are crystallized by the 'flux-fusion' method of growth and contain liquid 'feather' inclusions of the type with which we have become familiar in synthetic emeralds grown by Chatham and by Gilson. Chatham rubies have sometimes been based on small seeds of natural Burma rubies, which, of course, show the typical natural features of these well-known stones. The border line between the seed fragment and the flux-fusion growth can best be seen under dark-ground illumination under the microscope. Chatham has produced clustered groups of ruby and sapphire (both blue and orange) crystal types which are not found naturally, and the crystals of other manufacturers sometimes display forms and habits seldom seen in nature. Such crystals are readily recognizable; it is when they are cut that the gemmologists' problems begin. Present-day corundum crystal growers include Chatham, Kashan, Knischka, Kyocera, Inamori, Ramaura, and Lechleitner, among others. They use fluxes which include lead oxides and/or fluorides, and boron oxides, plus, of course, the necessary alumina (Al_2O_3) and its colouring agents such as chromium oxide. Temperatures such as 1300°C are countered by the use of platinum crucibles. These fluxes are commonly included within the fabric of the ruby or sapphire crystals and give a very useful clue as to the method of manufacture. The flux may take the form of fingerprints of feathers which resemble those seen in natural stones, but on close examination these feathers, or tubes or tiny dashed lines are seen to be filled with solid flux, sometimes with gas in the interstices of the cavities, but not with liquid. The flux may be white in colour and yellow or orange flux is often seen in Kashan rubies. Flux-grown rubies (and sapphires) may show growth lines, and these may be very close together and possibly at angles which resemble the growth lines at 120° seen in natural corundum. These growth lines may require careful adjustment of the lighting to see them and fibre-optic lamps and shadowing techniques are most useful in identifying growth structures in suspect corundums. In some Ramaura rubies growth lines may be so close together that, at the appropriate angle of viewing, iridescent effects may be seen. Examples of inclusions seen in flux-grown rubies are shown in Figure 9.11 and Plates 6–8.

In some rubies grown by the Czochralski pulling technique the only discernible features may be faint growth striations. These very perfect rubies, used for lasers, may pose problems for the gemmologist. Properties such as refractive index, specific gravity, dichroism, and absorption spectra (in the visible wavelengths) are similar for synthetics and natural rubies, but gemmological laboratories with spectrophotometers may be able to detect differences in the ultra-violet or possibly the infra-red ends of the spectrum. Such sophisticated instruments are not available to the average gemmologist, and it may be some consolation to the puzzled gemmologist to know that very experienced laboratory workers sometimes disagree on the pedigree of an apparently clean ruby.

Recently a stone was only categorized by the presence of a single tiny surface inclusion which proved, on electron microprobe analysis, to be margarite, a mineral which could only occur in a natural ruby.

Gemmologists, however skilled, are now faced with some difficult problems in the distinction between natural and synthetic stones. One general fact may be worth remembering: in no synthetic stone does one find inclusions of 'foreign' minerals – that is, of minerals consisting of chemical constituents other than those belonging to the host mineral.

During the 1980s glass-infilled rubies (and sapphires) have made their appearance (see Plate 10). These are usually encountered on the surfaces of natural rubies and presumably are introduced to fill up small surface cavities which, if polished out, would mean considerably reducing the weight of a valuable ruby. They are easily detected (if looked for!) by the change in lustre at the glass/ruby junction and by the common presence of included gas bubbles.

The only natural stone resembling Burma ruby at all closely in colour is the red **spinel**, which is found in the same district. The colour in this case is also due to chromic oxide, but the tint is more brick- or orange-red than true ruby-red. In any case, the lack of dichroism in spinel and its single refraction (RI = 1.72), different absorption spectrum, and inclusions serve to distinguish it with complete certainty. When testing a parcel of rubies under the microscope it is surprising how obviously different in colour an 'intruder' spinel appears and how noticeable is the dichroism in ruby if the stone is turned in the tongs.

The colour of red **garnets**, apart from certain pyropes already mentioned, is closer to that of Thai ruby. Garnets of this type have a very characteristic absorption spectrum containing three broad bands crossing the yellow, green, and blue parts of the spectrum, respectively, and this, apart from other tests, provides a very rapid and sure means of identification. **Tourmaline**, the only other natural stone that need be discussed in this connection, has far lower refractive indices (1.62 and 1.64) than ruby, and a larger double refraction, which latter property will enable the practised observer, using a lens, to detect a 'doubling' of the edges of the back facets of the stone if viewed through the front at a favourable angle (see Chapter 23).

Red **glass imitations**, which generally owe their colour to selenium, are not very often encountered since the advent of synthetic ruby. Their lower, single refractive index, which is usually near 1.68, their lack of dichroism, and their softness, all serve to distinguish them from ruby. Red **doublets**, almost invariably composed of a thin slice of almandine garnet serving as the table facet fused to a red glass base, are still sometimes seen. The table facet usually gives a refractive index reading of about 1.79, and the back facets about 1.63. Careful scrutiny with a lens will reveal the junction between the two layers as a thin line girdling the facets of the crown just below the table. The line in itself might not be noticeable, but there is an abrupt change in lustre between the garnet and the glass surfaces at the junction between the two which is completely revealing when viewed at the correct angle. Though the garnet slice is very thin, the absorption bands of almandine can usually be detected in the spectroscope when light has passed through the stone. Detection of the fake is easy if the stone be viewed from the side while immersed in a liquid, against a white background. A highly refractive liquid such as 1-bromonaphthalene is best, but even an egg-cup full of water will serve the purpose very well. Under the microscope the doublet, if examined through the table facet, may show rod-like crystal inclusions in the almandine layer, followed by bubbles of air in the glass base and at the junction between the two layers.

Far more dangerous are the ingenious doublets of modern manufacture described in Chapter 9. In these, use has been made of thin pieces of poor-quality natural Australian sapphire, which are cleverly cemented to a base of synthetic sapphire or synthetic ruby and faceted so that the junction line is exactly at the girdle of the finished stone. When provided with a base of synthetic ruby and mounted in jewellery the effect is very much that of a natural stone, since the crudity of the colour of the Verneuil synthetic ruby is modified by passing through the sapphire crown, while the latter provides a convincing effect of silk and other natural corundum inclusions. Successful frauds have already been carried out by using these doublets, definitely built to deceive. The best defence (as prescribed for the defence of freedom) is eternal vigilance. Once suspicion is aroused, many factors (such as the red fluorescence shown under ultra-violet light by the base of these composite stones) can be marshalled and used as proof.

A list of red stones which might be confused with ruby, as far as appearance goes, is given in Table 13.1, together with the most important physical and optical properties which serve to distinguish between them.

Zircon is included in the list for completeness, though red zircon bears no close resemblance to ruby. **Topaz** is also given, as deep pink topaz might be confused with pale Sri Lanka ruby of the type which grades into the so-called 'pink sapphire'. **Almandine** garnet could only be mistaken for the Thai ruby. The dichroism of the latter is by no means strong, but distinction from almandine is easy if a refractive index reading is taken carefully in sodium light, since the double refraction of ruby will then be clearly seen. The absorption spectrum of the garnet is also easy to recognize.

Table 13.1 Identification of red stones

Stone	H	SG	RI	DR	Pleochroism
Ruby	9	3.99	1.76–1.77	0.008	Strong
Zircon	7½	4.69	1.92–1.98	0.059	Weak
Almandine	7½	3.9–4.2	1.76–1.81	None	None
Pyrope	7¼	3.7–3.9	1.74–1.76	None	None
Spinel	8	3.60	1.72	None	None
Topaz	8	3.53	1.63–1.64	0.008	Distinct
Tourmaline	7	3.04	1.62–1.64	0.018	Strong

14

The identification of sapphire

As in the case of ruby, the most effective substitute for natural sapphire is its synthetic counterpart, and methods for detecting the differences between the natural and the manufactured stone will be found below, although the whole subject of synthetics has already been dealt with in Chapter 9. The cobalt-blue forms of synthetic spinel must also be considered as a sapphire substitute, since this is their *raison d'être*. Glass imitations of sapphire are not often seen, since the synthetic stones are clearly more effective as a substitute. The old-fashioned garnet-topped doublet on a blue glass base is also seldom seen, but a much more effective type of doublet has been exploited in recent years, in which a crown of natural Australian sapphire is cemented to a base of synthetic blue sapphire; fuller reference to this type of doublet, which has also been discussed in the previous chapter on ruby imitations, will be found at the end of this chapter.

As for natural blue gemstones, blue spinel and blue tourmaline are not uncommon, and exceptional specimens of these minerals may be mistaken for sapphire, though the blue of spinel is commonly a greyish-blue, and indicolite tourmaline a deep greenish indigo-blue in colour. The rare gemstones kyanite, benitoite, and iolite resemble sapphire more closely so far as colour goes, but these are collectors' stones scarcely seen in commercial jewellery.

Far more important than these is the transparent blue variety of **zoisite** discovered in 1967 in Tanzania. Large clear crystals of this new and important gem variety have been mined from a locality in the Merelani Hills, some 40 miles east of Arusha in a vein of graphite schist, with calcite, sulphur, quartz, and apple-green grossular garnet. The crystals are found in pockets in a variety of colours, shades of brown being the most comon, but also stones of an almost amethystine blue. The stones are usually heat treated at fairly low temperatures after they have been cut, and are transformed into the lovely blue of a fine Sri Lanka sapphire. The name 'Tanzanite' was coined by Tiffany for these stones and seems to have gained universal currency.

Blue crystals as found have a remarkable degree of pleochroism. The fast ray, X, is red-violet in colour; the intermediate ray, Y, is deep blue, and the slow ray, Z, is a yellow-green. After heat treatment the pleochroism is much modified: X becomes violet-red and both Y and Z are deep blue in colour. A cautionary note may be added: tanzanite is peculiarly liable to damage if subjected to ultrasonic cleaning.

The SG of the mineral is very constant at 3.355, refractive indices are 1.692 and 1.701 for the lowest and highest readings, with 1.694 for the intermediate index, and double refraction 0.009. The hardness is 6½–7 on Mohs' scale (rather too soft to wear well as a ringstone) and there is no easy direction of cleavage: thus cutting presents no great problem. The composition of this zoisite is essentially calcium aluminium silicate, with 2 per cent water and 0.02 per cent vanadium to which the colour is probably due. In the crystals as found, three broad absorption bands are visible, which naturally vary in strength according to direction. The strongest of these is centred near 595 nm, the others being in the green (528 nm) and blue (455 nm). In heat-treated stones even the 595 nm band may not be strong enough to aid in identification.

During the 1970s, precious-stone dealers were startled by the appearance of handsome sapphire-blue varieties of both **beryl** and **topaz**, thus enlarging in an unexpected way the colour range of these well-known gem species and making the identification of sapphire by appearance alone a more difficult proceeding. The blue beryls, coming reputedly from Goias in Brazil, were found to have peculiarities previously noted only in beryls from the Maxixe mine in Minas Gerais, and thus conveniently described as 'Maxixe type'. The unusual features referred to in these beryls are, first, a series of striking absorption bands at the red end of the spectrum, the strongest centred at 699 and 655 nm (see Chapter 10) and second, an anomalous dichroism showing the extraordinary ray as nearly colourless and the ordinary ray displaying a deep blue colour – the reverse of normal aquamarines. The old Maxixe stones had been found to lose their fine blue colour on prolonged exposure to sunlight, and the same can unfortunately be said of these recently marketed stones, the fine sapphire-blue colour of which initially earned them an unusually high price. The colour was in some cases found to be restored by neutron irradiation, and all these attractive stones may have been irradiated before sale to enhance their colour. Incidentally, the stones originally found in the Maxixe mine had a notably high density for blue beryl (2.80). The modern stones have more normal values.

The advent of 'new' gem varieties that can so closely resemble sapphire makes it clear that the first step in testing a sapphire-blue gemstone (unless preliminary lens inspection has given a clear indication of its nature, enabling a 'short cut' to be taken) is to measure its refractive index or indices on a standard refractometer, using sodium light or a good colour filter. Where readings of 1.76–1.77 are visible the stone must be either a natural or a synthetic sapphire – and this will be the case in the great majority of stones of this colour which are tested.

Dealing with the natural stones first: the blue zoisite just fully described will reveal itself sufficiently in the refractometer test alone, though **kyanite**, fine blue specimens of which are occasionally cut, has indices only a little higher. The double refraction of kyanite (0.016) is perceptibly greater, and adds another distinguishing factor. The very easy cleavage of kyanite makes it difficult to cut satisfactorily, and tends to give it a characteristically flaky appearance. Kyanite is notable for the extreme difference in hardness between 4 and 7 on Mohs' scale, according to the direction in which a scratch is attempted – but in a cut gemstone this has, of course, no practical application.

Blue spinel gives single refractometer readings which are usually near 1.72, and thus perceptibly below the characteristic value near 1.728 of synthetic spinel. But it is wise to remember that zinc is nearly always present to some extent (replacing the magnesium of normal spinel), as well as iron, to which the colour is due. Zinc may raise the refractive index considerably, and values up to 1.735 are not uncommon in consequence. In some 'gahno-spinels', as the zinc-rich types have been called, something like 20 per cent zinc is present, giving an index of 1.75 and a SG of over 4.0. A useful feature for checking the identity of blue spinel lies in its absorption spectrum. There is a broad band in the blue, centred at 459 nm, which it is important not to confuse with the narrow band at 450 nm which is the most important feature of the absorption complex seen in natural sapphire. Fortunately, this is not all: there is a narrower band nearer the green at 480 nm and a further two bands in the green (550 nm) and yellow (592 nm) which are peculiarly distinctive. A further band in the orange (632 nm) makes this spinel spectrum a very 'rich'-looking one when well developed. These bands are also shown by purplish- or brownish-red spinels, and are useful ones to know.

Blue tourmaline – sometimes known as indicolite – has indices which vary little from 1.62 to 1.64 for the two edges, and a refractometer test alone gives sufficient identification. A rather narrow but indistinct absorption line between the green and the blue, at 497 nm, can, on occasion, prove a helpful feature.

Though its colour tends to be rather sombre or sad, and lacks the brilliance and richness of the best sapphires, **iolite** can be a very attractive gemstone, and is also hard enough to deserve more frequent use in inexpensive jewellery. Its refractive indices are very near those of quartz, though a careful study of the shadow edges while the stone is being turned on the refractometer will reveal that the mineral is biaxial, not uniaxial as is quartz. The astonishing pleochroisim of iolite is well known; when viewed in one direction it shows a pleasing violet-blue, while in another its colour appears to be pale smoky yellow. This property formed the basis for one of its alternative names 'dichroite' while 'cordierite' is yet another alternative. Colourless iolite has been discovered in Sri Lanka.

Another of the natural sapphire-blue stones to be mentioned is **benitoite**. Were it more plentiful, its beauty and hardness would earn for it a more prominent place among the gemstones. It has been found in California in one or perhaps two localities, and never in great quantity or large sizes, so it remains an item for the collector rather than the jeweller. Benitoite has refractive indices near those of sapphire, as Table 14.1 will show; however, it has a far larger double refraction of 0.047, by which it is easily distinguishable on the refractometer, though at the time of its discovery (1907) this was not an instrument in common use, and it was its dichroism which first gave rise to the suspicion that it was not sapphire as originally supposed. In addition to its bright blue colour (rarely pink or colourless), benitoite has an added attraction in its high dispersion, which is actually as high as that of diamond. This gives it a very lively appearance.

Blue topaz produced by irradiation has been available commercially for more than a decade, and the colour resembles those of some sapphires. By using Co-60 in a gamma cell a blue of low intensity is produced, but by using high-energy electrons in a linear accelerator or neutrons in nuclear reactors a more intense blue results. Gamma rays and high-energy electrons change the topaz to a brownish colour and the desired blue is produced by heating to about 200°C. Neutron-irradiated stones can be darker blue or have inky or steely-blue hues. By these various methods, topaz of a pale 'aquamarine' to quite deep blue is produced. By using the low-energy radiation such as that in a Co-60 gamma cell, no residual radiation effects are induced, but high-energy electrons and neutron activation induces radioactivity in some topaz which is unacceptable by safety standards. These stones may require a cooling-off period, depending on the dose received. For the gemmologist blue topazes are easily identified by their refractive indices, but a second crucial point is whether the irradiation treatment can be detected. The answer may be in the affirmative, but detailed spectrophotometric work is necessary at present.

Some irradiated stones develop cracks and partings parallel to the basal plane (as in normal topaz crystals) and a zoning parallel to the cut surface may be visible. This zoning may be blue around the rim, with a colourless intermediate zone with defects (cracks, etc.) and a colourless core. However, stones with prominent defects are likely to be rejected after irradiation, and, in any case, a totally enclosing blue rim is not easy to detect and is rarely seen. For the average gemmologist the best assessment of a blue topaz is to look at the colour. If it is deeper than a 'normal' aquamarine blue then the stone is virtually certainly irradiated and paler colours may very well be so treated. For those who maintain that permanently altered, fast colours need not be disclosed, blue topaz presents no problems. For the total disclosure fraternity it is possibly wise to state that nearly all blue topaz on the market today is treated. Even blue topazes mined as such have probably been irradiated naturally. There is no simple positive detection test at present (1989). These stones have refractive indices of 1.612–1.622 (DR 0.010) and a density of 3.56.

Let us turn now to the far more usual problem: the distinction of natural from synthetic sapphire. In distinguishing between natural and **synthetic sapphire**, the same general methods are used as with synthetic ruby. The refractive index and SG will not effect a separation, and the most reliable tests involve an examination of the internal features of the stone. With sapphire, mere inspection with the naked eye is sometimes sufficient if the stone be immersed in a liquid such as 1-bromonaphthalene and viewed against a white background. In fact an egg-cup full of water will often suffice to show the distribution of colour, which in Verneuil synthetic sapphire is always in the form of curved bands (see Plate 12), whereas in the natural stones such banding is straight in zones parallel to one or more of the trigonal growth faces in the original crystal (Figure 14.1). However, straight growth-lines are no longer the hallmark of natural sapphires and rubies; they also occur in the flux-grown synthetics made by Chatham and others. The curved bands in synthetic sapphire are more easily detected than in synthetic ruby, since they occur in broad swathes (Figure 9.3) of colour, whereas in ruby they are closely spaced and difficult to detect with the naked eye. The gas bubbles, which in groups, swarms, or as isolated individuals are a typical sign of all synthetics made by the Verneuil process, are also usually to be found in synthetic sapphire (see Figure 14.2).

Natural stones, on the other hand, almost invariably carry within them clear signs of their slow crystallization in an environment consisting of chemically complex liquors and mineralizing agents. It is only to be expected that some of the substances surrounding the growing crystal should become impounded and remain preserved within the sapphire as frozen witnesses for our interrogation

Figure 14.1 Straight colour bands in natural sapphire

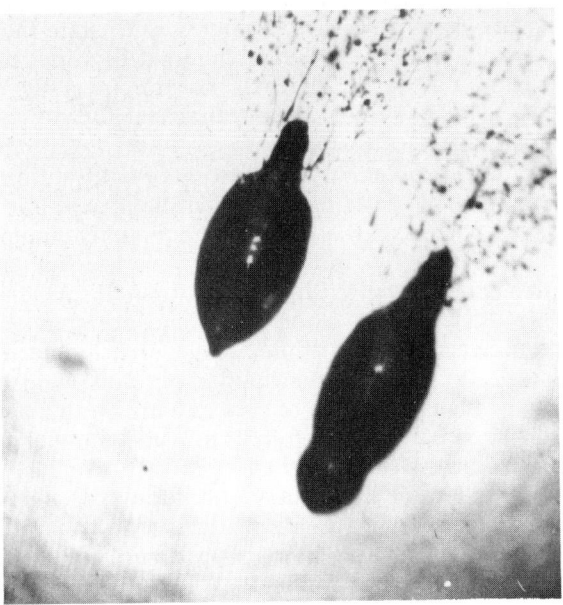

Figure 14.2 Bubbles in synthetic sapphire

aeons later. These positive signs, included in natural sapphires, apart from the straight zones of colour already mentioned, vary somewhat according to the locality where the sapphire was mined.

Thus, Sri Lankan sapphires contain typical 'feathers' consisting of layers of minute crystalline or liquid inclusions which reflect light from one plane, often slightly curved, like a thumb or fingerprint (see Figure 14.3 and Plate 11). However, feathers of similar appearance occur in many of the flux-grown

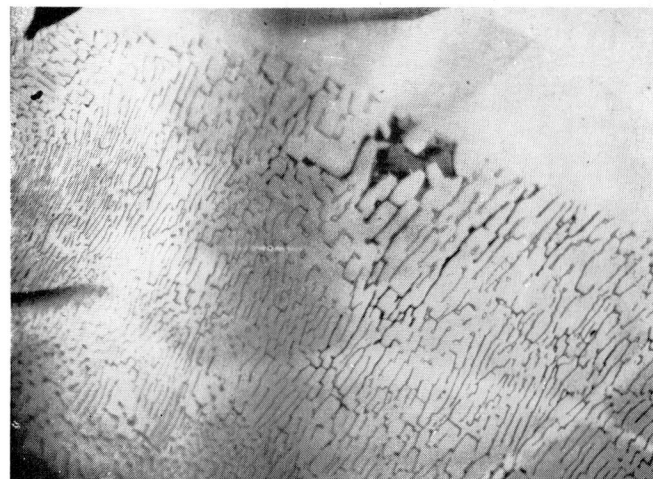

Figure 14.3 Liquid 'feather' in Sri Lankan sapphire

synthetic sapphires. Whereas in natural stones the 'droplets' of the feathers contain liquid (and possibly a gas and a solid crystal), synthetic feathers consist of particles of solid flux with or without gas in the cavities. Examination by reflected light assists in the elucidation of the nature of the droplets and may also reveal the presence of shiny crystals of platinum which are sometimes seen in the flux-grown material. Other typical features of Sri Lankan sapphires are rounded crystals of zircon, showing high relief and surrounded by crack-like tension haloes (Figure 14.4). Three-phase inclusions containing liquid, a movable gas bubble, and well-shaped crystals of hematite or other minerals are also not uncommon in Sri Lankan sapphires (Figure 14.5). A form of long fine silk (rutile needles) is also common in Sri Lankan stones (Figure 14.6), but these and Burma stones may show both coarse and fine and long and short crystals. These inclusions often show a hexagonal arrangement which is crysallographic-ally oriented. Sapphires (and rubies) of basaltic origin (Thai, Cambodian, Australian, and some other corundum) may contain crystals of pyrochlore,

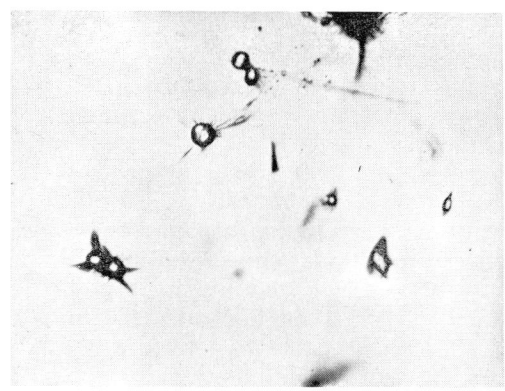

Figure 14.4 Typical zircon inclusions in Sri Lankan sapphire

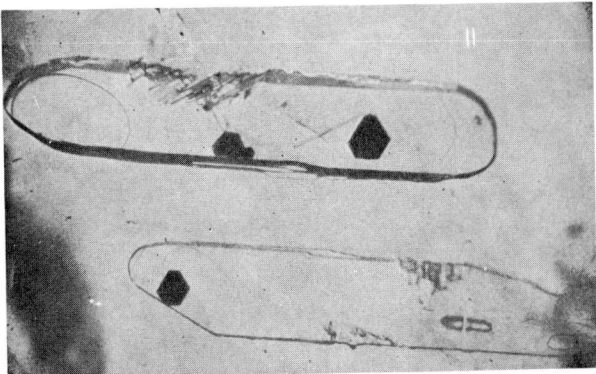

Figure 14.5 Three-phase inclusions in Sri Lankan sapphire

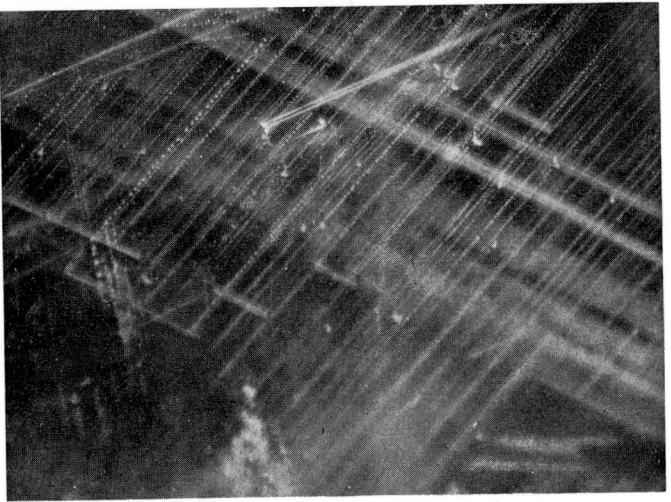

Figure 14.6 Form of long 'silk' often seen in Sri Lankan sapphire

garnet, and plagioclase feldspar and twin lamallae are seen in Thai and Cambodian (Pailin) stones. Thai sapphires resemble Thai rubies in frequently containing opaque crystals (some are pyrochlore) surrounded by a disc-shaped lacy feather. Hexagonal growth structures (lines) occur in many sapphires, but those from Australia and Montana show them well; it must be remembered, however, that these hexagonal structures have also been seen in the synthetics grown by Chatham and others.

Kashmir sapphires owe their attractive milkiness to minute liquid or exsolved inclusions often in hazy bands and nebulous clouds. The corundum from the Umba River area of East Africa may be characterized by the presence of boehmite (an aluminum oxide mineral) which often decorates the lines of twinning planes as fine fringes. Apatite crystals are also a feature of Umba stones, but apatite is possibly the most widely distributed of all accessory minerals in sapphires.

One test, which is often very valuable in checking on the natural origin of a sapphire in which clear signs are lacking, is to examine, by means of a small prism spectroscope, light which has passed through the stone. The manner in which this can best be done has been explained in Chapter 10. Most natural sapphires show a narrow absorption band in the deep blue of the spectrum, and sometimes a group of three bands, the strongest being at 450 nm. In the past, the 450 nm band alone was considered to be the hallmark of a natural sapphire, but times change. Nowadays the blue flux-grown sapphires by Chatham may show a faint band at 451 nm (near enough for confusion!) and often may be present. However, the presence together of the 450, 460 and 470 nm bands is still a reasonable indication of nature's own handiwork, especially if accompanied by suitable inclusions. Australian sapphires invariably show all three bands strongly, while Montana, Thai, Kenya, Kashmir, and Burma stones show them in diminishing strength in that order. In some Sri Lankan stones even the 450 nm band is virtually absent, but these frequently contain a trace of chromium, and then show the bright red fluorescence doublet of ruby by scattered light (Figure 14.5). (As a warning, it should be said that some synthetic sapphires also show a red fluorescence.)

A short-wave ultra-violet lamp such as the 'Mineralight' provides another useful aid in distinguishing natural from synthetic sapphires, though, as with all fluorescence tests, it must be used with discretion and not relied upon absolutely. The titanium in synthetic blue sapphires gives rise to a whitish or dusty green fluorescence on the surface, which requires almost complete darkness and dark-adapted eyes for its examination. The effect is most clearly seen when the illuminated stones are viewed on edge. Natural Sri Lankan sapphires sometimes show a similar effect when almost free from iron, but the majority of natural stones, unless heat treated, are inert under the rays. If the fluorescing synthetics be examined with a lens, the curved structure lines may often be seen clearly on the surface of the stone (more clearly than they can be seen under normal lighting), and where seen this, of course, makes the test a positive one. The type of situation in which a short-wave fluorescence test is particularly valuable (as pointed out in Chapter 13) is where one or two synthetic sapphires have been detected under the microscope in a hoop ring or a line bracelet in which the stones are for the most part natural. If the suspected stones show the typical fluorescence of synthetics this acts as a useful confirmatory check and reassures the operator that no other synthetics have been overlooked.

Besides synthetic sapphire, **synthetic spinel** of a deep royal blue colour is sometimes used to simulate sapphire. This fake can usually be detected by mere inspection, since in artifical light or sunlight red reflections from some of the back facets can be seen, due to the presence of a considerable amount of red in the cobalt-blue which is the colouring agent. This also causes such stones to appear deep red under the Chelsea colour filter – a similar effect being noticeable in **blue pastes** when these are coloured with cobalt, and in **blue doublets** in which the table facet is garnet and the remainder a cobalt-blue glass.

A new and most deceptive type of sapphire doublet was mentioned at the beginning of this chapter and has caused a considerable amount of trouble. A thin layer of natural Australian sapphire, typically greenish blue in colour, is made to form the crown of the faceted stone, and is cemented at the girdle to a base of blue synthetic sapphire. The join is so well contrived that parcels of these doublets in the loose state were accepted as natural sapphires for some time without question. The crown of the stone shows straight-zoned banding and natural inclusions, giving a strong impression that the stone is natural, and through the spectroscope a distinct absorption complex can be seen in the blue, apparently confirming this judgement. As always, when dealing with a doublet the first and most important step is to suspect that the stone may be a composite one. If, in this case, the setting makes it difficult to detect the curved lines or bubbles in the base of the stone, a strong clue is provided by the whitish-blue or greenish fluorescence under short-wave ultra-violet light.

Some glass imitations of sapphire are not coloured by cobalt but by iron, etc., and these are not to be detected by the filter. Blue synthetic spinel is, of course, also easily distinguished from sapphire by its lack of dichroism and by its single refractive index of 1.727. A warning note should perhaps be entered here. Sri Lankan sapphires frequently contain a trace of chromic oxide, and these will tend to appear red under the Chelsea filter. Effects seen through the filter must in fact never be accepted blindly as complete proof of the nature of the stones concerned, but only as valuable indications (see the full account in Chapter 4).

Heat treatment

During the 1970s it became known that greyish milky sapphires known in Sri Lanka as 'geuda' could be heat treated to produce good blue sapphires. It was also found that rubies with blue areas could be heated to eliminate the blue and that the colour of many blue sapphires could be improved. These changes are permanent, and there are those who maintain that it is not, therefore, necessary to disclose the treatment to a customer; others contend that all should be told. Since the temperatures involved in heat treatment may reach 1900°C it follows that some inclusions, such as albite, calcite, pyrite, and others become molten and lose any crystal form. As a result of the increase in volume with temperature the body of the host crystals bursts (locally) and a tension halo forms.

With larger inclusions the host may disintegrate under the strain. Negative inclusions containing fluids crack and any liquid leaks out and may help to partially re-heal the tension fissures which have formed. These and other features can be detected by careful microscopic examination. Melted inclusions surrounded by tension halos are a characteristic feature of many treated rubies and sapphires.

Table 14.1   Identification of blue stones

Species	H	SG	RI	DR	Pleochroism
Sapphire	9	3.99	1.76–1.77	0.008	Strong
Benitoite	6½	3.67	1.75–1.80	0.047	Strong
Kyanite	4–6	3.69	1.71–1.73	0.016	Strong
Synthetic spinel	8	3.63	1.727	None	None
Spinel	8	3.60	1.72	None	None
Topaz	8	3.56	1.61–1.62	0.008	Moderate
Zoisite	6½	3.35	1.69–1.70	0.009	Strong
Tourmaline	7	3.10	1.62–1.64	0.020	Strong
Beryl	7½	2.70	1.57–1.58	0.006	Strong
Iolite	7	2.59	1.53–1.54	0.009	Strong

Diffusion treatment

By heating corundum in alumina (Al_2O_3) powder with varying proportions of colouring agents (iron and titanium for blue sapphires, chromium for ruby, and iron with chromium for orange sapphires) it is possible to diffuse extra colour into a pre-formed or faceted stone. The temperatures and time involved may be in the 1800°C region for 24 hours.

Penetration of colour is very shallow and care must be taken in repolishing or the coloured layer will be removed. Diffusion-treated cut stones often show facets with different shades of colour; concentration of the deeper colours around the girdle (see Plates 13 and 14) when immersed in fluids and penetration of colour along any cracks help in the identification of such stones.

Table 14.1 shows the constants of the various blue stones.

15

The identification of emerald

The variety of beryl known as emerald shares with ruby and sapphire the highest esteem in the popular imagination, and, as with the corundum gems, its rarity and costliness have served to stimulate man's ingenuity in providing artificial substitutes. Just as the red of ruby and the blue of sapphire cannot properly be matched by any other natural mineral, so is the pure emerald green unequalled by any other transparent natural gemstone.

Beryl is a silicate of beryllium and aluminium; the colour of emerald is mainly due to traces of chromium replacing to a small extent the aluminium ions in the crystal lattice of the hexagonal beryl crystal. It is a feature of this colouring agent, which also causes the red in ruby and in spinel and the betwixt-and-between colour in alexandrite, that even when it produces a green colour it transmits a proportion of deep red light. Chromium is also the cause for the red fluorescence that can be seen in all these gemstones, when suitably stimulated. It is for these reasons that the majority of emeralds appear red or reddish when brightly lit and viewed through the Chelsea colour filter, which transmits only deep red light and a narrow band in the yellow-green which in emerald is partly absorbed. The majority of glass *imitations* absorb red light strongly and thus appear green through the filter, and so does heat-treated green tourmaline, which is not unlike emerald in appearance. Hence the appearance of a doubtful 'emerald' under the filter can be a useful guide. However, this simple test must be used with great discretion, since there are many exceptions to the simple rule that red through the filter is a sign for genuine emerald and stones which remain green must necessarily be something other than emerald. In the first place, synthetic emeralds appear strongly red under the filter: in fact their ruby red colour is hardly matched by the very finest Colombian stones and furnishes a hint of danger. Then there are other natural green stones which assume a reddish tinge under the filter.

Another consequence of the presence of chromium in emerald is the absorption spectrum it displays, which is distinctive enough to enable the stone to be distinguished from any other gem species, and also serves to act as a guide where there is some doubt whether one is justified in calling a stone emerald or merely green beryl – since there are some green beryls which owe their colour mainly to iron.

In recent years there has been growing recognition of the part played by vanadium in inducing bright green colours in several of the gem minerals, including tourmaline, grossular garnet, and beryl. The absorption spectrum of vanadium lacks the sharp absorption and fluorescence doublet in the red and accessory narrow lines produced by traces of chromium in a mineral, but the absorption maxima of the two elements are in very nearly the same position (in the orange-yellow near 600 nm), and it is this that plays the main part in determining the colour of the stone. In beryls from most of the famous emerald localities traces of both elements are present, but chromium has the dominating influence on quality, and its presence, readily ascertained by means of a hand spectroscope, is accepted by most authorities as an indicator that the term 'emerald' can be applied to the stone. This has been agreed by the Gemmological Association of Great Britain, the National Association of Goldsmiths of Great Britain and Northern Ireland, the Gem Testing Laboratory of Great Britain, and the Nomenclature Committee of CIBJO, an international organization representing the jewellery trade, gem dealers, and other groups.

The absorption spectrum of emerald has been already described and illustrated in Chapter 10. In the ordinary ray, in addition to the strong doublet at 683 and 680 nm in the deep red there is a line at 637 nm of almost equal strength. In the extraordinary ray there are two weaker bands at 662 and 646 nm with a narrow region of high transmission beside each, giving a curious and characteristic effect. The only other stones of approximately an emerald green that show a chromium spectrum at all resembling that of emerald are green jadeite and chalcedony stained with a chromium salt. In none of these is the doublet in the deep red nearly so sharp as in emerald, and being polycrystalline materials there is no variation in the spectrum when the stone is rotated in polarized light.

As for the other distinctive properties of emerald, the refractive indices and SG vary perceptibly according to the locality where it was mined, due to slight variations in chemical composition. Traces of iron and, more notably, the presence of the heavier alkali metals such as caesium tend to raise the value of these constants. The inclusions, also, are often distinctive for each locality, and gemmologists will do well to familiarize themselves with these things if they wish to distinguish between natural emeralds and the **synthetic emeralds**.

Figures for the SG of Lechleitner type of synthetic emerald, described in Chapter 9, have no great significance since they vary according to the type of natural beryl used for the faceted 'seed' on which the hydrothermal synthetic overgrowth is grown. Even the refractive index of the coating varies a little from sample to sample.

Of course, there are slight variations within each locality, but the figures given may be found a useful guide.

Inclusions are not easy to tabulate (see Table 15.1), and a fuller indication of the most characteristic types can be briefly given. In Colombian stones, which constitute a high proportion of fine emeralds used in jewellery, the most constant features are flat cavities with upper and lower margins jagged like a coarse saw, containing liquid, a bubble of gas, and a little cube of rock salt. Sometimes these three-phase inclusions may have only a single 'spike'. The cube of rock salt is often elongated in one direction, giving it a rectangular

Table 15.1 Inclusions and other characteristics of emeralds

Source	RI		DR	SG	Inclusions, etc.
	Min.	*Max.*			
Synthetic (flux melt)					
Chatham Gilson Inamori Lennix Russian Zerfass	1.558–61	1.564–7	0.003–5	2.60–5	Wispy veils; occasional phenakite crystals (Chatham, Gilson, Inamori, Zerfass); zoning, two-phase and spear-like inclusions (Lennix)
Gilson 'N'	1.574	1.580	0.006	2.70	Non-fluorescent; absorption band at 427 nm
Synthetic (hydrothermal)					
Biron/Pool	1.569	1.573	0.004–5	2.68–71	Wispy veils (two-phase); nail-head spicules, phenakite crystals
Lechleitner	1.566	1.587	0.005–6	2.68–80	Naturals on rear unpolished facets; crazing on surface layer
Linde/Regency	1.566–70	1.576–8	0.005–6	2.67–70	Sparse nail-head inclusions; strong fluorescence
Russian	1.573–9	1.580–6	0.006–7	2.68–70	Step-like growth lines; colour zoning; angular growth patterns

Natural

Brazil:

Carnaiba	1.566–75	1.572–82	0.006	2.67–72	Brown mica laths; 'stars' of particles with one- and two-phase fillings
St Terezinha	1.580–86	1.588–93	0.006–8	2.70–76	Dolomite, pyrite and chromite grains
Itabira/Belmont	1.580–2	1.589–90	0.007–8	2.70–74	Biotite, pyrite, chromite grains; multi-phase inclusions
Colombia:					
Chivor	1.571–4	1.577–80	0.006	2.69–71	Three-phase, jagged inclusions; pyrite crystals
Muzo	1.576–80	1.582–6	0.006	2.71–2	Three-phase, jagged inclusions; calcite rhombs
Tanzania: Lake Manyara	1.578	1.585	0.007	2.74	Mica flakes; liquid-filled cavities
USSR: Urals	1.581	1.588	0.007	2.74	Actinolite blades; flaky discs parallel to basal plane
Zambia	1.581–2	1.588–91	0.007–9	2.68–74	Elongate two-phase inclusions; tremolite/chrysotile fibres, mica
Habachthal	1.584	1.591	0.007	2.72–6	Actinolite fibres; mica flakes
India	1.585	1.595	0.007–10	2.72–4	Rectangular cavities with bubble, parallel to c-axis
Sandawana	1.583–8	1.590–6	0.004–6	2.74–5	Curved tremolite fibres; deep colour
Transvaal	1.583–6	1.593–4	0.006–7	2.75–6	Biotite mica flakes; sparse molybdenite
Pakistan	1.588–93	1.595–600	0.007	2.75–8	Curved fluid inclusions; mica flakes; dolomite crystals

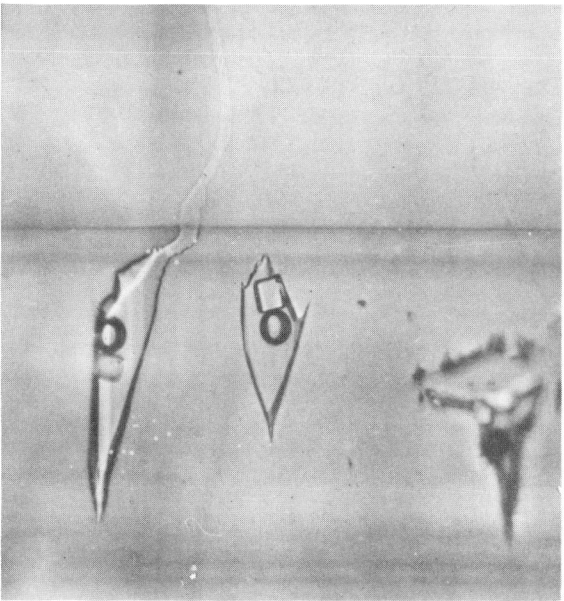

Figure 15.1 Three-phase inclusions in Colombian emerald

outline, or it may be seen obliquely, in which case it presents a lozenge-shaped outline. These familiar inclusions are most usually oriented parallel to the main axis of the original crystal. They are common to emeralds from both the Chivor and Muzo mines (see Figure 15.1). Distinctive for Chivor emeralds, when seen, are beautifully formed little crystals of iron pyrites, recognizable by their metallic brassy lustre and 'pyritohedral' form. Muzo stones, on the other hand, may reveal included rhombohedral crystals of calcite, and occasionally little pinkish inclusions of the rare-earth mineral parisite, which gives a strong rare-earth absorption spectrum. Emeralds from the Ural mountains (Siberian or Russian emeralds) have quite a different occurrence, and this is reflected in their inclusions. Flakes of mica may be seen, often broken or resorbed, but the most distinctive features are blades of green actinolite; cracks across the length of these gives them a 'bamboo shoot' appearance (see Figure 15.2). Nothing resembling these green blades is seen in other emeralds, with the exception of the masses of tremolite fibres seen in some of the Sandawana, Zambian (see Plate 15), and Habachtal stones. Another feature seen in Siberian emeralds is a development of thin, disc-like cavities parallel to the basal plane. These show a silvery lustre by reflected light, but in some directions may appear black, due to total reflection. In Transvaal emeralds mica is usually a major inclusion, while in Indian emeralds are found hexagonal cavities ('negative crystals') parallel to the main axis of the crystal. In profile these have a rectangular outline, often with a small projection from one corner, and they enclose a small bubble of gas in a liquid – that is, they are two-phase inclusions. Mica is also often present (Figure 15.3).

Emeralds from the 'Sandawana' emerald mine in Zimbabwe have played quite an important role in the trade since their discovery in 1956. These are rich in chromium and thus a very deep and vivid green, which is seen at its best in

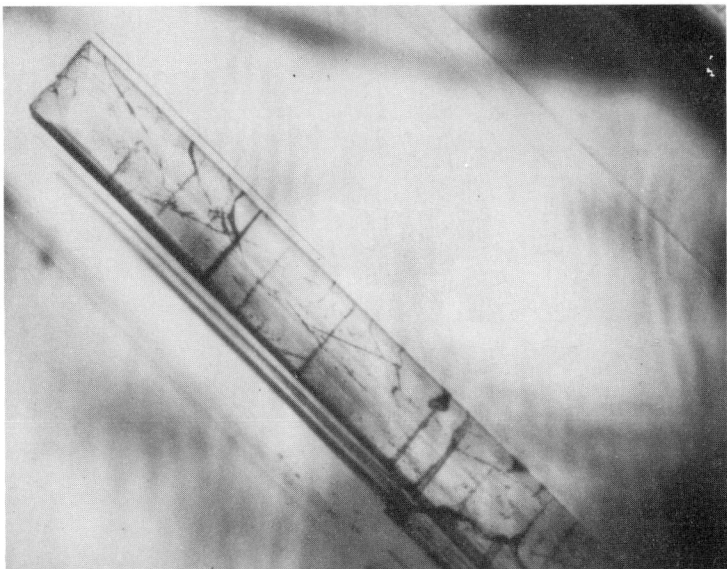

Figure 15.2 Actinolite crystal in Siberian emerald

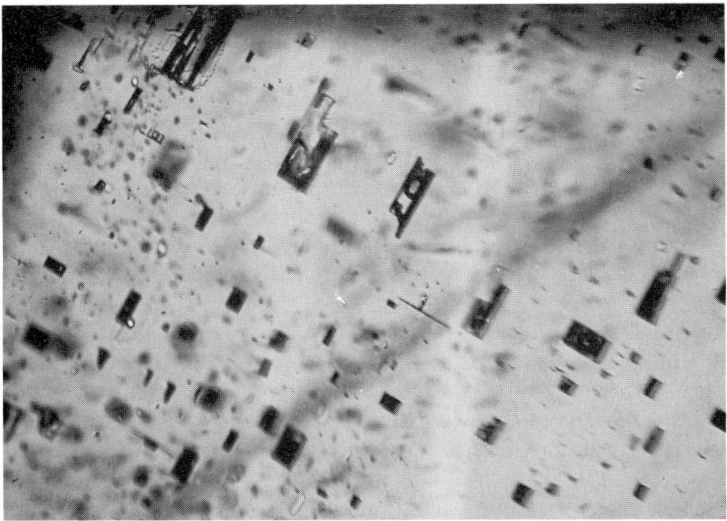

Figure 15.3 Negative crystals, containing liquid and bubble, in Indian emerald

small stones. A typical feature is the presence of numerous fibres of tremolite which are often bent and intersecting (Figure 15.4). Similar masses of fibres are encountered in the far less valuable emeralds from the old Habachtal mine in the Tyrol, which is still worked sporadically on a small scale.

Dark green emeralds have recently emanated from Pakistan, and these contain rather indeterminate inclusions, among which flakes of mica and small crystals of phenakite and dolomite could be recognized.

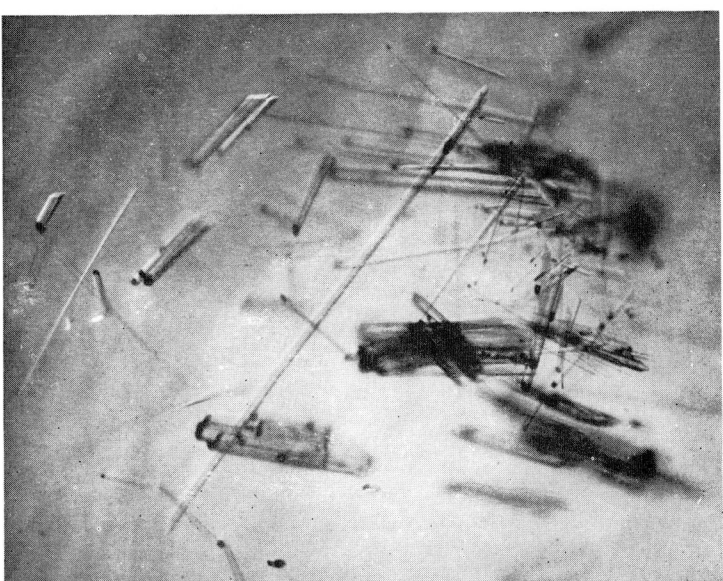

Figure 15.4 Tremolite inclusions in Sandawana emerald (photo: R. Webster)

Another discovery of emerald was near Lake Manyara in Tanzania. The occurrence is in what has been described as an actinolite schist containing the necessary chromium to result in the formation not only of emerald beryl but also of the still rarer alexandrite variety of chrysoberyl. These Tanzanian emeralds have rather high constants, the SG averaging 2.74 and refractive indices 1.578 and 1.585 for the extraordinary and ordinary rays. The inclusions are mainly mica and liquid-filled cavities. At their best, the colour of these emeralds is said to resemble that of stones from Sandawana and Colombia – than which there can be no higher praise. It is interesting to note in this connection that no vanadium is present.

Synthetic emeralds, for many years the prerogative of the San Francisco chemist Carroll Chatham, are now manufactured by a number of producers and by several different processes. A full description of these has already been given in Chapter 9 (on synthetic stones), but the main points will be summarized here.

The process used by Chatham, Gilson, Inamori (Kyocera), Lennix (for some of the Russian stones), and Zerfass is the 'flux-fusion' method, involving crystallization of the beryl constituents from a melt in some suitable solvent such as lead molybdate. The resulting stones tend to have characteristic veil-like feathers which are well illustrated in the photomicrographs shown in Chapter 9, and in Figure 15.5. Such stones also tend to have refractive indices, birefringence, and SG markedly lower than those for natural emeralds and to have a stronger red appearance under the Chelsea filter, and stronger red fluorescence under crossed filters and ultra-violet light.

By addition of some iron to his formula Pierre Gilson produced synthetic emeralds having rather higher constants (RI = 1.574–1.580; SG = 2.70) and with very little fluorescence. These, however, show an absorption band at 427 nm in the violet due to iron which is similar to that shown by aquamarine and some

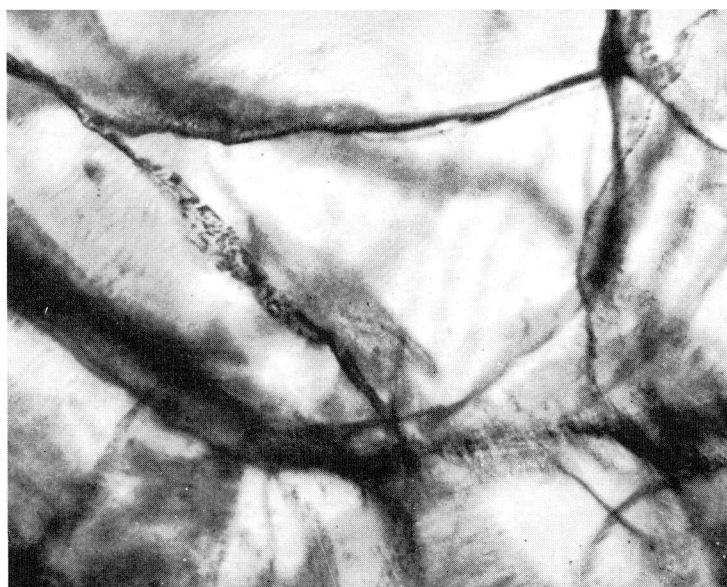

Figure 15.5 Twisted feathers in Zerfass synthetic emerald

natural emeralds from Zambia. The twisted feathers remain a feature of these rather 'awkward' synthetics.

In addition to these flux-fusion stones there are the hydrothermally grown types. First came the ingenious Lechleitner process in which an overgrowth of synthetic emerald was deposited on a pre-faceted pale beryl of small value. The hydrothermally grown stones were usually lightly polished on the crown but allowed to retain their crystalline structure on the base; they were characterized by crack-like markings in the surface layer (see Plate 16), and the layer itself could be seen as a thin, deep green edging to the stone when this was immersed in a matching liquid such as bromobenzene. Refractive indices were variable and in the range for natural emerald, while the SG also varied according to the type of beryl used as the basis for the overgrowth.

The firm of Linde took over the Lechleitner process by agreement, but soon afterwards produced entire emeralds made by a hydrothermal process of their own devising. These Linde stones have a fine colour and properties very near those for natural stones. They lacked the feathers typical of emeralds grown by flux-fusion, and had small 'nail' and 'brush-mark' inclusions of their own. Their most distinctive feature was their strong red fluorescence under all radiations including visible light. The Linde process has now been taken over by Vacuum Ventures Inc., who market Regency synthetic emeralds. Other hydrothermal synthetics are now manufactured by Biron and Pool (Australia) and the Russians. These hydrothermal stones often show growth features such as wavy lines near seed junctions and step-like and 'mountain-peak' features (see Plates 17 and 18). Colour zoning and nail-head spicules with tiny phenakite crystals are also seen. In view of continued modifications of this and other methods resulting in altered characteristics one has to admit that the gemmologist is often faced with a difficult proposition in having to distinguish between natural and synthetic emeralds.

Even when dealing with **doublets** the gemmologist can now be faced with difficult problems since, not content with producing 'beryl-on-beryl' composites with an emerald-green layer to provide the required colour, ingenious manufacturers have now assembled 'emerald-on-emerald' stones. To make things even more confusing, these are sawn in the plane of the girdle of the already faceted stone and reassembled in the same position with only the wafer-thin green layer between the upper and lower portions. This means that under the microscope any inclusions are seen to continue without apparent break throughout the whole stone, giving an appearance of homogeneity. In a mounted stone with the girdle concealed this presents a very awkward challenge.

Obviously, if the sale of doublets *as such* is to be profitable, the manufacturer must use natural emeralds of very inferior colour; and this provides a clue for the wide-awake gemmologist, since the chromium lines are barely discernible through the spectroscope which would not be the case were the stone truly of so fine a colour as it appears to be. Moreover, when viewed from the correct angle through the spectroscope, absorption bands in the orange reveal the presence of the artifical colouring matter in the junction plane.

Compared with such 'teasers' as the above, the testing of the old quartz soudé doublets (see Plate 19) and the newer synthetic spinel doublets used to imitate emerald is a very simple matter. Refractometer readings will in themselves establish the nature of the fake in each case.

Glass imitations of emerald are also frequently seen and can be deceptive to the eye. These are usually lead-glass pastes of refractive index between 1.60 and 1.66, with SGs between 3.40 and 4.00. Often the deception is made more plausible by the presence of 'feathers' which, under the microscope, are seen to be rafts or chains of bubbles. A rarely encountered imitation is a fused beryl-glass coloured by chromium. This is unusually hard for a glass, but its SG (2.42) and refractive index (1.52) are much lower than for crystallized beryl.

Of natural stones which resemble emerald in appearance, there are only three which deserve serious mention. One is the emerald-green **fluorite** from Namibia (and presumably from other localities, since material of this kind is sometimes met with in carved Chinese images and *objets d'art*). This has been mentioned briefly already as showing a reddish residual colour under the Chelsea filter. A further similarity between green fluorite and emerald may sometimes be noted under the microscope, since three-phase inclusions with gas bubble and cube of rock salt may be found in green fluorite resembling rather closely the well-known inclusions in Colombian emerald. Other tests, however, make distinction between the two an easy matter for the gemmologist. On the refractometer, for instance, fluorite gives a single shadow edge at the low figure of 1.434, and, being singly refracting, fluorite has no dichroism. With loose stones, the higher SG of fluorite (3.18) is quickly revealed by a trial in bromoform, in which it sinks decisively, whereas emerald rises to the surface with equal alacrity. Also, under ultra-violet rays (365 nm), the fluorite displays a violet fluorescence. Fluorite ranks as the standard 4 on Mohs' scale of hardness, against emerald's 7½ – and this, together with its low refractivity, gives it a far lower lustre than beryl.

The second natural stone to be mentioned for its colour resemblance to emerald is a variety of **tourmaline** which also, by a curious chance, comes from Namibia. As found, it is a deep blue-green in colour, but by suitable heat treatment it assumes a paler and purer green shade, not far removed from that of emerald, and the strong dichroism of the untreated tourmaline is no longer noticeable. The recent discovery of finely coloured, green tourmalines in Tanzania should be noted. Some of these apparently owe their colour to vanadium, some to chromium, and others, probably, to both these elements, as so often happens in emerald. The chrome-green tourmalines show a narrow absorption line in the red, but not the clear-cut group of lines so distinctive for emerald. The chrome tourmalines also show red through the filter, and this fact may cause an error of judgement unless the stones are carefully tested. Here again the refractometer provides the surest single test, since the values of 1.620 and 1.638 given by tourmaline are completely distinctive. The expert eye is, in any case, unlikely to be deceived in this instance, as the colour of the tourmaline is definitely not quite 'right' for emerald.

The third stone which might on occasion be confused with emerald is **jadeite**, which also owes its colour to chromium, and occasionally achieves an almost emerald green. Jadeite is very rarely transparent, however, so that the confusion would only arise between this and emerald of a quality hardly suitable for use in jewellery. The slight similarity between the absorption spectra of the two species has already been mentioned earlier in this chapter. The appearance under the microscope and the higher refractive index of jadeite (with cabochon stones the 'distant vision' method can be used) will help to remove any doubts. Another distinctive feature of jadeite which may be mentioned is its slightly dimpled or shagreened surface when polished, due to slight local differences in hardness, although this may not be seen in modern diamond-polished material.

Of the other natural green stones, **demantoid garnets** often contain traces of chromium and then have a fine colour – more yellowish-green, however, than with emerald. The brilliant lustre and 'fire' of demantoid in themselves are distinctive – as also its 'horsetail' fibrous inclusions, single refraction, and absorption spectrum. **Green sapphire** and **green zircon** need hardly be mentioned as serious rivals to emerald, though their properties are included in Table 15.2 for completeness. The same can be said for **peridot**, which has a distinctive yellowish-green colour of its own. Recent studies of Roman cameos and intaglios have revealed that many stones described as 'prase' are chrome chalcedony. Their rich colour is such that they could easily be mistaken for emerald but refractive index measurements and examination of the pattern of chromium absorption lines will enable distinction to be made.

One last note of warning may be added before leaving the subject of emerald and its identification. A disturbing feature of the trade in emeralds, although it is seldom mentioned in books, is the practice of 'oiling' flawed stones to conceal those cracks that reach the surface of the finished stone. After immersion for some days in a suitable fine oil, perhaps with gentle warming or even with the assistance of a vacuum, surface cracks will have virtually disappeared from view and the appearance of the stone (and, of course, its apparent value) will be very much upgraded. Unless green colouring matter has been added to the oil the practice is not, unfortunately, considered illegal; but the eventual result to purchasers and probably to the jewellers who supplied them with the jewels containing such stones can be pretty disastrous. It only needs a careless immersion of the oiled stones in water containing a detergent (as in washing up) for the oil to be leached out and the flaws to make their presence very apparent. Worst of all is the effect when the jewel has been subjected to ultra-sonic cleaning.

Manufacturing jewellers should be on guard when purchasing parcels of emeralds for mounting, and should inspect any flaws that reach the surface. Gentle warming under a desk lamp will tend to make the oil (if present) 'sweat' out of the flaw and be visible under lens inspection at the surface. The oil used may also fluoresce yellow when the stone is bathed in long-wave ultra-violet light. If colouring matter has been added this can be seen if the stones are gently wiped with a piece of white blotting paper. It should also be mentioned that much emerald from Brazil (and elsewhere) is being treated with various colourless resins for the same reasons as oil is used. The effects are similar in that flaws are made less visible.

Other gemstones, notably ruby and sapphire, may have colour and appearance improved in similar fashion where the stones are marred by surface-reaching flaws. It will be understood that it is the entry of a film of air into such flaws that makes them so painfully visible, due to the reflection of light at the stone/air interface, and the entry of oil prevents this effect only so long as it remains in position.

Table 15.2 shows the constants of various green stones.

Table 15.2   Identification of green stones

Species	H	SG	RI	DR
Emerald	7½	2.71	1.57–1.58	0.006
Zircon	6½	4.0[a]	1.82[a]	0.01[a]
Sapphire	9	4.00	1.76–1.77	0.009
Demantoid	6½	3.85	1.89	None
Peridot	6½	3.34	1.65–1.69	0.037
Jadeite	7	3.33	1.65–1.67	–
Fluorite	4	3.18	1.43	None
Tourmaline	7	3.05	1.62–1.64	0.018
Beryl glass	7	2.42	1.52	None
Chrome chalcedony	6½	2.60	1.54	None

[a] The constants for green zircon are very variable.

16

Aquamarine and alexandrite

The beautiful sea-green or sea-blue beryl so appropriately known as aquamarine is one of the most popular among the gemstones, and a few notes on its identification may be useful to the jeweller. As far as colour goes, there are only two natural species which closely resemble aquamarine – **blue zircon** and **blue topaz**.

The former of these can easily be recognized by its 'fire', lustre, and strong double refraction, but topaz, in the rare cases when it is found naturally (or as is more often the case with the irradiated material) with a distinctly blue tint, can hardly be distinguished from aquamarine by eye alone. In addition to these two natural minerals, there are artificial substitutes which are likely to cause trouble. These include **glass imitations** of various types, **synthetic blue spinel** similar to that used in simulating zircon, and, possibly, **doublets**.

The trained eye can usually 'spot' a synthetic spinel at sight: the blue is a little too gaudy and the lustre and fire are a little too bright to belong to a beryl, in which the lustre is decidedly glassy and the dispersion low. Without the use of instruments, **glass imitations** are often difficult to identify. Under a lens, glistening spherical bubbles may sometimes be detected, and when seen provide conclusive evidence that the specimen is a glass, since no natural mineral shows bubbles of this kind, and in synthetic spinel any bubbles will be 'profilated' or of so small a size that a microscope is needed to recognize them for what they are.

Glass being a bad conductor of heat, these imitations will feel warm to the touch compared with a genuine aquamarine. The stone should be cleaned, allowed to attain room temperature, then gently touched with the tip of the tongue, using tongs to hold the stone (unless it is mounted in jewellery). A similar test carried out on a piece of rock crystal and a piece of ordinary glass will provide a helpful comparison. It is said (we hope untruly) that the jeweller still sometimes reaches for a file when testing for a paste – but not only are some 'aquamarine' glasses very hard, but also it should be realized that even a piece of paste jewellery may be valued by its owner, who may not welcome its return with a large file mark across it!

Synthetic spinel, apart from its too-handsome appearance, can be separated from aquamarine at once by holding the stone under a good light and looking at it through a Chelsea colour filter. The blue synthetic will show a bright orange or red, while aquamarine appears a very distinct green, since it cuts out the deep red very effectively. Synthetic blue spinels which contain no cobalt have been made, and these do not show red through the filter. Stones of this type, however, are practically never seen in jewellery.

Doublets made to simulate aquamarine are hardly ever met with, but the possibility of a doublet should always be borne in mind by the careful jeweller when examining any gemstone.

From what is written above, it should be clear that only the following are likely to be confused with aquamarine: zircon, topaz, synthetic spinel, and glass imitations, all of the appropriate colour. Of these, examination with a lens will eliminate zircon (strong double refraction) and many glasses (bubbles), while the Chelsea filter will detect synthetic blue spinel. If these positive signs are lacking, however, we are still left in doubt as to whether our stone is indeed an aquamarine, a blue topaz, or some form of glass imitation free from obvious bubbles. Once more, the scientific tests recommended and described in this book make it possible quickly to arrive at a definite decision.

First, let us turn to the refractometer. Aquamarine will give a reading near 1.58, with the small double refraction of 0.006. Any stone giving readings below 1.56 or above 1.60 can be definitely rejected as not true aquamarine. The reading for topaz is decidedly higher (1.62), with double refraction 0.01, while the majority of glass imitations have their single refractive index in the 1.50 to 1.51 region or are lead glasses with refractive index near 1.64. However, there are some glasses in which both refractive index and SG are nearly the same as for beryl. Using sodium light, or a spinel refractometer, the distinction would, of course, be simple, since aquamarine would then show its distinctive double refraction, impossible in a glass. However the double refraction is too small to be seen in white light on a standard refractometer, and thus, should a reading near 1.58 be obtained, it will still be wise to examine the stone under the microscope, or with a dichroscope, to confirm that it is a genuine stone.

The dichroscope, perhaps, provides the simpler test, and can easily be applied to a stone even when in a setting, providing that it has an open back so that light can pass through it. Aquamarine is decidedly dichroic; with a green specimen, one image in the dichroscope will appear a deeper green than the specimen as a whole, while the other image will be practically colourless; with a blue specimen one image will again be colourless, and the other a pronounced blue. With Madagascan stones in particular this blue image is a magnificent colour; unfortunately (a note for the interest of advanced readers) the blue colour belongs to the 'extraordinary' ray, and thus, however the stone be cut, it is always diluted with the colourless 'ordinary' ray when viewed with the naked eye. This is not true for the 'Maxixe' type of blue beryl, described later and in Chapter 14, in which the deeper blue colour of the stone is found in the ordinary ray.

In testing for dichroism, one must remember to view the stone from several angles to ensure seeing the full effect. Also, it is better to use light reflected from a white wall or cloud, as a sunlit sky is in itself slightly dichroic, and may lead the observer to credit this slight effect to the specimen.

The microscope, which is always invaluable when it comes to distinguishing natural and artificial gems by its revelation of their internal features, can also be used to help us here. Glass imitations will usually show a bubble or two, even if these be too small to be detected with a pocket lens – or they may show 'swirl marks' as a result of improper mixing of the constituents. Aquamarine on the other hand, being a natural mineral, will commonly show layers of small crystals ('feathers') or thin, needle-like crystal inclusions, or thin tubes containing liquid and perhaps a bubble of gas. These lie parallel to the main (hexagonal) axis of the original crystal (see Figure 16.1).

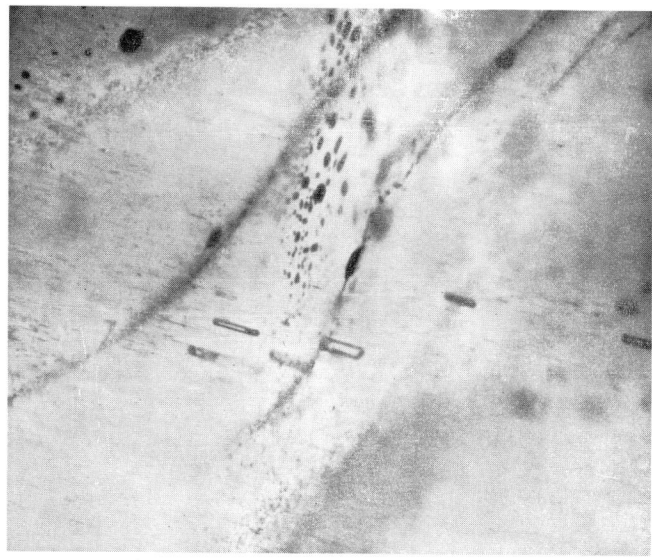

Figure 16.1 Inclusions in aquamarine. 'Rain' and two-phase inclusions parallel to main axis and layers of discs at right angles to these

If no refractometer is available, and the stone can be obtained free from its setting, it may be tested in a heavy liquid. Here one can separate topaz, zircon, synthetic spinel, or the heavier pastes from aquamarine simply by immersing the stone or stones in pure bromoform (2.9) or di-iodomethane (3.3); aquamarine floats while the others sink. Any pale blue specimen, in fact, which floats in these liquids and is also dichroic can be accepted as genuine aquamarine. Even without such aids, dealers accustomed to handling aquamarines can usually detect the intrusion of a blue topaz when weighing these stones individually, since a topaz of this colour is more than 30 per cent heavier than an aquamarine of the same size.

Stating the matter as briefly as possible, one may say that the only single test which will in itself provide a positive determination of aquamarine is a careful refractometer reading in sodium light (or with a spinel refractometer in ordinary light) when not only the mean index but also the double refraction can be measured. Failing this, either a rough refractometer reading or a SG test may be made, combined with a dichroscope observation.

Before concluding, it may be well to state here that although aquamarine can be synthesized it has not yet appeared *on the market*. Specimens so called will either be synthetic blue spinels or pastes. A glass made from fused beryl, and thus having the same *chemical composition* as aquamarine, is indeed sometimes seen, but, not being crystalline, its properties are quite different from those of beryl, as can be seen from Table 16.1, in which the physical constants of all the stones and imitations discussed in the chapter are given in a form suitable for reference. The constants are given for a special hard glass imitation of aquamarine.

Table 16.1   Identification of aquamarine

Species	H	SG	RI	DR
Aquamarine	7½	2.70	1.574–1.580	0.006
Zircon	7½	4.69	1.926–1.985	0.059
Synthetic spinel	8	3.63	1.727	None
Topaz	8	3.56	1.610–1.620	0.010
Hard glass	6	2.37	1.50	None
Beryl glass	6½	2.42	1.52	None

Lead glass imitations may vary from SG 2.63, RI 1.57 to SG 3.85, RI 1.64 or higher. Their hardness is only about 5 on Mohs' scale.

'Maxixe type' beryls

Early in the 1970s beryls of a strking cobalt-blue colour appeared as attractive cut stones on the market in New York and London, and fetched high prices. These could not properly be described as aquamarine, but it seems appropriate to mention them here. In addition to their unusual colour these stones had uncommon optical properties. The deeper of the two dichroic rays belonged to the ordinary and not to the extraordinary ray – the reverse way round to the case of aquamarine, and a fact that enabled the best use of the deep colour by cutting stones with the table at right angles to the optic axis. There were also (in the ordinary ray only) a series of strong absorption bands to be seen in the deep red, where there were strong bands centred around 697 and 655 nm, extending to weaker bands towards the yellow.

These properties were reminiscent of blue beryls found in the Maxixe mine in Minas Gerais some fifty years previously, and like these attractive stones the new ones showed a marked tendency to fade in bright sunlight. The stones were thoroughly investigated by Dr Kurt Nassau and others. Nassau found that exposure to neutrons or other penetrating radiations often restored the deep colour to faded stones. According to another account the original colour of the stones was pink, and the colour induced by some unspecified treatment. In view of the above, any dark-blue beryls showing the optical properties of 'Maxixe' type as indicated above should be treated with caution as being liable to fade after prolonged exposure to light.

Alexandrite

Alexandrite is a variety of the mineral *chrysoberyl*, which also provides stones of a clear greenish-yellow, of a brown colour not unlike tourmaline, and honey-coloured cat's-eyes showing a narrow silky ray when cut *en cabochon*.

Although alexandrite has not much intrinsic beauty, the curious change in colour from shades of green in daylight to raspberry red in artificial light has its attraction, and this, combined with the rarity of good specimens, places it among the more costly gemstones.

As far as appearance goes, no natural gemstone is likely to be confused with alexandrite, except perhaps some colour-change garnets and the uncommon **andalusite**, which has reddish gleams mingled with the green body-colour of the stone, due to its strong pleochroism. So-called 'synthetic alexandrites', however, are often met with, which are generally **synthetic corundum**, but sometimes may be **synthetic spinel**.

True **synthetic alexandrite** has now been added to the growing list of man-made gemstones. Specimens of this rarest form of chrysoberyl have been grown by the flux-fusion and floating-zone methods and by the 'pulling' technique developed by Czochralski. The colour change shown by these man-made stones may, as might be expected, be more pronounced than that found in natural examples, except perhaps for the best Siberian stones, now exceedingly rare and almost invariably flawed, and the recently discovered Brazilian stones. The green or bluish-green in daylight is quite intense and in electric light it shows the 'raspberry' red so much sought after in this stone. The traces of chromium added to form the synthetic display these clear colours despite the fact that iron is also frequently present.

Those stones grown by flux-fusion contain the twisted veil-like liquid feathers associated with this method, together with an occasional metallic inclusion with a triangular or hexagonal outline. Stones made by the 'pulling' and the floating-zone methods may show curved striae of a kind. In each case the clear-cut absorption lines due to chromium are seen, and a strong red fluorescence often occurs under ultra-violet light.

Far more frequently seen, of course, are the so-called 'synthetic alexandrites' that are really synthetic corundums coloured by traces of vanadium. These are a peculiar greyish-blue in daylight and in electric light a strong purple. These stones are frequently bought by the more gullible members of the public, blissfully unaware of the extreme rarity of the natural stone, which would make a necklet of large clear examples an impossible occurrence. The clear-cut narrow line in the blue at 475 nm seen in these imitations is a convenient proof that can be relied upon: no natural stone shows such a line. Alexandrite-coloured synthetic spinels are seldom seen, but as far as appearance goes they are more convincing than the corundum counterfeits.

There are several simple tests which can be made to distinguish true alexandrite with certainty from synthetic corundum and synthetic spinel.

First, since they belong to a different mineral species, with different properties (see Table 16.2) they can be quickly distinguished on the refractometer.

The refractive index of alexandrite being not far below corundum, a careful reading is needed for a decisive distinction between the two minerals; but the synthetic spinels are very easily distinguished by their decidedly lower, and single, refraction.

Under the microscope, the synthetic corundums will reveal the usual spherical bubbles; if these are hard to find, the curved structure lines are invariably present, and can be seen clearly if the stone is viewed from the correct angle. The spinel imitations may not show bubbles, but the lack of double refraction will be evident under the microscope by the failure to show any doubling of the back facet edges when viewed in sharp focus through the front of the stone, while between crossed 'Polaroids' it will show a typical 'tabby extinction' effect.

Finally, the appearance through the dichroscope and spectroscope of light transmitted by genuine alexandrite is distinctive, especially with the latter instrument. Careful observation from different angles with the dichroscope will reveal three differently coloured rays in alexandrite: green, purple, and orange – though, of course, only two of these can be seen together at any one time in the two adjacent images visible through the eyepiece of the dichroscope. The corundum counterfeit shows brownish and mauve while the spinel, being singly refracting, has no dichroism at all.

Alexandrite owes its peculiar balance of colour between red and green to its absorption spectrum, and this in turn is due to the presence of small amounts of chromic oxide in this variety of chrysoberyl. The spectrum may be recognized by the two strong dark lines very closely together (known as a 'doublet') in the deep red, two weaker lines in the orange-red, and a broad absorption band covering the yellow and some of the green. Synthetic 'alexandritic' corundum owes its colour partly to vanadium, and has none of the above characteristics, but a tell-tale narrow absorption line at 475 nm in the blue, which proclaims it at once for the fraud it is. It also shows a distinctive mustard-coloured glow under long-wave ultra-violet light. The spinel counterfeits may show a vague cobalt spectrum with three broad bands in the orange, yellow, and green respectively.

All in all, a careful refractometer reading is the best test for alexandrite. If the reading is different by more than a few integers in the third place of decimals from the average figures 1.745 and 1.754 for the lower and upper shadow-edges, the stone is not an alexandrite chrysoberyl. If the indices are right for chrysoberyl there is still the possibility that the stone may be a true synthetic alexandrite. Where this is so, the observation of an exceptionally clear colour change must put one on one's guard.

Siberian and Brazilian alexandrites show the most decided colour change, though possibly rivalled by those found near Lake Manyara in Tanzania. Those from Sri Lanka, though inferior in this respect, are obtainable in larger and more flawless pieces. In a case where there is so little colour change that it is doubtful whether one should term the stone chrysoberyl, or whether one is ethically

Table 16.2   Identification of alexandrite

Stone	H	SG	RI	Pleochroism
Alexandrite (natural and synthetic)	8½	3.71	1.745–1.754	Strong
Synthetic corundum	9	3.99	1.761–1.770	Strong
Synthetic spinel	8	3.63	1.727	None
Andalusite	7½	3.15	1.635–1.645	Strong
Garnet	7½	3.75–3.99	1.740–1.770	None

entitled to use the better selling name of alexandrite, the spectroscope may possibly be called in as arbiter. If the chromium lines in the red can be seen distinctly, the stone may have a right to the title 'alexandrite'.

Alexandrite also shows a red fluorescence between crossed filters, that is, when a powerful beam of light is directed onto the stone through a copper sulphate solution, which is then viewed through a good red filter. Any chrysoberyl not showing this fluorescence under such conditions cannot be classed as alexandrite. Laboratories are sometimes faced with such distinctions, as the name 'alexandrite' has a powerful sales attraction, while plain 'chrysoberyl' means little to the average jeweller or to the public.

The properties of alexandrite, andalusite, and the synthetic counterfeits are shown in Table 16.2.

17

Zircon

Although zircon (or 'jargoon' as it was frequently called) has been known as a gemstone for centuries, it did not become really popular with the public until some sixty years ago, when the magnificent appearance of the blue heat-treated stones from Indo-China excited admiration when they first reached the Western market. Golden-brown and clear, colourless zircons came from the same source, and their beauty was greatly enhanced by skilful cutting with symmetrical and optically flat facets which had been carried out by German lapidaries working in Bangkok.

The rough crystals and pebbles, found in several deposits near the Mekong river and its tributaries in Indo-China, are an uninteresting brown colour when found, but can be changed into blue, golden, or white by heating in primitive stoves in a reducing or an oxidizing atmosphere. Unfortunately the sky-blue, heat-treated stones often fail to retain their glorious colour, and only their fine appearance and relatively low cost enable them to retain favour for use in jewellery. White zircons also tend to become discoloured, though with these and with the blue stones the colour may often be restored by careful reheating.

Before the influx of these handome forms of zircon, Sri Lanka was practically the only source for the stone as a gem, and it is by far the commonest precious mineral in the famous gem gravels ('illam') from the favoured island. The range of colour here runs from yellow and orange through brown, greenish-brown, to olive green. Even in early times the superficial similarity between well-cut colourless zircons (probably heat treated) and diamond was realized and the term 'Matura' or Matara diamond was applied to these.

Zircon is essentially a silicate of zirconium, though up to 4 per cent of the zirconium can be replaced by the closely related element, hafnium. Compounds of uranium and thorium are also isomorphous with those of zirconium, and this accounts for traces of these two radioactive elements which are found in the majority of zircons. Their presence in zircon renders it slightly radioactive, and may have far-reaching effects on its structure and properties, as will be described later.

Zircon is slightly harder than quartz, but has a curious brittleness or friability which is shown in the frequent appearance of 'rubbed' facet edges on cut specimens. Most dealers are aware of this peculiarity, and if a number of zircons are assembled in a packet, take the precaution of wrapping each stone individually in a screw of tissue paper. The SG (4.68) of normal zircon is higher than that of any other common gemstone, giving it a distinctly heavy 'feel' in the hand. It is sometimes useful to remember that the weight of a brilliant-cut diamond is almost exactly three-quarters of that of a white zircon of equal spread. The refractive indices are also high, being 1.926 and 1.985 for the ordinary and extraordinary rays. These values are, of course, beyond the range of the standard refractometer, on which zircon (except for a few 'low' types, described later) does not display any shadow edges. Both the double refraction (0.059) and the dispersion (0.038) are high enough to be noticeable features of the stone: the double refraction makes a 'doubling' of the back facets easily observable when viewed through the front of the stone with a lens, while the dispersion gives a degree of 'fire' which approaches that of diamond.

In order to prevent leakage of light through the back of the stone a series of eight extra facets is often added around the culet of a brilliant-cut zircon – this is the so-called zircon cut.

A distinctive feature for which zircon is justly famed is the remarkable absorption spectrum which it usually displays. This consists of evenly spaced narrow lines covering the whole spectrum; the strongest of which are some twelve in number. These lines were first observed by A. H. Church in 1866 and correctly ascribed by Sorby to the presence of uranium. There are very few gem zircons which do not show at least the strongest of these lines, which is seen in the red at 653.5 nm. Broadly speaking, all the white, blue, and golden heat-treated zircons which are those most used in commerce, reveal the 653.5 line and its weaker comapnion at 659 nm when viewed through the spectroscope by a skilled observer. With these, the spectrum lines can be seen more strongly by internally reflected light than by transmitted light. The stone is simply placed table facet down on a piece of black cloth, a powerful beam of light directed on to the specimen, and the light reflected from the inside of the table facet picked up with the spectroscope.

Of the other gem zircons, Sri Lankan stones of yellow, brown, and green shades practically all show ten or more lines in varying degrees of sharpness. Anomalies shown by certain 'low' type green zircons are discussed later under 'metamict zircon'. Burma zircons, which are usually shades of green or brown, are recognizable by their very intense spectrum, in which many faint lines between the prominent main bands can be discerned.

Only in some red, orange, or pale brown untreated zircons can none of these lines be detected, and certainly the zircon spectrum as a rule provides the quickest positive means for its identification.

Except for blue zircon, in which, for some unexplained reason, the effect is strong, the mineral shows little or no dichroism. This lack of dichroism, however, can prove useful in distinguishing zircon from sphene or cassiterite, which rather closely resemble it, but which are both quite strongly dichroic.

Many zircons are luminescent; white zircons in particular always showing a yellowish glow under long-wave ultra-violet light or under X-rays. An exciting and distinctive feature of this fluorescence is revealed when it is viewed through the spectroscope, since this resolves it into bright lines forming two groups in the yellow and blue regions. A 'line' luminescence of this kind is typical of certain rare-earth elements and in this case has been ascribed to europium. As a practical test, however, it should be used with caution, as white zircons when exposed to ultra-violet rays tend to revert to their original brown colour. Forunately, heating to dull redness will restore the stones to their desired colourless condition. If the heating process is carried out in the dark on a hot plate, thermo-phosphorescence will be seen, and through the spectroscope this also shows the same line spectrum noted above.

One other strange fact about zircon, of no practical significance, is that its refractive index rises when its temperature is raised. Diamond and beryl are among other gemstones which are known to show this remarkable property.

Metamict zircon

Sri Lanka was for hundreds of years the main source of gem zircons, and from the nineteenth century onwards they became the object of considerable study because of peculiarities in their behaviour. In particular it was found that there were major differences in their SG and refractive index which could not be accounted for by any variations in their chemical composition. At the top of the range there were zircons with SG 4.69 or very near that figure, and these had a refractive index of 1.926 for the ordinary ray and 1.985 for the extraordinary ray, giving a birefringence of 0.059. At the bottom there were stones with a SG as low as 3.95 and a single refractive index of about 1.78. In between these extremes there was a whole range of zircons having intermediate properties, and it was noticeable that the lower types were almost invariably green. Zircons from other localities mostly showed fairly constant figures near the top of this range, which can be considered as characteristic of 'normal' zircon.

For more than fifty years these extraordinary variations in the properties of Sri Lankan zircons formed one of the mysteries of minerology, and formed the subject of many scientific papers. The solution came in 1936, when Chudoba and Stackelberg, using the methods of X-ray crystal analysis, established that in Sri Lankan zircons there can be a gradual breakdown in the tetragonal crystal lattice due to prolonged bombardment by alpha particles shot out by radioactive atoms of uranium and thorium which are almost always present in the mineral in small amounts. In Sri Lanka the zircons are from ancient rocks, and the breaking-down process may have been going on for some 800 million years. In the lowest types, only an almost amorphous mixture of silica and zirconium oxide remains. This inevitably lowers the SG and refractive index of the stone considerably, and the double refraction naturally dwindles to nothing when there is no crystal structure left.

Any minerals containing radioactive atoms, either as an essential constituent or as an 'impurity', and which in consequence have lost their original crystal structure, are known as 'metamict' minerals; and 'low' zircons in which this process is complete or nearly complete can correctly and conveniently be described as metamict zircons. The rare mineral 'ekanite', discovered in the Sri Lankan gem gravels, is also a metamict – due in this case to its large thorium content. Ekanite is, in fact, known only in the metamict state, and is quite strongly radioactive and potentially dangerous to the lapidary.

Metamict zircons, as mentioned above, are almost always some shade of green. Low zircons of an unattractive brown shade are, however, not uncommon, while occasionally zircons of quite a pleasing orange colour are found to be metamicts. Not all green zircons are metamict: those from Burma, particularly, are fully crystallized 'normal' zircons.

The nature and position of the absorption bands in minerals is very largely influenced by their crystal structure, and this is well shown in the case of zircon. In Sri Lankan zircons the degree of internal decomposition towards the metamict state is reflected not only in a gradual lowering of SG but also in a lessening of the sharpness of the characteristic absorption bands. In the completely metamict state, zircons only show a rather vague broad band in the red, with sometimes a narrow band in the green at 520 nm, which is not part of the normal zircon spectrum.

Heating to redness intermediate zircons showing a woolly version of the normal spectrum bands has the effect (after cooling) of sharpening the bands considerably, and prolonged heating will cause the SG to rise. With low metamict zircons in which the normal spectrum has disappeared the effect of heating to a bright red heat (800°C) is quite different and most surprising. A strong series of absorption bands makes its appearance showing quite a different pattern from the bands in the normal spectrum, though the general appearance is much the same. In place of the usual strong line at 653.5 nm in the red, there are three strong bands in this region, centred at 687, 668.5 and 652.5 nm. Of the seven or so other bands in the anomalous spectrum, none quite corresponds with those in the normal spectrum. Particularly noticeable are two powerful bands in the blue at 473.5 and 451 nm. The two spectra can be seen for comparison in Figure 17.1. This astonishing change in the spectrum after heat

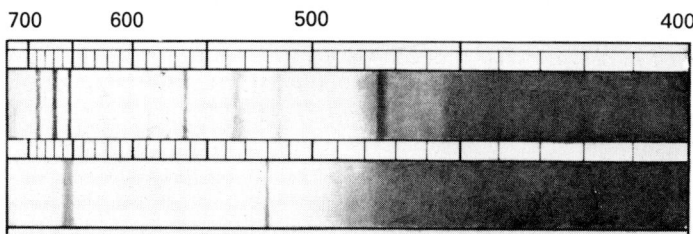

Figure 17.1 (*Top*) Anomalous spectrum of heat-treated metamict zircon; (*bottom*) typical spectrum of metamict zircon before heat treatment

treatment of low metamict zircons is accompanied by a *decrease* in the already low SG and refractive index. In terms of SG the dividing line between zircons showing a rise in the properties on heating and those which sink still further and develop the anomalous spectrum, is about 4.02. However, when heated for a considerable time at still higher temperatures (over 1100°C) even the lowest types of zircon tend to revert to normal.

X-ray investigations of zircons showing the anomalous absorption spectrum have been carried out by Vance and others. These indicate that this spectrum is produced by the presence of uranium in a lattice of tetragonal ZrO_2, which is the crystal phase produced when low-metamict zircon is strongly heated. It seems logical that in the few cases where the anomalous spectrum is seen in gem zircon the stones have in fact been subjected to high temperatures either by man or (less credibly) in nature.

Recognition of zircon

The properties of zircon have now been sufficiently described; it remains to consider how this popular species can be recognized and how it may be distinguished from other stones which somewhat resemble it.

Probably no stone, apart from opal and diamond, can be so readily identified at sight or with a pocket lens, as zircon. Its peculiar lustre, part adamantine due to its high refractivity, and partly greasy or resinous due to some special reaction to the polishing process, added to a fairly distinctive colour range, often makes it recognizable at a glance. Closer inspection with a lens will reveal the rubbed facet edges (see Plate 23), a slight milkiness, and, above all, the marked 'doubling' of the back facet edges when viewed through the front of the stone (Figure 17.2). The ability to recognize this 'doubling' effect and to gauge approximately the amount of double refraction in a faceted gemstone gives the practical gemmologist an enormous advantage over even an experienced jeweller or dealer who has no gemmological knowledge.

The only natural stones having comparable double refraction and colours falling within the zircon range are sphene and sinhalite. Both these are rarer and more highly prized than zircon, and their recognition, even with instruments, needs some skill.

Sphene is usually seen in shades of yellow or sometimes green or brown. Its double refraction is twice that of zircon, which help one to identify it. It has an exceptional degree of 'fire' and unlike zircon of this colour, it is distinctly dichroic. As with zircon, its refractive indices (1.90–2.03) are beyond the refractometer range. Its absorption spectrum may show faint lines in the yellow

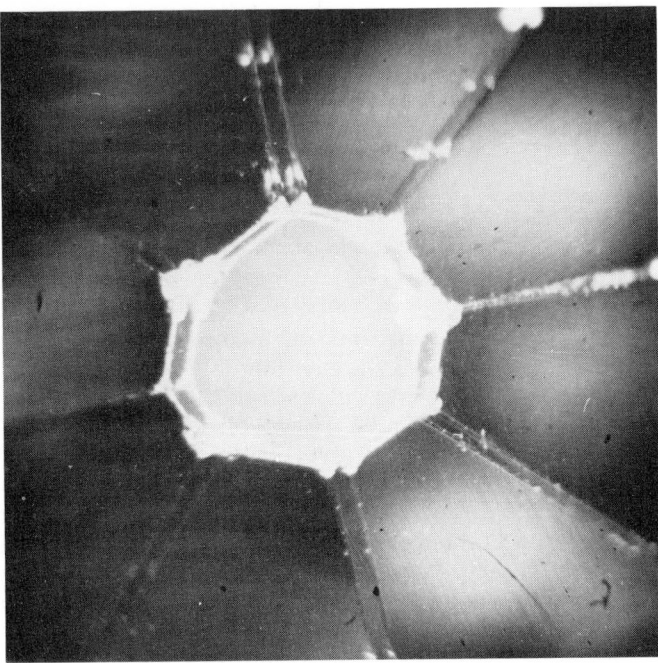

Figure 17.2 Double refraction seen through table facet of zircon (magnified 18 diameters)

due to the didymium metals, but there is nothing resembling the zircon spectrum. If the stone is unmounted, the lower SG of sphene (3.52) provides a useful test.

Sinhalite, from Sri Lanka, has only since 1952 been recognized as a separate mineral. For years, when its constants were measured, it was considered to be a brown or yellow form of peridot, richer in iron than the green variety. More often, specimens lurked unrecognized in parcels of chrysoberyls or of Sri Lankan zircons, where they had been placed on the basis of appearance only. Actually sinhalite is not a silicate at all, but a borate of magnesium and aluminium with a variable content of iron. Its colour varies from very pale straw yellow to an unattractive dark brown, the best stones being a golden or greenish-brown much resembling chrysoberyl or zircon, as indicated above.

The double refraction of sinhalite is 0.038 – that is, not so high as in zircon but quite strong enough to show marked doubling of the back facets in a stone of fair size. For sure identification a refractometer reading of approximately 1.670–1.708 will prove conclusive. A SG test with a loose stone will give a value near 3.48. To one familiar with the spectrum of peridot the bands seen in sinhalite must seem very similar, but an 'extra' band in the blue at 463 nm is a distinctive feature. Sinhalite has a hardness of 6½ and it shows little 'fire'.

Cassiterite, or tinstone, is very rarely transparent enough to be worth cutting as a gem, and is usually very dark brown in colour. However, due to its high lustre and strong double refraction (0.10), it can resemble brown zircon rather closely, which justifies its mention here. In pale stones its dispersion, which is higher than for diamond, gives it a lively appearance. When unmounted its phenomenal SG (6.9) makes it noticeably heavier in the hand even than zircon. Unlike zircon, it is dichroic; in fact, with dark specimens one ray may be so strongly absorbed that no doubling can be seen with the lens. There are no distinctive absorption bands in the cassiterite spectrum.

Peridot is another stone showing marked double refraction, though its colour and lustre are not those of green zircon. Where there is any doubt a refractometer test or observation with the spectroscope will at once settle the issue.

Demantoid garnet, with its high lustre and 'fire', can sometimes be confused with singly refractive green zircon, and the distinction is important since the garnet is far more highly prized. The 'horsetail' fibrous inclusions in demantoid, which are almost invariably present, makes it easy to recognize with a lens, and it has its own distinctive spectrum with a heavy band (which may be a cut-off) in the violet at 443 nm.

Of synthetic stones resembling zircon, **synthetic rutile** shows a far greater degree of 'doubling' and a superabundance of dispersion colours which is quite distinctive. **Blue synthetic spinels** have been made in colours to imitate blue zircon, but should not deceive the practised eye. There is, of course, no 'doubling' of the back facets in the spinel; it shows red through the Chelsea filter and exhibits a cobalt absorption spectrum.

Distinctions which are easy in stones of fair size become much more difficult in quite small specimens. Thus small zircons (sometimes rose cut) may represent diamonds in cluster surrounds rather deceptively, and, on the other hand, small white synthetic spinels have been (unintentionally) sold in goods advertised as 'zircon rings'. The spectroscope is most useful in such cases, as even the smallest white zircon will show the distinctive 653.5 line in the red by reflected light.

The properties of some individual zircons illustrating the entire range are given in Table 17.1. It will be noted that 'low' zircons are mostly green, and all from Sri Lanka. The last specimen on the list was loaned to Professor K. Chudoba, and used by him in establishing the reason for the strange variation in the properties of zircon in the research already mentioned.

Table 17.1   Properties of some individual zircons

Specimen	SG	RI	DR
Golden-brown, Burma	4.711	1.933–1.992	0.0588
Golden, Indo-China	4.697	1.926–1.985	0.0594
Brown, Burma	4.693	1.926–1.984	0.0580
Colourless, Sri Lanka	4.687	1.926–1.985	0.0590
Blue, Indo-China	4.690	1.927–1.986	0.0590
Dark red-brown, Sri Lanka	4.618	1.925–1.976	0.0510
Green, Sri Lanka	4.546	1.920–1.962	0.042
Brownish, Sri Lanka	4.472	1.910–1.938	0.028
Yellow-green, Sri Lanka	4.43	1.900–1.925	0.025
Green, Sri Lanka	4.30	1.870–1.888	0.018
Green, Sri Lanka	4.24	1.870–1.884	0.014
Green, Sri Lanka	4.12	1.832–1.849	0.017
Orange, Sri Lanka	4.008	1.823	None
Green, Sri Lanka	3.99	1.830	None
Greenish brown, Sri Lanka	3.98	1.818	None
Green, Sri Lanka	3.965	1.792–1.796	0.004

18

Topaz and other yellow stones

In times past the name 'topaz' has been applied, with and without prefixes, to yellow gemstones of several different species. In particular there has been an obstinate tendency in the trade to use the terms 'topaz', 'topaz-quartz' or 'quartz-topaz' as good selling names for the yellow quartz or citrine which so closely resembles it in appearance.

During the last few decades there has been a worldwide movement to establish an international nomenclature of precious stones based on sound principles and aimed at ensuring that to each gemstone is ascribed a correct name which is unequivocal and which cannot be applied to a stone of any other species. In Britain, at least, all argument is over since, in accordance with the Trade Descriptions Act 1968, it is now illegal to use the name 'topaz' in connection with any stone other than what is popularly known as 'Brazilian' or 'precious' topaz.

Let us now turn to a consideration of the characters of topaz itself. In nature this mineral is most often colourless, though frequently found in shades of yellow, and less commonly of blue and pale green. Sherry-coloured types from Ouro Preto, Brazil, when carefully heated, assume an attractive shade of pink, and it is these heat-treated stones which provide the majority of the pink specimens on the market, although natural pink stones have been found in Brazil and Pakistan. Topaz is one of the hardest of gems, being the standard 8 on Mohs' scale.

The refractive index and SG vary slightly, and in opposite senses, for yellow and pink stones, on the one hand, and white and blue, on the other, due to the presence of more fluorine and less hydroxyl in the latter. Yellow and pink stones have refractive index 1.63–1.64, double refraction 0.008, and SG 3.53, while for colourless and blue stones the corresponding figures are 1.61–1.62, 0.010, and 3.56. Dichroism is distinct in the pink stones, less strong in the yellow. Topaz takes an exceptionally high polish, and some have claimed that a cut topaz can be recognized by its characteristically slippery feel. The great Victorian gemmologist, Sir Arthur Church, relates in his privately printed memoirs how he was able to extract the topazes from a mixed bag of stones offered at half-a-crown (12½p) apiece in a junkshop by inserting his hand into the bag and applying 'the inner aspect of a well-educated thumb' to the stone.

Sherry-brown or golden-yellow topaz from Ouro Preto in Brazil contains traces of chromium and shows an orange fluorescence under long-wave ultra-violet light or between crossed filters. In the pink heat-treated stones derived from these the fluorescence is rather stronger and redder, as the chromium has then entered the crystal lattice. When this fluorescence is studied with the spectroscope a weak line at 682 nm can be seen, different in appearance and position from the clear-cut bright line seen in the fluorescence spectrum of sapphire.

When dealing with a large parcel of yellow stones, or yellow stones in mounted jewellery, this fluorescent property of topaz can prove very useful, but it must always be remembered that yellow sapphire from Sri Lanka shows an apricot fluorescence and synthetic yellow or orange sapphires (which are often used to represent topaz) may show a red chromium fluorescence under ultra-violet light or crossed filters.

As for the stones which may be confused with topaz, undoubtedly the **yellow quartz** (citrine), as mentioned above, is the most likely substitute. Relying on apperarance alone, it must be admitted that the two can often not be distinguished with any certainty, though the colour of topaz is usually richer and more subtle than that of quartz and it takes a higher polish. To emphasize how entirely different in every respect except appearance these two minerals are, let the reader glance at the characters of each in Table 18.1.

Table 18.1

	Composition	System	H	SG	RI
Topaz	$Al_2(F,OH)_2SiO_4$	Rhombic	8	3.53	1.630–1.638
Quartz	SiO_2	Trigonal	7	2.65	1.544–1.553

The figures given in the table for topaz apply to the yellow or pink types used in jewellery. The difference in form between topaz and quartz crystals can be seen in Figure 18.1, which also shows the typical appearance of water-worn pebbles of white topaz, which are plentifully found in Nigeria and other parts of Africa. These are often mistaken for diamond by hopeful amateur prospectors: their considerable hardness and their SG lend some plausibility to this idea.

If the stone is unset, the considerably higher SG of topaz may be perceptible when the stone is balanced in the hand; it is far more satisfactory, however, to make really certain by slipping the stone into either bromoform or di-iodomethane, when topaz will sink whereas quartz will float. Quartz indeed has so constant a SG, whatever its colour, that it is worth keeping a bromoform solution diluted with a suitable solvent to an exact SG of 2.65, controlled by the suspension of a specimen of rock-crystal quartz as an indicator. Then amethyst or colourless or yellow quartz can all be identified with virtually complete certainty in the liquid, as they too will remain suspended or rise or fall very slowly, as the indicator does. Incidentally, such a liquid will also serve to distinguish between quartz and chrysoberyl cat's-eyes and in the identification of synthetic emerald.

Figure 18.1 Topaz (*front, left to right*): crystals, cleavages and rolled pebbles, crystals; quartz (*back row*): crystals and rolled pebble

Should it be desired to test the stone in its setting, the refractometer, as usual, provides the most satisfactory, or rather, the *only* satisfactory means, though a fluorescence test can be very helpful, and will also serve to separate topaz from other stones or counterfeits of similar appearance. Apart from quartz, these are most likely to be yellow specimens of **sapphire, tourmaline, beryl**, and (less probably) **chrysoberyl** or **zircon**. Where pink topaz is in question, pink **tourmaline** or **sapphire** have to be considered, while blue topaz is often mistaken for **aquamarine** (which it closely resembles) or even, in the deeper blue types now available, for **sapphire** (see Chapter 14).

Colourless topaz is of small commercial importance, though a well-cut specimen may be quite attractive. It is not easy to differentiate at sight from colourless specimens of quartz, tourmaline, beryl, or corundum. To the above list must be added the counterfeits – **pastes** and **synthetic sapphires** of appropriate colour. Doublets representing topaz are seldom met with, though they should be borne in mind as a remote possibility.

Topaz and all the above-mentioned stones of like appearance with the exception of zircon give positive readings on the refractometer, and zircon, both on account of its distinctive appearance and its strong double refraction as seen with a lens, is hardly likely to cause difficulty in any case.

As may be seen from the figures shown in Table 18.2, **tourmaline** is the only natural mineral (disregarding such rarities as apatite and danburite) resembling topaz which gives shadow-edges in the same region as the topaz. Since the double refraction of tourmaline is double that of topaz, the two can be clearly separated by a careful reading of the maximum and minimum readings. With tourmaline, when the stone is rotated to the position where the two edges are most widely separated, they can be seen at two separate edges even in ordinary light on a standard refractometer. With topaz under these conditions the two edges are so close that they merge into one another, though a slight variation of the confused edge should be observable as the stone is rotated on the instrument.

Table 18.2   Identification of topaz

Species	H	SG	Mean RI	DR	Pleochroism
Topaz	8	3.53	1.63	0.008	Distinct
Zircon	7½	4.69	1.95	0.059	Weak
Corundum	9	3.99	1.76	0.008	Weak
Chrysoberyl	8½	3.72	1.74	0.008	Distinct
Paste[a]	5	3.7	1.63	None	None
Apatite	5	3.20	1.64	0.002	Strong
Tourmaline	7	3.10	1.63	0.018	Distinct
Danburite	7	3.00	1.63	0.006	Weak
Beryl	7½	2.68	1.57	0.006	Weak
Quartz	7	2.65	1.55	0.009	Weak[b]
Orthoclase	6	2.56	1.53	0.005	Weak
Brazilianite	5½	2.98	1.613	0.21	Weak

[a] Different pastes vary greatly in the RI and SG. This is one typical value.
[b] The majority of yellow quartzes sold owe their colour to heat treatment. These have practically no dichroism. Untreated citrine or smoky quartz has quite distinct dichroism.

Pastes of similar refractive index are not uncommon, and, having a higher dispersion than topaz, give a rather sharper and single shadow-edge. To be really certain of distinguishing between these three possibilities it is advisable to use sodium light (salt in a gas burner in a darkened room will do) or a strong light combined with a good colour filter to get critical readings. The special spinel refractometer made by Rayner is ideal for measurements in this particular region, giving clearly defined shadow-edges in ordinary light for the natural stones, and a colour fringe with paste due to its higher dispersion (the reverse effect, be it noted, to that seen in the standard refractometers using a lead glass hemisphere or prism). Assuming only an approximate reading of 1.63 can be obtained on the refractometer, examination of the stone under the microscope may assist the jewellers. Topaz is usually very 'clean', while tourmaline is seldom without characteristic flaw-like, liquid inclusions; the presence of spherical gas bubbles or swirl marks, on the other hand, will prove the specimen to be a paste.

Though not used commercially, **danburite** and **apatite** can both provide transparent yellow stones, and, since they have very similar indices to those of topaz, they should be briefly mentioned for the benefit of students. With a spinel refractometer danburite can be seen to have a distinctly lower birefringence (0.006) than topaz (0.008), while the very small double refraction of apatite (about 0.003) is entirely typical for the stone. Such nice distinctions are not easily made on a standard refractometer, even in sodium light, as the scale is more cramped. Where doubt exists, resort may be had to the spectroscope or to heavy liquids. Both danburite and apatite differ from topaz in floating in di-iodomethane, while in 3.06 liquid only the danburite would float. Yellow apatite shows quite a strong didymium spectrum (see Chapter 10), while danburite shows a similar spectrum, but so faintly as to be hardly visible. Brazilianite is another stone encountered occasionally. It is greenish-yellow in colour with refraction indices of 1.602–1.623.

The refractometer will clearly distinguish **beryl**, yellow **sapphire**, and **chrysoberyl** from topaz. Yellow chrysoberyl or yellowish-green chrysoberyl is a lovely, bright stone which is curiously neglected in jewellery. It often contains crystalline 'feathers', and shows a broad absorption band in the blue-violet which serves to distinguish it from corundum should the refractometer leave any doubt (Chapter 10).

Before attempting a summary of the above tests, it should perhaps be mentioned that topaz has a strong tendency to cleavage parallel to the base of the natural crystal, and this fact makes it important to handle these stones with a certain amount of care, since any sharp concussion may cause unsightly flaws to develop along cleavage planes, or even cause the stone to break. The writer has seen long drop-shaped topaz ear-rings which have broken across a cleavage plane, merely by being dropped on to the plateglass surface of a dressing table.

A brief résumé of the simplest methods by which topaz may be identified can now be given. The stone most frequently confused with yellow topaz is quartz (citrine). Where the stone is unset, a simple trial in bromoform will at once decide this particular issue, since quartz will float and topaz sink in the liquid. To distinguish topaz, whether yellow, pink, blue, or colourless, from all other substitutes it is only necessary to take a careful reading of the refractive index of the stone on a refractometer. If the reading is not in the 1.63 region the stone cannot be topaz; if it *is* near 1.63 then it may be topaz, tourmaline, or a paste. If two edges are discernible, showing a double refraction of about 0.02, the stone is a tourmaline. The sure distinction on the refractometer between topaz, with its small double refraction of 0.008, and paste (singly refracting) needs either monochromatic light, a good colour filter, or a spinel refractometer. If dichroism is present, or the specimen is found to scratch a polished quartz surface, it cannot be a paste. Under the microscope, paste will usually show a bubble or two, or swirl-like striae. Topaz, even if free from crystalline inclusions, should show enough doubling of facet edges as seen through the stone to be detected under the microscope. As a quick check for yellow topaz, its fluorescence between crossed filters should be remembered. If there is no fluorescence the stones cannot be topaz. A weak red or orange glow points to topaz with synthetic or natural sapphire as a possible alternative.

In Table 18.2 the rare gems danburite (pale yellow or colourless), apatite (yellow or blue), brazilianite, and yellow Madagascan orthoclase are included for completeness.

Other yellow gemstones

In the above discussion on yellow gemstones, topaz held the centre of the stage and other yellow stones were considered only in relation to this best-known representative of stones of this particular colour. However, there are several important yellow gems that certainly should not be thought of as topaz substitutes, and this is an appropriate place to describe these as individuals.

Yellow sapphire can be a beautiful gemstone; it has, of course, the great hardness (9 on Mohs' scale) of all corundum gems and at its best a most pleasing colour. Stones from Sri Lanka are probably those most commonly seen. Frequently these are too pale in tint to be valuable, but the deeper coloured specimens are, to the discerning eye, probably the loveliest of their kind. These usually have plentiful inclusions in the way of liquid 'thumbprint' feathers, zircons showing expansion cracks or 'haloes' indicating their metamict nature, and so on. Under long-wave ultra-violet light they exhibit an apricot-yellow fluorescence and virtually no sign of the absorption band at 450 nm seen in all the natural sapphires containing iron. Yellow sapphires from Thailand and from Australia, on the other hand, invariably contain iron and have a strong absorption complex in the blue-violet and *no* fluorescence. These latter varieties have a stronger colour than the stones from Sri Lanka and can be very handsome. In recent times strongly coloured yellow to yellowish-brown sapphires have appeared on the market which have neither the apricot fluorescence nor the line at 450 nm. These are heat-treated stones and the most common inclusions are dust-like particles arranged in definite directions.

During the past decade or so many yellow-orange sapphires have been irradiated and their distinction from natural and heated stones is far from easy. R. W. Hughes has described a fade test which may help. Sapphires are exposed to the radiation from a 150-watt spotlight for up to one hour, the stones being brought to within 1 cm of the light source. Some stones may darken when hot and return to the original colour on cooling, others may fade and some may not change at all. This test involves some risks and permission should be sought from the owner beforehand. Hughes has found the results set out in Table 18.3.

Table 18.3   Fade test on yellow-orange sapphires

Origin	Reaction during fade test
Natural Sri Lankan yellow-orange sapphires	No change in most. Some may show some fading
Natural or heat-treated yellow-orange sapphires from Thailand, Australia, Tanzania	No change
Irradiated Sri Lankan yellow-orange sapphires	The colour fades within one hour
Heat-treated Sri Lankan yellow-orange sapphires	The colour temporarily becomes darker and more brownish. As the stone cools to room temperature, the colour reverts to its original state

Synthetic yellow sapphires are notorious among practising gemmologists for showing virtually no completely conclusive signs of synthesis. Though grown by the Verneuil process the boules are developed slowly and only show curved striae with difficulty even by the photographic method recommended in Chapter 9. The entire absence of fluorescence, absorption bands, or inclusions is a strongly suspicious sign, as is the crude machine cutting, usually with thick girdles, conforming with exact millimetre diameters or lengths.

It should be remembered that pale Sri Lankan sapphires assume a darker and richer tint of yellow after a few minutes' exposure to X-rays. The colour, however, fades rather rapidly in sunlight, and the process is thus not considered commercially legitimate. Unfortunately only a fading test will prove the fact that a yellow sapphire has been treated, and this procedure is never welcomed by an owner and would-be vendor.

Yellow chrysoberyl, when of fine quality, is another magnificent gemstone. Chrysoberyl, with its hardness of 8½ on Mohs' scale, is hard enough to take a fine polish and to retain it indefinitely under normal wear. In colour it varies from pale straw yellow through the desirable golden tint to stones of green or brown colour, iron being the key to these variations. Although its refractive indices (1.747–1.756) are only a little lower than those for corundum and its birefringence is nearly the same, a careful refractometer reading should leave one in no doubt as to the identity of each species. The density of chrysoberyl is fairly constant at 3.72, and another very useful identification feature of the stone is the broad absorption band centred at 444 nm in the violet. This is noticeably different from the sapphire spectrum in consisting of a single band and not a complex of three bands stretching into the blue. Chrysoberyl of this colour shows no fluorescence, as the iron content precludes it. Small, pale chrysoberyls of indifferent quality, probably coming from Russia, are not infrequently met with in Victorian jewellery, and in past days were known as 'chrysolites'. The finest specimens today of yellow chrysoberyl are those from Brazil and Sri Lanka. Incidentally, when pure, chrysoberyl is quite colourless and it is worth noting that specimens of this type are sometimes found in Burma and in association with diamond in alluvial deposits.

Yellow beryl is another stone which, at its best, has great beauty, and sufficient hardness to take and retain a good polish. In common with beryls other than emerald, it has a peculiarly limpid quality that is most attractive. Exceptional transparency is needed to enable a stone to show this effect. Pale yellow beryls are rather insipid; those of golden colour to which the name 'heliodor' has been applied are most prized. Yellow beryl has no particularly striking characteristics. A refractometer reading is the easiest means of identification, 1.570 and 1.575 being typical values for the two indices, the stone being optically negative. The density is commonly 2.68, one of the lower values for a gem beryl. It should perhaps be mentioned that the feldspar **labradorite** is sometimes found in transparent pale yellow pieces and has a density and refractive indices very close to those of beryl. The double refraction, however, is twice as large and the intermediate ray is almost half-way between the other two, so that its biaxial character should be noticeable if a careful reading is taken. The colour of the feldspar is insipid, and its is a collector's item rather than a commercial gemstone.

Yellow tourmaline is sometimes seen, and can easily be identified on the refractometer. Both the refractive indices and the density are rather higher than most of the transparent tourmalines, typical values being 1.623–1.643 for the refractive indices, uniaxial negative, and 3.10 for the density.

Sphene is typically yellow and **zircon** often so; each shows strong doubling of the back facets and each has refractive indices higher than 1.81. Sphene has a density of 3.53, only slightly higher than diamond, whereas yellow zircons are usually in the higher range for this mineral, with density between 4.65 and 4.70. Zircon can be recognized by its absorption spectrum, though in heat-treated types this may consist only of the 653.5 nm line accompanied by the line at 659 nm. In sphene the only absorption lines seen will be faint didymium lines in the yellow.

Yellow apatite will show the didymium lines as two groups in the yellow and in the green much more clearly, and its refractive indices of 1.634–1.637 with the low birefringence of 0.003 are distinctive.

Yellow danburite, with refractometer readings of 1.630 and 1.636, is close at hand, and being a calcium mineral may also show faint didymium lines, so a density test (apatite 3.21, danburite 3.00) may be necessary in confirmation. These last two stones are collector's items and are hardly ever used in jewellery.

19

Pink, mauve, or lilac stones

The simple methods by which colourless, red, blue, green, pale blue-green, and yellow stones can be identified have now been considered and we can turn to the pink, mauve, or lilac-coloured gems which are occasionally met with in jewellery. These include **kunzite** (lilac-coloured spodumene), **morganite** (rose-coloured beryl), **pink topaz** (see Plate 21), **pink tourmaline, pink sapphire**, and pale mauve specimens of **amethyst, scapolite** and **spinel**. **Synthetic corundum** and **spinel** of appropriate colour, **pastes**, and **doublets** must also be considered in this group.

Some of these are very difficult to identify by sight alone, though the following considerations may assist in forming a decision if proper tests are not available.

Kunzite is more lilac than pink, and usually rather pale in tint. On examination with a lens it should reveal distinct 'doubling' of the back facet edges, as the double refraction (0.015) is moderately strong. It is a difficult stone to cut, on account of its ready cleavage, and should be handled with care for the same reason. In the hands of a skilful lapidary it takes a good polish, and forms bright and attractive stones. Under long-wave ultra-violet light, kunzite shows a distinct orange glow. The dichroism is unusually strong for so pale a gemstone.

The attractive pink beryl called **morganite** is more rose-coloured than lilac and red beryl is now found in Utah, USA. As with kunzite, the colour is commonly pale, and, in common with other pale beryls, has a curious 'liquid' clarity. This variety is usually very 'clean' and the double refraction is not strong enough to be detected with a lens, save in very large stones. Morganite tends to have a higher refractive index and markedly higher SG than other beryls. **Pink topaz** of the deep colours found in Pakistan or that produced by heat treatment of certain brownish yellow Brazilian stones, can be very attractive; it has a more pronounced colour than either kunzite or morganite, takes a very high polish, and has the characteristic 'slippery' feel of all cut topaz. It is commonly cut in elongated oval shape, in accordance with the shape of the original crystal.

Pink tourmaline closely resembles the topaz, and is cut in elongated shapes for the same reason; it is seldom quite 'clean' and can often be distinguished by its typical feathery, flaw-like inclusions (Figure 23.2 in Chapter 23). The stronger double refraction (0.018) may also be noted, causing a more pronounced 'doubling' than with topaz. Pink corundum, usually called **pink sapphire**, since it owed its colour chiefly to a trace of chromium, may in one sense be considered merely a pale form of ruby. But almost all true pink sapphires come from Sri Lanka, and as such have a hint of blue in their make-up which results in a truly pink hue which, even when quite deep in tone, is distinct from the true ruby-red. The decision whether to name a given borderline specimen 'pink sapphire' or 'ruby' must often depend on subjective opinion. Judgement must always rest upon its appearance in daylight. Unfortunately, such a decision may not be merely a talking point but of some commercial importance, due to the prestige value of the name 'ruby'. Similar considerations make the distinction between 'emerald' and 'green beryl' and 'alexandrite' and 'green chrysoberyl' sometimes both difficult and commercially significant. As far as a scientific test can help to decide the issue it may depend in each of these cases upon the chromium content of the specimen, and the spectroscope may be helpful. Pale **amethysts** are not often cut, since those of a deeper tint are so plentiful, but some specimens are very similar to kunzite in hue, though less brilliant. The colour is often unevenly distributed in these pale amethysts, being located in zones or patches. Mauve to **purple scapolite** from East Africa has now appeared on the market; it closely resembles amethyst at first sight. The refractive indices (1.532–1.539) are lower than quartz, and the optic sign is negative in contrast to positive quartz. The SG is also lower at 2.59.

Synthetic corundum is manufactured in shades of pink – usually in an attempt to simulate pink topaz, for which it looks 'too good to be true'. It may prove hard to be certain that a very clean pink sapphire is synthetic. The laboratory worker is lucky in being able to note its continuing red glow after exposure to X-rays. **Synthetic spinel** when well cut forms a very bright and attractive gem indeed, and is not quite akin in appearance to any of the above, though natural spinels are sometimes found in tints of pale mauve or lavender, so that a refractometer test (see Table 19.1) is advisable. Synthetic pink spinel owes its colour (rather surprisingly) to ferric iron, and shows two distinct absorption bands: in the yellow at 580 nm and a stronger band in the green centred at 555 nm. Pastes and doublets are not uncommon in shades of pink, and may well deceive at first sight, though usually not after careful scrutiny by a practised observer. Pink **cubic zirconia** may be encountered but its high lustre, dispersion, and negative reading on the refractometer should serve to identify it.

Now let us consider the simplest tests by which these similar-looking stones can be separated and identified with certainty. On this occasion the table of constants, in which the stones are listed in order of diminishing SG, will be given first.

Glancing through Table 19.1 one may see at once that, if the stones are unset, successive trials in bromoform and di-iodomethane will separate morganite and amethyst from all the others, and kunzite and tourmaline from topaz, spinel, cubic zirconia, and corundum. If, as has already been recommended, a special bromoform solution matched for quartz is kept handy, amethyst can be identified outright by this means.

Once more, however, a careful refractometer reading will provide the surest single test, and one that can be applied without removing the stone from its setting. Where monochromatic (sodium) light, or white light in combination

Table 19.1   Identification of pink, mauve, or lilac stones

Species	H	SG	Mean RI	DR	Pleochroism
Pink corundum	9	3.99	1.76	0.008	Weak
Synthetic spinel	8	3.63	1.727	None	None
Natural spinel	8	3.59	1.715	None	None
Pink topaz	8	3.53	1.63	0.008	Distinct
Kunzite	7	3.18	1.67	0.015	Distinct
Pink tourmaline	7	3.04	1.63	0.018	Distinct
Morganite	7½	2.80	1.59	0.008	Weak
Amethyst	7	2.65	1.55	0.009	Weak
Scapolite	6	2.59	1.54	0.007	Distinct
Cubic zirconia	8½	5.65	2.17	None	None

with a good colour filter can be used, the refractometer test alone is practically decisive for all these stones. Natural and synthetic corundum give identical readings, and may need to be examined under the microscope for a sure distinction, but synthetic spinels have a perceptibly higher index than the pale mauve natural spinels, and show anomalous double refraction between crossed polars.

If a reading is given in the 1.63 regions care must be exercised to separate tourmaline, with its large double refraction of 0.018 from topaz (only 0.008) and from paste, which has none and thus gives only a single edge. Doublets will almost invariably give a shadow edge near 1.79, representing the almandine garnet which forms the table, while the back facets, if accessible, will give some such value as 1.62, according to the type of glass used for the main body of the doublet. Kunzite has sufficient separation of the shadow edges to be clearly distinguishable from paste, the only competitor in this region of the scale for specimens of pink colour. Morganite gives higher readings and larger double refraction than other beryls, and its SG is likewise higher, due to the presence of rare alkali metals.

The figures given in Table 19.1 for morganite are representative but not invariable, so that jewellers must not be unduly troubled if their readings do not quite correspond in this instance; they will in no case overlap those of any other pink stones. As for other simple tests, the great hardness of sapphire, and the softness of paste, may be remembered. The differences in strength of dichroism in pale-coloured specimens are difficult to assess, but if perceptible dichroism is seen one may at least be sure that the specimen is not a paste, doublet, or spinel of any kind.

In summing up, one may say that of all these pink stones, kunzite and amethyst show the greatest tendency to lilac or violet tints. Without recourse to other tests, *kunzite* can usually be distinguished by its depth of cutting, distinct dichroism, and considerable double refraction, *tourmaline* by its typical inclusions and double refraction, *synthetic spinel* and *cubic zirconia* by their bright, clean appearance, *pink sapphire* by its silk and feathers, and *pink topaz* by its long oval shape and slippery surface. The most decisive test for all these stones is provided by the refractometer, with subsequent microscopic examination where necessary to separate natural from synthetic corundum. Where stones are free from their setting, a trial in heavy liquids will assist in deciding the issue, while a fluorescence test may either act as a useful guide or provide a short cut to a complete identification.

20

Brown and orange stones

In the previous chapters on practical testing methods applied to particular cases we have already dealt with gems of almost every colour. Brown or orange stones, however, form a category which has not yet been covered, and since there are several gem species quite commonly found in these colours it will be as well to devote a chapter to their consideration. Practically the only gemstone of true *orange* colour is the fire opal; those best described as *orange-brown* include hessonite garnet or 'cinnamon stone', once sold as 'jacinth', and zircon closely resembling it in colour, to which the term 'jacinth' more properly applied; while the *brown* stones most commonly seen are tourmaline and chrysoberyl, apart from quartz and topaz, which can be distinguished by the methods described in Chapter 18. Rarer stones, spessartine and sinhalite, sphalerite, and sphene add an alliterative flavour to the chapter, while andalusite, axinite, cassiterite, enstatite, idocrase, kornerupine, and brown diamond can come under this colour category.

As few of these stones are very costly, counterfeits are not often met with, but a synthetic corundum of peculiar brownish-orange tint, to which the name 'padparadscha' (variously spelt) has been applied, should be mentioned, though whether intended to represent garnet, topaz, or zircon is something of a mystery. The author once encountered a complete necklace of these things, faceted and mounted in gold. Natural corundums which might be described as orange are only very occasionally found.

Unlike other opals, **fire opal** usually shows no play of colour; it has little brilliance to recommend it, and thus its chief claim to beauty lies in its magnificent flame-red colour. Again there is dispute over the terminology, with one school adhering to this definition while the other says 'fire opal' must have a play of colour, and if there is no play of colour the stone should be called 'orange opal'. The best specimens are almost completely clear, though a slight milkiness is often apparent. The cutting most usually adopted is a step or mixed cut with a slightly domed table facet. Being amorphous, fire opal shows no dichroism or double refraction, and in this it tallies with the imitations sometimes made in **glass** or one of the plastics such as **bakelite**.

If a stone is under suspicion it is best to remove it from its setting and test its refractive index and SG, both of which, in true fire opal, are fairly constant and characteristic. It may be possible to obtain a refractive index reading for fire opal even if it has a domed table by using the 'distant vision' technique, though this is usually more easily carried out with stones having a surface with steeper curvature. Under the microscope, glass will usually reveal bubbles or striae, and any of the plastics will peel under the blade of a pen-knife, carefully applied – the degree of toughness varying according to the type of plastic (see Chapter 27).

The commonest plastics vary in SG from 1.25 to 1.45, far lower than fire opal (2.00), and, conversely, they have a higher refractive index – 1.49 to 1.65 as against 1.45. Such imitations are seldom seen except in costume jewellery. Other tests for fire opal will be found in Chapter 21.

The two stones to which the unsatisfactory name 'jacinth' was once commonly applied, namely, orange-brown **hessonite** garnet and **zircon** or 'jargoon', are very similar in appearance. The cinnamon garnet (the jeweller's 'jacinth') can usually be recognized by its multitude of small inclusions, which give it a peculiar granular appearance (Figure 22.3), while zircon has its own characteristic resinous-adamantine lustre and large double refraction. The dichroscope is of little value here, as garnet is a cubic mineral and therefore non-dichroic, and no zircon except the blue variety shows perceptible dichroism. On the refractometer, hessonite normally gives a reading near 1.743, but the presence of the almandine molecule may raise the index. Almandine-rich hessonites have a distinctly reddish tint, and show the easily recognized almandine absorption spectrum.

A distinct possibility is that a seeming hessonite may turn out to be another member of the large garnet family: **spessartine**. Fine specimens of this manganese garnet are scarce, though not so scarce as formerly, and command relatively high prices among discriminating collectors. The refractive index and SG of spessartine are very near those for almandine (e.g., 1.79 and 4.10), but the colour is quite different, being yellow, orange, or orange-brown, instead of deep, purplish red, and bands due to manganese may be seen in the blue and violet of its absorption spectrum (Chapter 10). Sometimes mixed garnets containing hessonite and spessartine or almandine and spessartine molecules may be encountered. **Zircon**, with its high refractivity, gives a 'negative' reading on the ordinary refractometer – that is, a shadow extending right up to the edge representing the contact liquid used. This, if considered in conjunction with the appearance of the stone, is virtually conclusive, but brown **sphene** should be borne in mind as a possibility, since this also will give a negative refractometer reading, and shows even stronger double refraction, though this is sometimes masked, as far as doubling of facet edges is concerned, by the complete absorption of one ray. The dichroscope here will settle any doubt, since brown sphene is strongly dichroic while zircon of this tint is virtually non-dichroic. The absorption spectrum of zircon, if seen, provides conclusive evidence.

Brown specimens of **peridot** are rarely seen in jewellery, although among the mainly green pebbles recovered in Arizona deposits brownish iron-rich samples are not uncommon in the rough. By chance, tests on two such stones gave similar high figures for refractive indices and density (quoted below). There can easily be confusion here with the mineral **sinhalite**, which was first recognized as a new mineral in 1952. This was the second occasion on which a new species was initially established from the examination of faceted gemstones: taaffeite (discovered in 1945) was the first case of such an unexpected happening. Sinhalite is a magnesium–aluminium borate and thus completely different in composition from the peridot group, which consists of silicates of magnesium and iron. However, the (orthorhombic) structure of the two minerals is very similar, and this explains the close relationship that exists in their physical properties. The constants for sinhalite are, in fact, far less liable to variation than those of the peridots since there is less tendency to alterations in chemical composition – in particular, in the amount of iron present.

The lowest values for sinhalite, found in a golden-brown specimen of 17½ carats, were 1.667, 1.697, and 1.705 for the three critical refractive indices, with density 3.465. The highest, in a brownish-green stone, were 1.676, 1.704, and 1.712, with density 3.52. The lower set of figures compares closely with those for the brown Arizona peridot specimens mentioned above, which had 1.668, 1.674, and 1.704 for the three indices, and density 3.43. In such a case, separation by means of refractive index readings depends upon the amount of movement seen in the shadow-edge representing the higher refractive index: in sinhalite the value for the intermediate 'beta' index is much nearer the maximum or 'gamma' edge than the lowest or 'alpha' edge, whereas in peridot the value for the 'beta' index is nearly half-way between the greatest and the least index. Fortunately, a study of the absorption spectra of the two species will aid materially in coming to a decision in the few cases where doubt exists. Peridot has three evenly spaced bands in the green and blue centred at 493, 473, and 453 nm, whereas in sinhalite the absorption bands are centred at 493, 475, 463 and 450 nm; the 'extra' band at 463 nm is the tell-tale feature.

At its best, sinhalite provides fine golden- or greenish-brown stones, clean and often of considerable size. After the announcement of its discovery specimens were found in parcels of chrysoberyls and of zircons of similar colour. Only its low hardness of 6½ on Mohs' scale prevents its competing favourably with better-known species.

If we consider every possibility, **diamond** and **cassiterite** are two other minerals which give negative refractometer readings and can fall in this colour category. Brown diamond, with its brilliant adamantine lustre, single refraction, and incomparable hardness, should be easy to distinguish, though the colour may have been induced by particle bombardment followed by suitable heat treatment. Methods for detecting treated diamonds are discussed in Chapter 12. Cassiterite, valuable as an ore of tin, as a gemstone must be considered a freak. As can be seen from Table 20.1, it has a SG far higher even than zircon.

The two other common gemstones which come under this colour group, brown **tourmaline** and brown **chrysoberyl**, are often very difficult to distinguish at sight. Both are more sheerly brown than the stones already discussed, and are indeed not very attractive. Tourmalines of this type absorb a great deal of light, and for this reason are very 'dead' in appearance. The doubling effect which can be detected with a lens in most tourmalines cannot be seen here, since one of the two doubly refracted rays (the 'ordinary' ray) is completely absorbed within the stone. The dichroscope image corresponding to this ray is therefore practically black, in contrast to the other image, which is brown. The refractometer as usual provides the most positive test for these two minerals; tourmaline showing two shadow-edges near 1.62 and 1.64 at their widest separation, while chrysoberyl has much higher indices, near 1.74 and 1.75, with a double refraction too small to be distinguished unless monochromatic light be used. Here again, an exciting but unlikely possibility exists for the wide-awake jeweller. The mineral **andalusite**, sometimes cut as a gem, has properties close to those of tourmaline, and may present a very similar appearance, though it is an entirely distinct mineral. Tourmaline belongs to the trigonal (rhombohedral) system of crystal symmetry, while andalusite is orthorhombic – a fact which provides the mineralogist with quite conclusive optical means for distinguishing the two. However, a simple refractometer reading, if carefully taken, will also enable a definite distinction to be made, since tourmaline of 'andalusite' colour (brown or brownish-green with hints of red) will always show a double refraction nearly twice as great as that of andalusite. The matter is given space here because dealers have sometimes attempted to satisfy collectors' inquiries for an andalusite with tourmalines of suitable appearance from their stock, not realizing that the collector is a discriminating person, and desires an andalusite *qua* andalusite, and not simply for its appearance!

Other brown or orange stones occasionally cut for collectors are **axinite**, **blende** (sphalerite), **enstatite, idocrase, kornerupine**, and **scheelite**. Axinite is usually dark brown in colour and exhibits intense pleochroism. Cuttable material has come from France and Baja California, Mexico. Blende is a cubic mineral having a refractive index nearly as high as that of diamond, and its dispersion is considerably greater, but its softness and its ready dodecahedral cleavage make it difficult to cut to good advantage and incapable of standing any wear. Its high single refraction distinguishes it from everything save diamond (and possibly cubic zirconia), and, once seen, its appearance is unlikely to be forgotten. The colour shown by transparent pieces is a magnificent golden-brown. Occasionally it displays an absorption spectrum with one or more

Table 20.1   Identification of brown and orange stones

Species	H	SG	Mean RI	DR	Pleochroism
Cassiterite	6½	6.9	2.04	0.10	Distinct
Scheelite	5	6.0	1.93	0.017	Distinct
Zircon	7½	4.69	1.95	0.059	Very weak
Spessartine	7	4.16	1.80	None	None
Blende	3½	4.09	2.37	None	None
Corundum	9	3.99	1.76	0.008	Weak
Chrysoberyl	8½	3.71	1.75	0.009	Distinct
Hessonite	7¼	3.65	1.74	None	None
Topaz	8	3.53	1.63	0.008	Distinct
Sphene	5	3.53	1.96	0.12	Strong
Diamond	10	3.52	2.42	None	None
Sinhalite	6½	3.48	1.69	0.038	Weak
Idocrase	6½	3.38	1.70	0.005	Distinct
Kornerupine	6½	3.31	1.67	0.013	Distinct
Axinite	6½	3.30	1.69	0.011	Intense
Enstatite	5½	3.25	1.66	0.008	Distinct
Andalusite	7½	3.15	1.64	0.010	Strong
Tourmaline	7	3.07	1.63	0.018	Strong
Quartz	7	2.65	1.55	0.009	Weak
Fire opal	5½	2.00	1.45	None	None

narrow bands in the red. This is mentioned lest the beginner might confuse these with rather similar bands seen in zircon.

Idocrase at its best can be an attractive gem and is sufficiently hard to take and maintain quite a good polish. Stones from the Laurentian Mountains in Canada are of a clear golden-brown. They may be distinguished by their refractive index of a little above 1.70, with a very small double refraction (0.005). Some specimens may show a faint didymium absorption spectrum.

Enstatite, a pyroxene mineral, is commonly seen as cut stones in shades of brown, but colourless facetable material has now been found in Sri Lanka. Green chrome-bearing material originates from the diamond pipes of South Africa. Enstatite shows a strong absorption line in the green at 506 nm and this line is present even in almost colourless stones.

Kornerupine comes from the gem gravels of Sri Lanka in shades of yellowish to greenish-brown and as brownish cat's-eyes. Bright green chrome-bearing material has been found in Kenya. Kornerupine has refractive indices in the 1.67–1.68 region and care must be taken not to confuse it with the pyroxene minerals diopside, enstatite, and even spodumene.

Finally, **scheelite**, an important ore of tungsten, should receive mention here, since transparent pieces of yellow or orange hue are sometimes cut for collectors, and can present a magnificent appearance. Its low hardness invalidates it for serious use in jewellery. The double refraction is marked enough to be visible with a lens, and this distinguishes it from diamond, while it is not strong enough to be confused with zircon. The absorption spectrum of scheelite often shows the group of fine lines in the yellow characteristic of the rare-earth metals 'didymium' – an indication that it is a calcium mineral. Its distinction from **synthetic scheelite** is discussed in Chapters 9 and 12.

The properties of this and other brown or orange minerals can be found summarized in Table 20.1, in which the stones are arranged in order of decreasing SG.

Plate 1 Rough diamonds fluorescing under ultra-violet light (photo: R. Webster)

Plate 2 Star stones. (*Top row*) Natural sapphire and ruby; (*bottom row*) synthetic ruby and sapphire; (*right*) star diopside

Plate 3 Naturally coloured cut diamonds (0.57–1.93 carat)

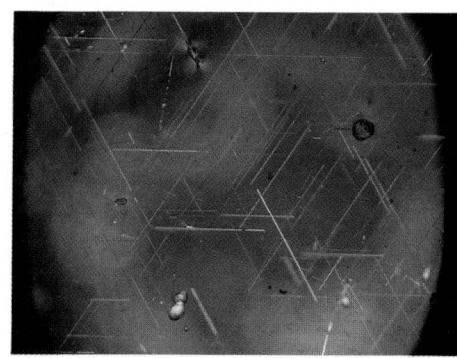

Plate 4 Sri Lankan ruby showing long rutile needles, rounded crystals, and zircon haloes

Plate 5 Thai ruby showing twinning planes by transmitted light and their outcrop at the reflected surface

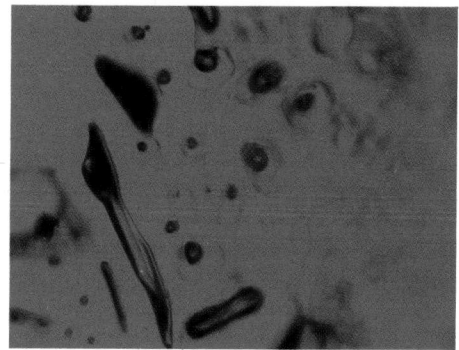

Plate 6 Bubbles in high relief surrounded by low-relief flux remnants in Knischka synthetic ruby

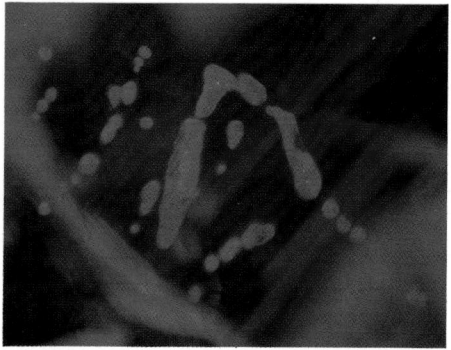

Plate 7 Orange-yellow flux remnants in Ramaura synthetic ruby

Plate 8 Hexagonal platinum plates in Chatham synthetic ruby

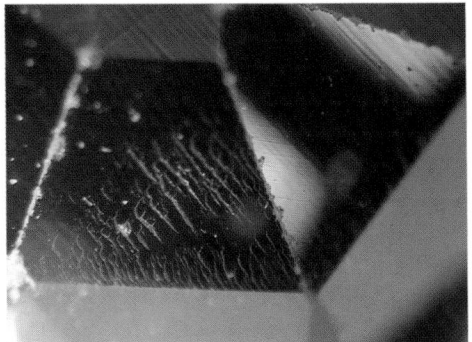

Plate 9 'Fire marks' and fine parallel surface polishing marks in Verneuil synthetic ruby

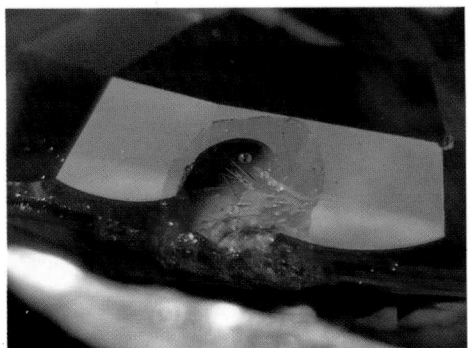

Plate 10 Glass-infilled cavity in natural ruby. Note the lustre contrast

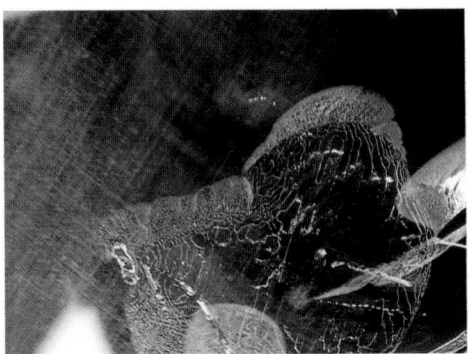

Plate 11 'Feathers' and fine rutile needles in Sri Lankan sapphire

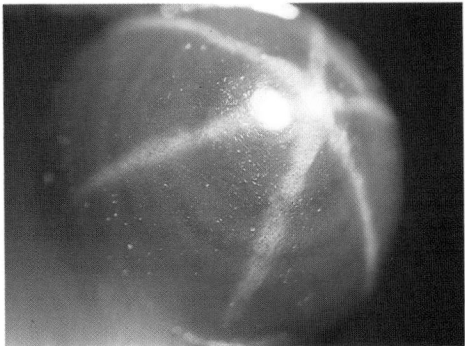

Plate 12 Verneuil synthetic star sapphire showing curved growth lines

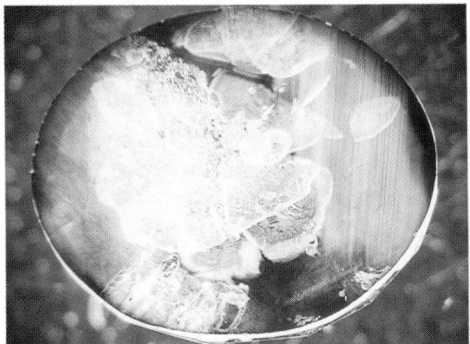

Plate 13 Diffusion-treated sapphire showing colour-enhanced girdle area

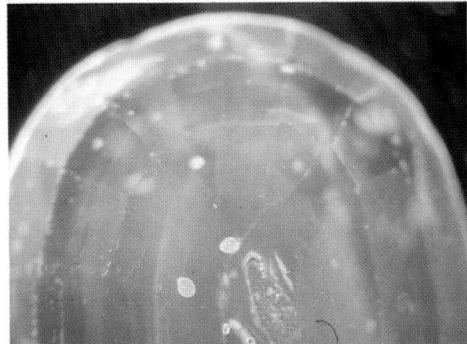

Plate 14 Diffusion-treated corundum showing some facets with coloured surface layers polished away

Plate 15 Two-phase inclusions and tremolite needles in Zambian emeralds

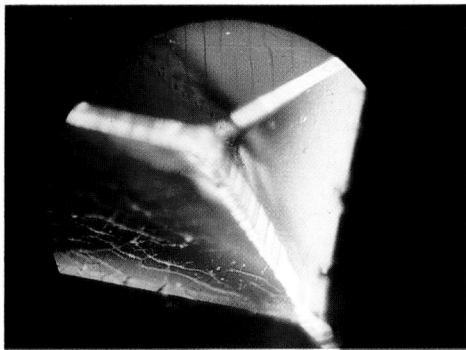

Plate 16 Surface cracks in Lechleitner coated emerald

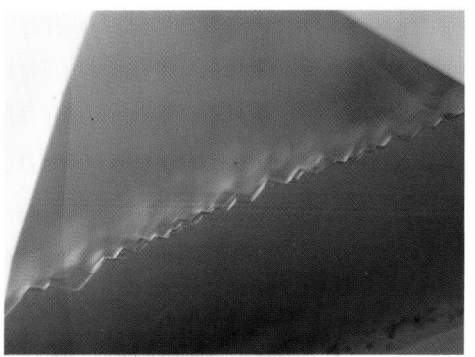

Plate 17 Typical 'mountain peak' growth pattern in Biron hydrothermal synthetic emerald

Plate 18 Parallel growth bands and 'arrow head' growth patterns in Russian hydrothermal synthetic emerald

Plate 19 'Soudé emerald' showing dark green central layers with colourless quartz base and crown

Plate 20 Imitation emerald crystal made from a quartz crystal with a green plastic coating with embedded mica and tourmaline crystal

Plate 21 (*Left*) Topaz (24 carats); (*right*) morganite (45 carats) and spinel (21 carats)

Plate 22 (*Centre*) Fibrolite; (*left*) kyanite; (*right*) andalusite (Al_2SiO_5 polymorphs)

Plate 23 Zircon showing wear on facet edges

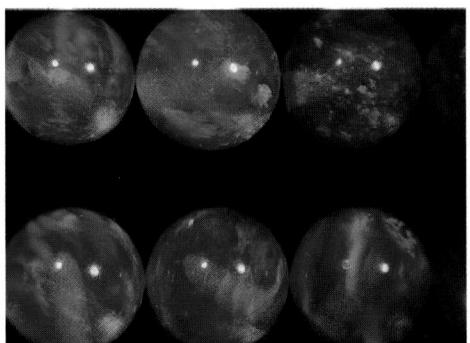

Plate 24 Gilson synthetic opal beads showing columnar structure

Plate 25 Gilson synthetic black opal showing 'lizard skin' structure and crenulate margins

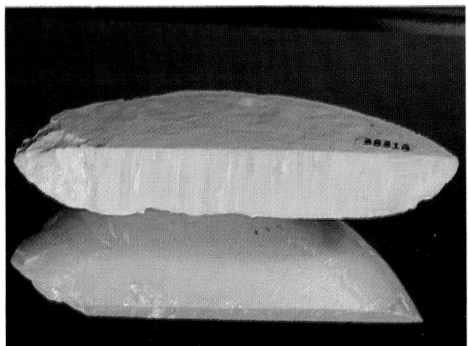

Plate 26 Sectioned saucer-shaped mass of Gilson synthetic opal showing columnar structure

Plate 27 Weathered gray charoite cut to show internal purple colour with a vase

Plate 28 Opal simulants. (*Front row, left to right*) Opal doublet and triplet, treated opal, Slocum stone; (*back row*) Gilson synthetic opal (2). 'plastic opal', Slocum stone

Plate 29 Chalcedony group. (*Top, left to right*) Sardonyx, Mocha stone, plasma, cornelian necklace (with sard and bloodstone); (*centre*) cornelian, onyx, chrysoprase, agate; (*bottom*) botryoidal chalcedony, chrysoprase vein

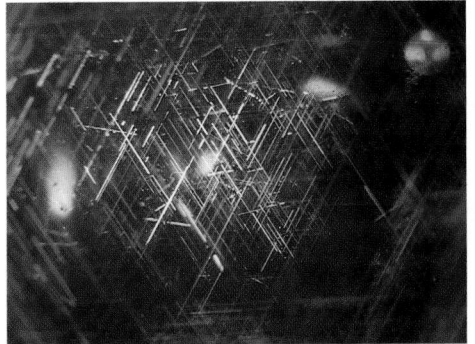

Plate 30 Rutile needles in almandine garnet

Plate 31 'Mantilla shawl' veils in spessartine garnet

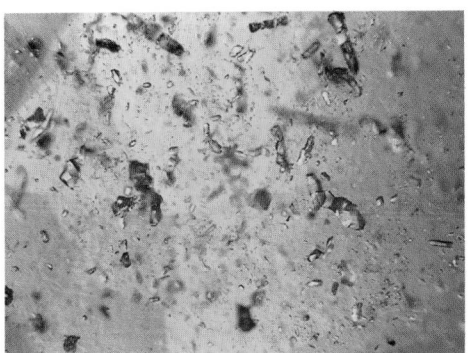

Plate 32 Granular apatite and calcite inclusions in hessonite garnet

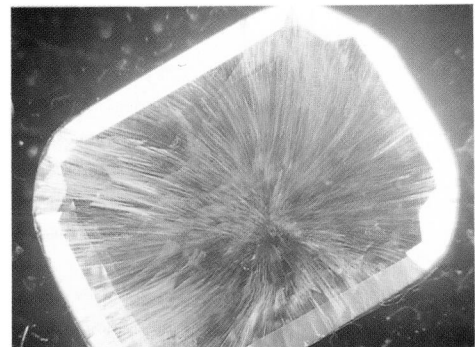

Plate 33 'Horsetail' asbestos (byssolite) inclusions in demantoid garnet

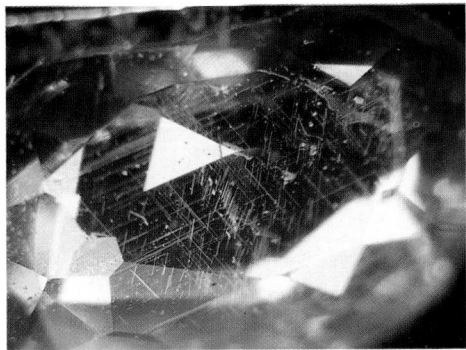

Plate 34 Garnet-topped doublet showing pink garnet with rutile inclusions and bubbles at junction with underlying greenish-blue glass

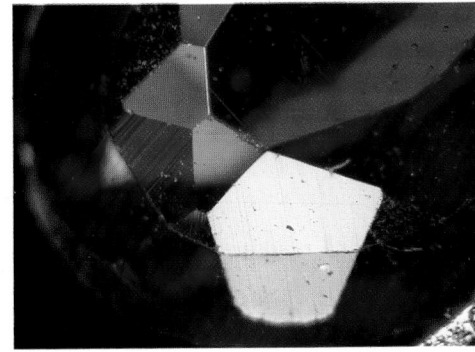

Plate 35 Garnet-topped doublet showing the higher surface lustre of the garnet top

Plate 36 Colour range in pyrope–almandine garnets

Plate 37 Colour range in grossular garnets

Plate 38 Colour range in tourmaline

Plate 39 Range of British ambers – mostly water-worn pebbles and boulders as collected

Plate 40 Andradite garnets: green demantoid, yellow topazolite and black melanite

Plate 41 'Water lily' pad in peridot

Plate 42 Colour range in Burmese spinels

Plate 43 Moonstones showing blue and white sheen

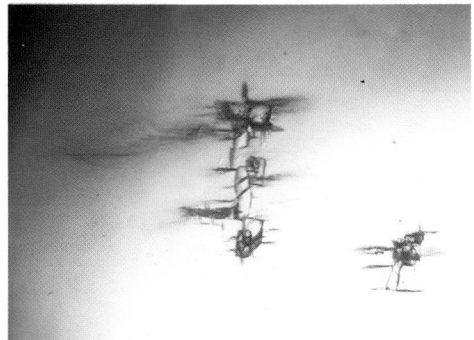

Plate 44 'Centipede' inclusion in moonstone – transmitted light

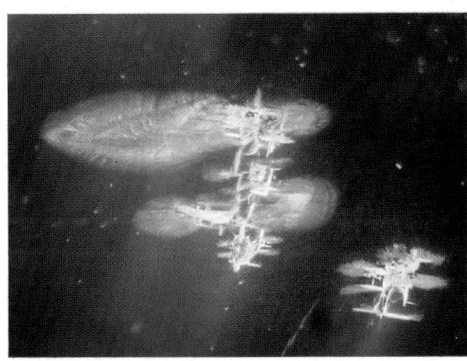

Plate 45 'Centipede' inclusion in moonstone – dark-field illumination

Plate 46 Sectioned jadeite boulder showing fibrous granular structure

Plate 47 Jadeite cabochon showing magnified dimpled surface after polishing

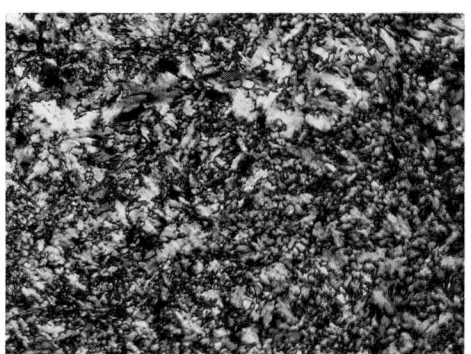

Plate 48 Thin section of nephrite between crossed polarizers showing interlocking fibres

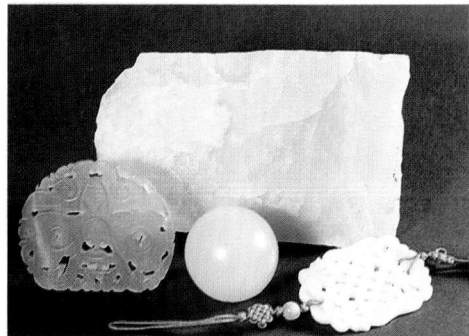

Plate 49 Bowenite serpentine – fashioned samples

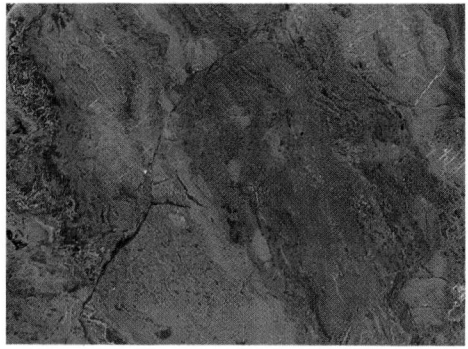

Plate 50 Polished surface of ruby-verdite – the ruby is often absent

Plate 51 Rough and worked aventurine quartz. The small darker-brown oval is aventurine glass

Plate 52 Nephrite vase and saussurite

Plate 53 Some mussels used for pearl culture. *Hyriopsis schlegeli* (*top left*); *Pinctada martensi* (*bottom left*), and *Cristaria plicata* (with juvenile specimens)

Plate 54 Chrysoberyl (*left*) and tourmaline cat's-eyes

Plate 55 Rose quartz exhibiting di-asterism

Plate 56 Scapolite cat's-eyes from Burma

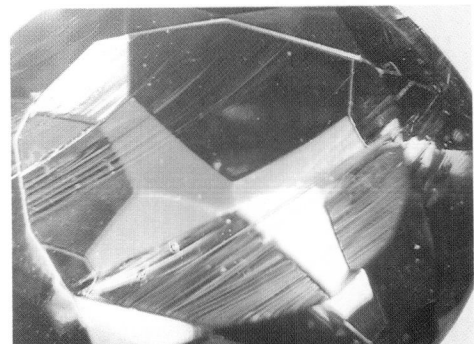

Plate 57 Swirl marks and bubbles in glass

Plate 58 Anomalous interference colours in glass

Plate 59 Conch pearl (20 mm) showing 'flame structure'

Plate 60 Fly and larger insect in Baltic amber

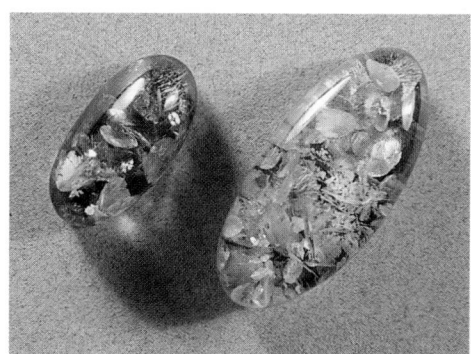

Plate 61 Spangles (stress figures) in treated amber

Plate 62 Heat-treated amber cut to show almost colourless interior

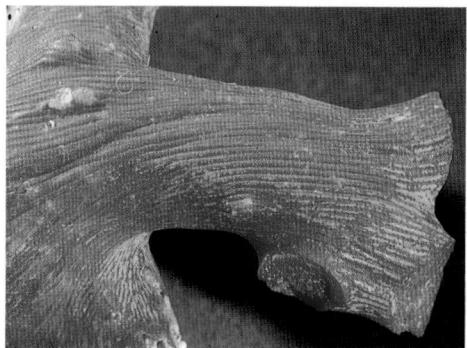

Plate 63 Red coral branch showing ridged longitudinal canals

Plate 64 Polished red coral showing longitudinal lines and cross section

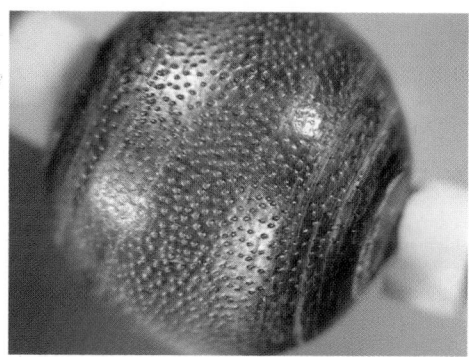

Plate 65 Golden coral bead

Plate 66 Magnified shell cameo showing linear structures almost at right angles in successive layers

Plate 69 Degraded natural pearl, pre-1600

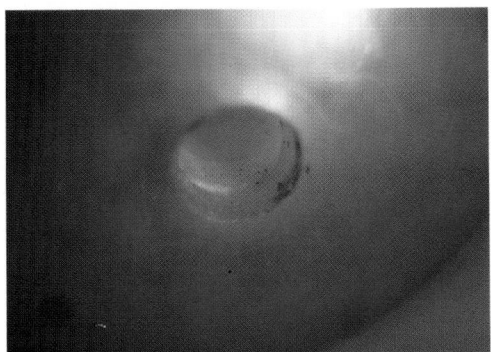

Plate 67 Cultured pearl showing junction within the drill hole

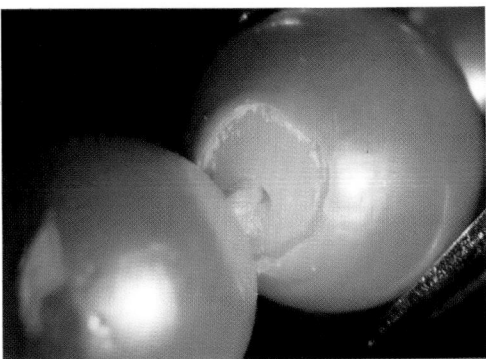

Plate 68 Poor-quality cultured pearls with thin nacreous layer

Plate 70 Naturally coloured pearls

21

Quartz, chalcedony and opal

Quartz is the commonest but nevertheless one of the most beautiful of all minerals. It is composed of pure silica (silicon dioxide: SiO_2), a compound which, either alone or combined in the form of silicates, accounts for nearly 60 per cent of the earth's crust. Quartz is also a very durable mineral both chemically and physically, so that when rocks become weathered and broken down after long ages of exposure, grains of quartz remain and form the bulk of the river gravels and the sands of the seashore and of the desert.

Jewellers know quartz in many forms, though they may not realize that all these belong to the same mineral species. The quartz minerals fall into two categories:

1. Crystalline quartz, occurring as individually visible trigonal crystals – typified by rock crystal, amethyst, citrine, rose quartz, and brown quartz.
2. Cryptocrystalline quartz, typified by chalcedony, agate, chrysoprase, cornelian, sard, bloodstone, jasper.

Stones belonging to the second group are composed of tiny grains or fibres of quartz too small to be distinguished as individuals. The varieties in these two groups are too numerous to allow for detailed discussion of each, and from the point of view of identification this is, in any case, unnecessary. The crystalline forms of quartz used in jewellery are for the most part transparent. Their properties are remarkably constant, and are as follows: hardness – the standard 7 on Mohs' scale; SG – 2.651; refractive indices – 1.544 and 1.553.

If readers have provided themselves with a bromoform mixture matched exactly with a rock-crystal indicator, as suggested earlier in the book, they will be able to identify any pure specimen of the crystalline quartzes (provided it is free from its setting) by a trial in this liquid. If a refractometer is available and the stone is faceted the refractive index will also be found a reliable guide.

Quartz has a spiral internal structure, which is often revealed externally by small extra faces on the crystal, which serve also to show whether the crystal is a 'left-handed' or a 'right-handed' one. In itself this has no bearing upon gem testing and will not be enlarged upon here. But the spiral structure also has a special effect upon rays of polarized light passing through the stone in the direction of the optic axis, rotating the plane of polarization. This, in turn, results in quartz having an 'interference figure' which is unique among gemstones, and where this can be seen the figure provides a beautiful and practical test for quartz.

In general terms, to see an interference figure in a doubly refracting mineral the specimen has to be viewed in the direction of its main axis in convergent light and between crossed polars. Where a parallel-sided slab of the material is available, correctly oriented, all that need be done to view the figure is to hold the piece, sandwiched between two 'Polaroid' sheets in the 'crossed' position, quite close to the eye, and look through the 'sandwich' at the window or at a well-lit sheet of white paper. The rays reaching the eye at such a short focal distance are necessarily convergent, and the figure should be plainly visible without the use of any lens. An easy way to try this out is to take a sheet of muscovite or biotite mica, which is easily held between crossed polaroids in the manner described and should show the two 'eyes' of a biaxial interference figure very nicely.

With faceted stones this simple technique is not usually possible. With these, figures can often be seen by turning the stone in tweezers between the crossed polars of an open polariscope with built-in lighting such as the Rayner versions described in Chapter 3, but a pocket lens or a small glass bulb filled with di-iodomethane must be held immediately above the specimen to give the necessary convergent light. (A strain-free sphere of glass at the end of a glass rod or a built-in lens serves the same purpose.) This is tricky and needs practice. But with spheres of quartz or spherical quartz beads (which are often used in necklaces) the interference figure is very easily seen simply by turning the beads between crossed Polaroids. No lens is needed, since the shape of the bead itself causes the rays to converge. All one has to do is to turn the bead into the correct position. What is seen is a dark cross with a hollow coloured centre, enclosed in what is the first of a series of faint rings. Figure 21.1 shows the effect produced on two quartz necklace beads in the manner just described. Where the effect is unique for quartz is in the hollow centre – a tourmaline or an aquamarine bead, for instance, would show the cross meeting at a dark circular centre without a gap. For those interested, details and pictures of interference figures can be found in any textbook of mineralogy. However, the techniques suggested will

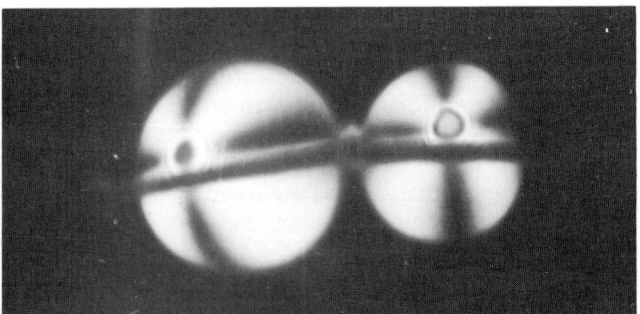

Figure 21.1 Two spherical beads on a necklace, between crossed 'Polaroids' showing the unique interference figure of quartz

be those involving a microscope and not really suitable for the practical gemmologist.

Before mentioning particular varieties and the other minerals with which each may be confused it will save needless repetition if the distinction of any of them from **glass** imitations is dealt with first, *en bloc*. Quartz gemstones, on account of their low refractivity, have a very glassy lustre, and are thus difficult to distinguish from glass counterfeits by inspection alone. One simple test which is remarkably effective when properly applied arises from the low heat conductivity of glass. If a specimen of quartz and one of glass be touched with the tongue, quartz will feel perceptibly the cooler, and the same will prove true for any of the crystalline gemstones, since they all conduct heat away from the spot touched by the tongue at a rate faster than glass does. Amber and other resins conduct heat even less than glass, and thus feel still warmer to the touch. Naturally, certain precautions are necessary in carrying out such a test. Hygiene demands that the specimens should be in a clean condition, and they must not be warm from recent contact with the hands. Each stone should be held in corn tongs for the test, and a direct comparison made with a known specimen of quartz, since glass also feels cold to the tongue and is only warm in comparison with true crystal. Other features by which glass can be recognized – bubbles, swirl marks, etc. – have already been described in Chapter 9.

Some of the crystalline varieties of quartz can now be individually mentioned and a brief reference made to stones with which they may be confused. The colourless quartz called **rock crystal** is too insipid to make a satisfactory gemstone, but it is a beautiful material for carvings, beads, crystal balls, and the like. Glass is the only likely substitute. Real crystal balls of some size and without flaws are rather valuable. It is a simple matter to ensure that it really is crystal by observing the 'doubling' when placed above print or the sharp edge of a card. The yellow quartz properly known as **citrine**, though commonly sold under the misnomer 'topaz', is easily mistaken for that more precious mineral. SG and refractive index tests each provide a clear separation. Most citrines have been produced from amethyst by heat treatment, the colour change being permanent. Such stones show practically no dichroism, whereas natural yellow quartz is distinctly dichroic. Another common feature in 'burnt amethyst' citrines is a banding of their colour. Citrine grades into the darker brown quartz (formerly known by the now discouraged name of cairngorm) or the smoky **morion** or smoky quartz. A clear green form of quartz has recently been marketed, for which the name '**prasiolite**' has been proposed. This is derived from certain types of amethyst found in Minas Gerais, Brazil, by careful heat treatment. The stones have some similarity to beryl or tourmaline in appearance, but have all the ordinary distinguishing features of quartz. **Amethyst** varies in tint from pale mauve to the deepest violet. Patchiness, and zoning of colour will distinguish it from paste, as also will doubling of the table facets, when viewed with a lens. A curious 'tiger-stripe' inclusion is a unique feature often seen in amethyst – see Figure 21.2. The curious amethyst/citrine quartz (sometimes known as 'Ametrine') may be mentioned here. This is produced by heating and/or irradiating sectored amethyst from Brazil and Bolivia and some material may have been irradiated naturally. The distinction of synthetic amethyst has been discussed in Chapter 9. Purple sapphire (miscalled 'Oriental amethyst') is sometimes seen in similar colours, but is a far brighter, harder stone, with the higher refractive index and SG of corundum. This type of sapphire shows the

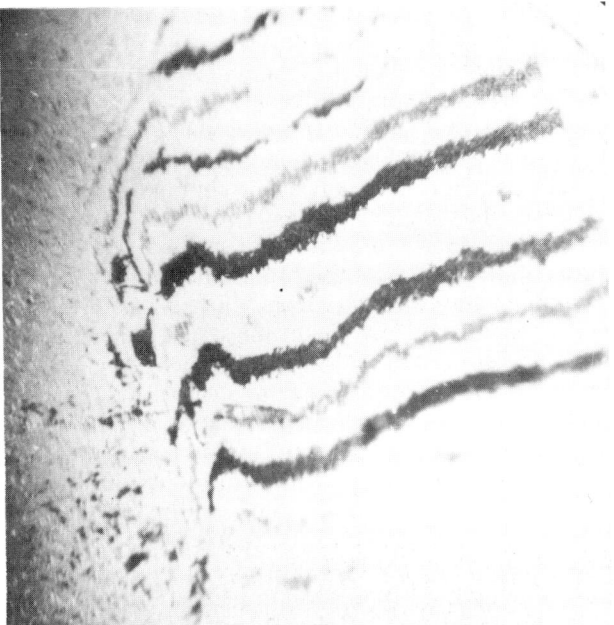

Figure 21.2 'Tiger-stripe' structure typical as an inclusion in amethyst

ruby fluorescence line when strongly illuminated. The purple scapolite from East Africa closely resembles amethyst in appearance and properties. Its identification requires careful determination of the optical properties – in particular, the optic sign (see Chapter 19).

Rose quartz is never quite transparent, but milky and often veined and flawed. When deep in tint it makes attractive beads and is a lovely material for small carvings or ornaments. If suitably cut, a six-rayed star is not uncommonly seen in rose quartz, though never very strongly marked. The effect has been ingeniously enhanced in cleverly made doublets in which a hemisphere of asteriated quartz is provided with a reflecting base. The resulting doublet presents a very striking appearance, with a brilliant star effect. This has been found to be due to extremely fine needles of rutile oriented in three directions at right-angles to the trigonal symmetry axis of the crystal.

It can be easily distinguished from star sapphire at sight, not only by the greater brilliance of the star but also by the fact that when viewed under an electric light a reflection of the electric light bulb can be clearly seen at the centre of the star – an effect not seen in untreated star stones. Later productions are more deceptive, though the rays are less well defined than in genuine or synthetic star sapphire, and the pink colour of the quartz can be seen when viewed at right-angles to its axis.

Quartz cat's-eyes contain parallel fibres or canals where fibres once were. Their usual colour is fawn-brown, greenish, or yellowish, and the ray, though distinct, is not so sharp or silky as with chrysoberyl cat's-eye because the reflecting fibres are coarser. A fine quartz cat's-eye is very similar in appearance to a poor chrysoberyl cat's eye, though the latter has a brighter surface lustre. If the stone is unset, a SG test in bromoform or di-iodomethane will provide a clear decision in such cases. When in a setting, the 'distant vision' method will distinguish between them on the refractometer. Impure cat's-eye quartz has been stained various colours, the products being sold as 'Hungarian cat's-eyes'. The magnificent golden-brown chatoyant **tiger's-eye** was originally a blue crocidolite asbestos, the fibres of which have since been replaced by quartz. The name crocidolite is often retained for the altered material. Fibres of reddish-brown rutile are often seen penetrating a colourless quartz in apparently random orientation. This rutilated quartz is known by such fancy names as 'Venus's hair stone' or 'flèches d'amour'. Continuing the list of quartzes containing inclusions, there is a green **aventurine quartz**, which owes its colour and spangled appearance to the presence of small flakes of green fuchsite mica. This is more properly classed as a variety of **quartzite**, being formed from a mass of small quartz grains. There is also a reddish-brown aventurine quartz containing spangles of an iron oxide mineral. The golden-brown **aventurine glass** so frequently seen has SG and refractive index near that of quartz, but is far more spectacular than any natural aventurine (see Figure 21.3).

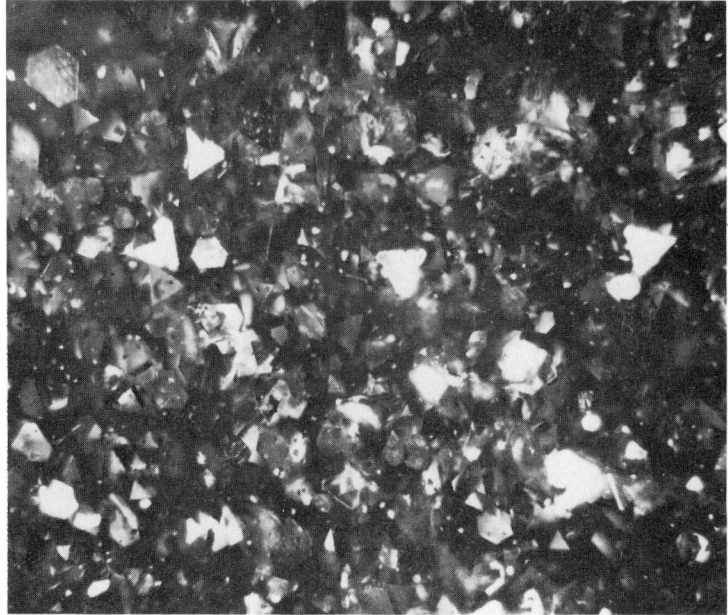

Figure 21.3 Copper crystals in aventurine glass

Rock-crystal beads are sometimes heated and cooled suddenly by plunging into water containing dyestuffs. The colour enters the cracks formed by the sudden chilling, and the resulting effect is quite pretty, though obviously 'faked'. **Jasper** is a very impure opaque multicrystalline quartz, usually some shade of brown in colour. Jasper artificially stained with Berlin blue is used as an imitation for lapis lazuli, and is known as Swiss or German lapis. Much jasper is really a crude form of chalcedony.

Passing now to the chalcedony group: here again we have a host of names to contend with for what is essentially the same material. The differences are chiefly in colour and types of banded structure (see Plate 29). Cryptocrystalline quartz when translucent and unbanded is known as **chalcedony**. The SG (about 2.60) and refractive index (about 1.535) are slightly lower than for quartz. A flat, polished surface of chalcedony, either banded or unbanded, will often show two shadow-edges on the refractometer, indicating a birefringence of about 0.006. This does not vary as the stone is turned on the instrument. The cryptocrystalline structure of chalcedony can be seen most clearly in thin section between crossed polars under the microscope. A chalcedony stalactite photographed under these conditions is reproduced in Figure 21.4. Unlike

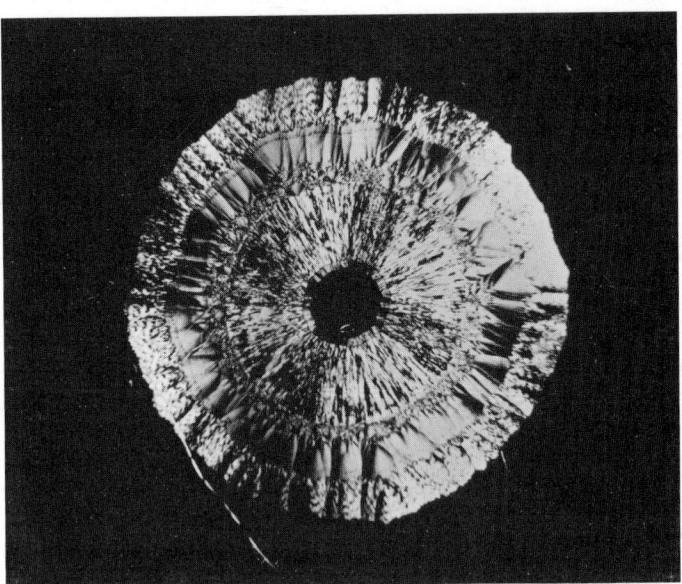

Figure 21.4 Chalcedony section as seen between crossed polars

crystalline quartz, the cryptocrystalline types are to some extent porous and thus susceptible to staining; full advantage is taken of the fact to 'improve' the dim and delicate colourings found in nature. Chalcedony itself is mostly stained in shades of blue or green. The blue is sometimes miscalled 'blue moonstone'; the green is produced by treatment with chromium salts and appears pink under the Chelsea colour filter and the stone usually shows a banded structure by transmitted light – thus distinguishing it from chrysoprase and from jade, for which it is sometimes wrongfully sold. Vague narrow bands due to chromium can be discerned in the absorption spectrum of this green chalcedony. A natural **chrome chalcedony** resembling cloudy emerald was used by the Romans for cameos and intaglios, and more homogenous material with a pleasing green colour has been recently found and marketed in Zimbabwe. This has a sharper one-band spectrum in the red compared with the woolly three-band effect seen in stained green chalcedony.

True **chrysoprase** owes its (natural) apple-green colour to a nickel compound. Though common enough in its poorer qualities, really fine pieces are comparatively rare and valuable. The absorption spectrum of chrysoprase shows no chromium lines, but a faint line due to nickel can be seen at 632 nm in the orange. **Agate** is the best known of the banded chalcedonies. The layers are curved and follow the shape of the rock cavity in which the silica was deposited in a series of rhythmic stages. Agate is extensively used industrially in the manufacture of balances, for pestles and mortars, etc., the natural colour being retained for such purposes. When required for ornament it is stained by a variety of chemical means to enhance the contrast of the various layers. Where the bands are straight, the names **onyx** (black and white), **sard** (red), **sardonyx** (red and white), are used. Cameos are often cut from such material, the carved relief being worked in the white layer with the underlying black or red layer acting as an effective background. **Black onyx** is produced by saturation of the natural greyish material with a solution of sugar or honey, and subsequent treatment with concentrated sulphuric acid, which deposits carbon in the pores of the stone. **Bloodstone** is a dark green chalcedony containing bright red spots; heliotrope is an old name for this.

Cornelian, a more correct version of the usual spelling 'carnelian', is a reddish or yellowish-red chalcedony which shows a banded structure by transmitted light. **Mocha stone** is the name given to black dendritic (fern-like) growths of manganese dioxide in a milky white to pale brown translucent chalcedony which is only faintly banded. **Moss agate** is a term used to describe chalcedony containing growths of variously coloured minerals which resemble plant life very closely indeed. The varieties Mocha stone and moss agate grade into each other. Such growths often mimic plant life to an astonishing degree. There are many other names given to members of the chalcedony group which have not been included above, but their use in jewellery is not at all extensive.

Opal

The brilliance and variety of the flashes of pure colour reflected from a fine opal are unequalled in the mineral kingdom, and its devotees may justly claim that it is the most beautiful of all gemstones. However, as opal is comparatively soft and easily damaged it is more suitable for use in brooches, pendants, and occasional jewellery than for a ring which is expected to withstand the hazards of everyday use. Moreover, to make the necessary impact in jewellery, an opal must be of fine quality. A mediocre specimen may be attractive enough when examined under a good light, but be quite ineffective when worn as a gem.

Unlike most other gemstones, opal is not a crystalline material but a solidified silica jelly. Although it has now assumed the state of a solid, it still retains varying amounts of water, which may amount to as much as 10 per cent in some cases. Common opal is often milky white and translucent, without any play of colour. A green variety, resembling chrysoprase, is sometimes cut as cabochons or for other decorative purposes. In recent years attractive pink and pale blue common opal from Peru has been marketed. It may be striped or contain black dendritic inclusions. A completely transparent variety is known as hyalite, and again has no commercial value. Precious opal is a general term for gem-quality opal showing a play of colour. The hardness of opal is about 6 on Mohs' scale; it is singly refracting, with an index of 1.45, and the SG varies from 2.00 for fire opal to 2.11 for black opal and white opal. The flashes of colour from precious opal are almost spectrally pure: the colours are in fact quite clearly due to interference of light rather than to selective absorption. The nature of the structures which cause some wavelengths of incident light to be reinforced and others suppressed has only recently been discovered, though for more than a century several theories have been confidently put forward. By using the enormous magnifications possible with the electron microscope, Baier in West Germany and Sanders in Australia were able to show that in precious opal there are areas of closely packed spheres of amorphous silica (Figure 21.5). When these spheres are all of one size and near a diameter similar to the wavelengths of visible light they give rise to diffraction, with the production of a play of colour. Where the spheres vary in size there can be no diffraction, and the result is milky or 'potch' opal (Figure 21.6). The longest wavelength of the flashes of ccolour was found to be about double the diameter of the spheres. Thus spheres of about 300 nm diameter can give red flashes, and also green, yellow, and blue, while spheres of smaller size can give rise to blue-flash opal only.

The types of opal most used in jewellery are categorized as black opal, white opal, fire opal, and water or jelly opal. In all categories (except possibly fire opal) a play of colour either on the surface or within the stone is an essential feature if the stone is to be classed as 'precious'. Dealers use special terms descriptive of the dimensions of the individual coloured patches, such as pinfire, flame, flash, and harlequin opal, but these terms have little importance to the gemmologist bent on determining the authenticity of a specimen. **Black opal** is the most highly prized variety, on account of its rarity and the outstanding beauty of the play of iridescent colours seen against a dark grey or black background, which

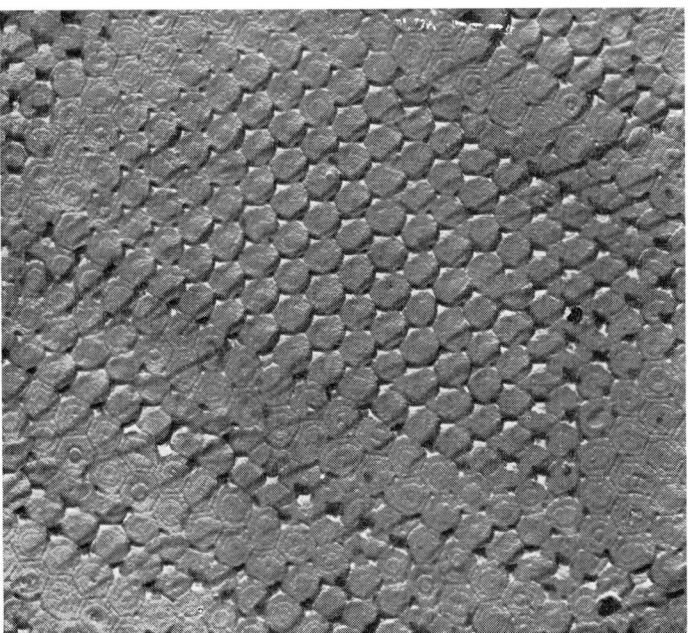

Figure 21.5 Structure of precious opal as revealed by an electron microscope (courtesy: CSIRO)

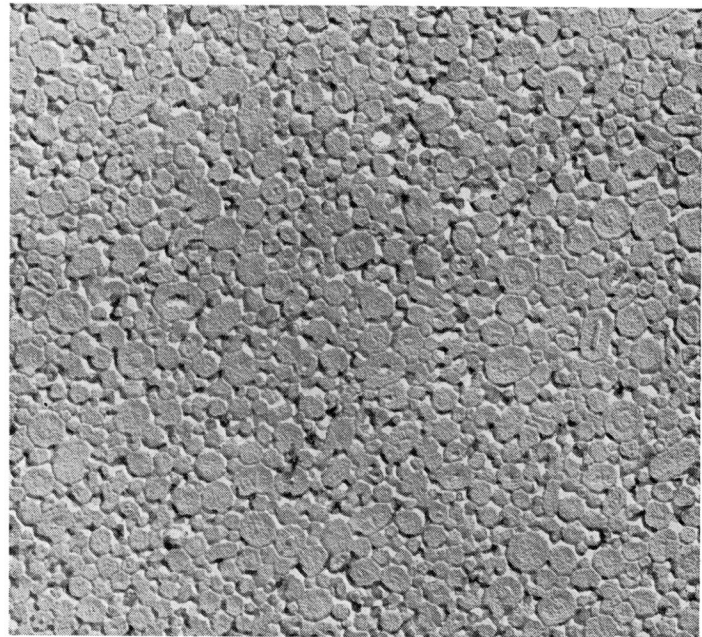

Figure 21.6 Structure of potch opal as revealed by an electron microscope (courtesy: CSIRO)

makes it one of the loveliest and most costly of all gems: each piece has its own individual beauty. In **white opal** the play of colour is seen against a white or whitish milky background and is thereby less spectacular. However, this was the traditional form in which opal had become famous in history, and for ten years or so after the first discovery of black opal at Lightning Ridge in New South Wales it was white opal that remained the more popular category.

Fire opal, chiefly found in Mexico, is transparent or only slightly cloudy. Though flashes of iridescent colours may be discerned emanating from inside the stone, the main attraction of fire opal lies in its tremendously strong orange colour – a colour, unfortunately, which does not blend with that of other coloured gems. It should be mentioned here that common opal (of which fire opal is an orange-coloured variety) may range in colour from colourless (hyalite) through pale yellow to orange, red, and deep red. Some authorities expect 'fire opal' to show a play of colour and describe orange and red material as 'cherry opal'. **Water opal** or **jelly opal** is a transparent form within which iridescent colours gleam with delightful effect, giving the appearance of solidified soap bubbles. It has been shown recently by H. Hanni that some 'water opal' from Mexico is, in fact, a natural glass of volcanic origin and composed of thin layers of the order of 2 μm in thickness. Interference at these thin layers gives rise to the iridescent effects in the glass in the same fashion as the films of oil on a wet road create brilliant colours.

In order to display their play of colour to best advantage, black or white opals are invariably cut in cabochon form, black opals usually with a nearly flat upper surface as the 'precious' layer is frequently thin. Only in water opals (or in crystal-topped triplets) is a high-domed cabochon effective. Only in fire opals are faceted forms employed, and even here the table facet is often curved.

Opal doublets (discussed later in the chapter) formerly presented the only serious problem when considering the authenticity of an opal specimen. Since the early 1970s this happy state of affairs has been significantly affected by the production of true synthetic opal (representing both black and white varieties) by Pierre Gilson of France, and by the appearance of plastic opal imitations manufactured in Japan and elsewhere. A further complication is provided by the ingenious form of unexpectedly plausible glass imitations made by John Slocum in Michigan, USA, who markets his product as Slocum stone (see Plate 28). Synthetic opal has already been described in Chapter 9, but a few extra remarks can perhaps be usefully added here. The Gilson stones are undoubtedly very beautiful and may often require a great deal of skill and experience to identify. As in all cases of synthetic stones, the gemmologist will be wise to obtain samples for study even at considerable cost. Careful inspection under a low-power microscope of synthetic opals will often reveal a 'lizard-skin' pattern on the parts of the surface showing the vivid flashes, which is not seen in natural opals (see Plate 25). Columnar structures are present in many specimens, and show especially in beads (see Plates 24 and 26). Another feature of white synthetic opals is that colour flashes from the underside of a cabochon will often all appear to be of the same colour. The black opals made by Gilson tend to be more translucent than the natural stones, and they show a predominance of red colour flashes which are rare in nature. They also transmit a greater proportion of long-wave ultra-violet light. The last-named effect can be demonstrated in the laboratory by placing suspected stones (together with comparison stones) on photographic printing paper in a glass dish of water in a dark or darkened room, exposing them for a few seconds to ultra-violet light, and developing the paper in the usual way.

Those Gilson black opals tested by the author have shown a slightly lower density than comparable natural opals, averaging about 2.06 compared with 2.11 typical of Australian stones. Though immersion tests are liable to spoil the appearance of opals where these are at all porous, the effect is unlikely to be permanently deleterious if they are cleaned in pure alcohol or distilled water containing a trace of detergent, gently wiped, and allowed to dry. If a suspected Gilson stone is placed in a bromoform solution which is diluted with benzene under contant stirring until the specimen is suspended, the fact that natural black opals are found to sink in this fluid provides a fairly convincing confirmatory test. Since some Gilson opals have been found to be noticeably porous, and even to stick to the tongue when gently licked, it would be wise to carry out a porosity test with clean water before embarking on immersion in any other fluids.

In the distinction between whhite opals and their Gilson counterparts a test under long-wave ultra-violet light may be useful. Australian white opals have a whitish fluorescence, which is followed by a rather prolonged phosphorescent afterglow. This afterglow is hardly perceptible in the Gilson synthetic white opals tested. It should perhaps be noted that the opals tested from volcanic rocks in the Rio Corrente area in Brazil do not show any reaction under ultra-violet light.

The plastic opal imitations, which possess a true precious opal structure, are very realistic and may pose problems when close set. The RI at 1.48–1.495 is higher than true opal, but the SG is much lower at 1.12–1.20, and this is obvious by the heft of loose stones.

Often the brilliant play of colour in white and especially in black opal exists in nature as very thin layers, and such layers are often sliced away from their matrix and cemented to a base of dark opal matrix or black onyx to form **doublets**, which can, of course, be sold at a relatively low price. Doublets can be superficially as beautiful as solid black opals, but are naturally far less valuable. In order to disguise their nature these are frequently mounted with the edge of the stone hidden by the closely fitting setting. Gem trade laboratories are often faced with the teasing task of deciding whether or not the stone is a doublet.

Warning signs of a doublet are an extreme flatness or very low curvature of the upper surface of the stone. If the lower portion is visible and seen to be black onyx, the stone is clearly a doublet. If it is of dark grey opal showing no trace of colour flashes it is merely highly suspect. Careful scrutiny with a lens round the margins of the stone where it borders the setting may reveal a point at which the 'join' can be glimpsed. Another worthwhile exercise is to examine the stone by overhead lighting, and focus with a lens, or, better still, a low-power microscope, through the translucent 'precious' surface of the gem in a search for bubbles at the intersurface where the opal layer is cemented to its base. Sometimes these bubbles are seen as a layer of flattened discs, in other cases as isolated gleaming spheres. Failing this, success may be gained by attempting to punch a powerful beam of light through the stone. Translucency will mean that the stone is almost surely a true black opal. In the case of doublets pinpoints of light may be seen which are revealed as bubbles under the lens or microscope.

Even where the stone is claw set and the edge is visible there may be argument as to whether the stone is a doublet or a solid opal in which the flashy layer changes abruptly to a layer devoid of colour play. In addition to the fact that the most clever cementing can hardly simulate the perfection of a natural transition from one layer to another, patient scrutiny of the margin in true opal will almost always reveal one or more places where a flash of colour dips into the less precious region. However, opal-on-matrix doublets have now been produced in which the adhesive is mixed with powdered matrix and the junction needs to be searched for carefully.

As well as the traditional form of doublet discussed above, a pretty form of opal triplet is on the market in which an opal-on-onyx doublet is capped with a cabochon of clear quartz. Such things are hardly intended to deceive, and provided one is aware of their existence, inspection with a lens should be sufficient to show their nature.

A further problem for the jeweller and gemmologist is provided by **treated opal**, in which Andamooka opal of poor quality has been 'improved' by impregnation of its interstices with carbon particles, greatly enhancing the showiness of the play of colour. The process is hardly new, as it is described in the 1823 edition of John Mawe's famous little book *A Treatise on Diamonds and Precious Stones*. Mawe's recipe is to warm the stone in oil or grease, which is afterwards burnt off. In modern times a similar soaking (in old sump oil) and burning has been described, and also heating, followed by impregnation with sugar solution and later 'carbonizing' with warm sulphuric acid – essentially the time-honoured process for producing black onyx.

These treated stones consist of true opal throughout and have the normal SG and refractive index. Fortunately, their appearance is extremely characteristic. They have a curious granular texture, precluding a really fine polish. The flashes of colour come from a multitude of small areas giving the whole a *cloisonné* effect.

Tumbled pebbles of treated opal can be purchased for a pound or two, and anyone dealing in opal should invest in a sample in order to ensure that he or she will not confuse it with the costly true black opal, in which the larger-scale ribbed flashes of colour are also highly characteristic under lens examination.

Another problem confronting the gemmologist in the last decade or so is the **plastic** or **resin-impregnated opal**. This treatment improves the visual continuity in the stone and is not always obvious, even under the microscope. The plastic may not be affected by solvents and very careful testing (but not touching) with a hot needle may reveal the characteristic smells of resins. A safer method is to subject the stone to testing in a laboratory by the infra-red spectrometer, which will readily reveal the organic nature of the impregnation.

Until recently the only plausible imitations of opal in **glass** have been those representing fire opal, in which the colour is the main source of attraction. The refractive index and density in the glass imitations are always distinctly higher than those in authentic opals. In the mid-1970s, however, experiments with special forms of glass carried out by John Slocum of Michigan have led to the large-scale commercial production of imitation opals of many varieties. These are astonishingly deceptive to the naked eye, though inspection under a lens or low-power microscope reveals at once the falsity of any impression that the stone might be some form of opal. The glass used by Slocum has a refractive index near 1.50 and a density range of 2.4 to 2.5. Further details of 'Slocum stones' are given in Chapter 9.

22

The garnet family

The name 'garnet' is given to a group or family of minerals which vary widely in their constitutents, but which all conform to the same general type of chemical formula, have the same internal structure, and crystallize in the same forms.

Garnets are typically formed under pressure, and thus occur in metamorphic rocks in ball-like forms consisting of rhombic dodecahedra (12 lozenge-shaped sides), icositetrahedra (24 trapeze-shaped sides), or combinations of these (Figure 22.1). They are also found as rounded pebbles in gravels derived from the weathering of such rocks.

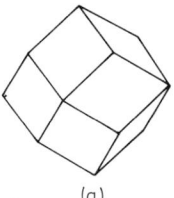 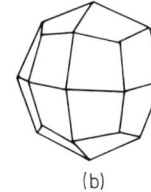 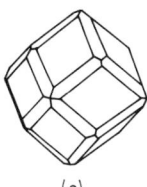

(a)   (b)   (c)

Figure 22.1 Three common habits for garnet crystals. (a) Rhombic dodecahedron; (b) icositetrahedron; (c) combination of the two forms

The general formula for the garnets can be written $3R''O.R'''_2O_3.3SiO_2$, where R'' may represent divalent magnesium, iron, manganese, or calcium and R''' trivalent aluminium, iron, or chromium. Mineralogists recognize six distinct members of this group, and five of these are known as gemstones; many other garnets are known. They are Pyrope (Mg, Al), Almandine (Fe, Al), Spessartine (Mn, Al), Grossular (Ca, Al), Andradite (Ca, Fe), and Uvarovite (Ca, Cr). Gem varieties of grossular and andradite are better known under the names of hessonite and demantoid, while uvarovite, though a fine chromium-green in colour, is too rarely found in transparent pieces to be used in jewellery.

The first three mentioned form a series conveniently known as 'pyralspite' since they are so close in structure as to be almost completely miscible. The remaining three form another intermiscible series 'ugrandite', in which the atoms are rather more widely separated in the crystal lattice, and there is only a limited mixing of these with garnets of the pyralspite group. Garnets corresponding exactly to their allotted chemical formulae are seldom found in nature, and the properties of these minerals vary widely according to the particular admixture of molecules present. Recently, however, pure samples of each garnet have been prepared, using high-pressure techniques, by J. Coes, and the constants of these have been measured by B. J. Skinner. These constants agree rather closely with figures calculated years ago by Ford, Fleischer, and others, on the basis of measurements on numerous analysed garnets found in nature. The figures now established for the pure garnets are listed in Table 22.1, not because they have any value as testing constants but because they enable the interested gemmologist to assess the effect of each garnet molecule on the properties of the mixed types one actually encounters in jewellery. Let us now leave the general for the particular and deal with each of the gem garnets in turn.

Table 22.1   **Properties of pure synthetic garnets**

Type of garnet	SG	RI	Atomic spacing (nm)
Pyrope	3.582	1.714	1.146
Almandine	4.318	1.830	1.153
Spessartine	4.190	1.800	1.162
Grossular	3.594	1.734	1.185
Andradite	3.859	1.887	1.205

Pyrope

Colour, SG, and refractive index usually enable a clear distinction to be made between the various gem garnets. Only between almandine and pyrope is the admixture so variable and continuous, and the appearance so similar, that one may feel the need of a term such as 'pyrandine' to represent intermediate types that cannot properly be designated either pyrope or almandine (see Plate 36). The name 'rhodolite' has, indeed, been used for an intermediate garnet having a rose-red colour, but the name is not a happy one, being so similar in sound to that of the manganese mineral rhodonite. One must admit that, in practice, jewellers do not here feel any need for exact nomenclature: if a given red stone can be established as a 'garnet' that is all they need to know.

Of the red garnets, pyrope has undoubtedly the finest colour, if one includes under this name only those comparatively 'pure' pyropes from Kimberley, Arizona, and the former Bohemia. All of these owe their colour to chromium, to which the red of ruby and spinel can also be credited. There are also gem-quality pyropes containing little chromium, the colour of which may be ascribed to iron or manganese, since 'pure' pyrope would be colourless. Those with iron show a rather weak almandine spectrum. Pyrope is seldom found in large sizes, and its colour, though fine, is often too dark to be appreciated. Normally, the single refractive index near 1.745 is sufficiently above that of almost all red **spinels** to make distinction easy, 1.730 being the usual upper limit for red spinel. However, one or two garnets have shown indices of 1.732, 1.731, and even 1.730, while an exceptional red spinel (mounted in a ring) was recently encountered which was sufficiently chrome-rich to give a reading of 1.7439.

Quite 90 per cent of the red spinels examined had refractive index 1.720 or lower, and there need be no question of pyrope in such cases. An additional check is also available in the absorption spectra of the two stones, though at first glance they are very similar. The most notable feature in each is a broad band near the centre of the spectrum. In pyrope this band is centred at about 575 nm, that is, in the yellow-green, whereas in spinel the band is nearer to the blue, being centred at about 540 nm. In pyrope, the strongest of the almandine bands can be detected in the blue-green at 505 nm and this, of course, is never seen in a spinel. A simple method for checking the position of the broad band in pyrope or spinel in order to distinguish the two is to interpose a very dilute solution of potassium permanganate, contained in a bottle or beaker, between the light source and the stone. When suitably diluted, the permanganate will show five evenly spaced absorption bands, centred at 570, 545, 524, 504, and 487 nm.

These act as a 'built-in' wavelength scale and by placing the stone in and out of the field, comparison with the permanganate bands makes it obvious whether the absorption of the stone is near 540 nm (spinel) and 575 nm (pyrope). Crookes glass (often used in anti-glare spectacles) can be made to function in the same way, since the glass shows clear-cut absorption bands due to didymium. The strongest of these is at 584 nm, and makes a useful standard for comparison.

Perhaps a simpler test which distinguishes red spinel from pyrope is its red fluorescence under ultra-violet light or under the more sensitive test of 'crossed filters'. Pyrope, due to its iron content, is normally non-fluorescent. The red glow of spinel when strongly illuminated in light which has passed through a flask of copper sulphate can be resolved by a prism spectroscope into a closely spaced group of narrow bright lines. This 'organ pipe' fluorescence spectrum is a very diagnostic test for red spinel. Some mixed red garnets have recently been encountered, having constants near those for pyrope, which surprisingly show a distinct red glow between crossed filters.

Almandine

Apart from pyropes coloured by chromium there is a whole series of red garnets, as previously mentioned, forming the almandine–pyrope range, in which the refractive index and SG become progressively higher as the proportion of the almandine molecule present becomes greater. The colour of true almandine is a peculiar purplish-red, so deep in tint in the heaviest types that stones appear almost black unless backed by a reflecting layer of foil. Such stones have often been cut as hollow cabochons, for which the name 'carbuncle' has been used for centuries. The highest figures for red garnets used in jewellery are about 4.20 and 1.81 respectively for SG and refractive index, corresponding to a garnet containing some 80 per cent of the almandine molecule.

The gemstones most likely to be confused with almandine are Thai **ruby** and purplish-red **spinel**, both of which may have a very similar colour. The spinel is easily separated by its lower constants, but those of the ruby are very close to certain almandines. Ruby is doubly refracting, and a refractometer reading will

Figure 22.2 Absorption spectrum of almandine

enable a clear decision to be made if carefully taken in sodium light. The dichroscope is also helpful here, although the dichroism of Thai rubies is weaker than that of rubies from Burma. A quick and certain test is provided by the spectroscope, since the almandine spectrum of three strong bands (Figure 22.2) is easily distinguished from the ruby spectrum of one broad band – accompanied by very narrow bands in the red and in the blue. Almandine typically contains rod-like crystal inclusions (see Plate 30) which are parallel to the dodecahedral faces of the original crystal, and thus intersect at angles of 70° and 40° (4 × 70° plus 2 × 40° equals 360°). These inclusions were proved by Mellis to consist of rutile needles, and are somewhat similar (though on a coarser scale) to the fine rutile needles, intersecting at 60°, that form the typical 'silk' in Burma rubies. The needle inclusions in garnet are often seen in the slice of almandine used to form the table facet of a **doublet** (see Figure 9.27) with a layer of bubbles immediately below, where the junction with the glass base occurs. Almandine may contain enough silk to show a rather feeble 'star' effect when suitably cut. A polished sphere of such material will show four six-pointed stars and six four-pointed stars by reflected light. As previously noted, the almandine bands may often be seen in garnets which have a preponderance of the pyrope molecule when these have insufficient chromium to mask the effect by a broad absorption region in the green.

Spessartine

The rare and beautiful manganese garnet spessartine has already been mentioned under 'Brown and Orange Stones'. The name spessartite' is used by petrologists for a type of rock first named from the Spessart locality in Bavaria. The terms 'spessartine' and 'almandine' were the names originally given to the minerals, and by the laws of priority a species once named in print should retain that name unless subsequent research or confusion decrees otherwise.

By the same token, names should not be coined for new species or new varieties of a mineral or gemstone which could be confused with well-established mineral names. A particularly bad example is the term 'ammolite', which refers to an iridescent form of aragonite in fossil shells and can be confused with the term 'ammonite', a type of extinct mollusc. This term is doubly unfortunate when we learn that 'ammonites' are preserved in 'ammolite'.

Spessartine can range in colour from yellow through orange to flame-red. Often it closely resembles hessonite garnet in appearance though it lacks the granular inclusions and treacly swirls seen in the commoner garnet. Where the almandine molecule is present to a considerable extent the colour may deepen to a port-wine hue.

The purest spessartines have a refractive index near 1.80 and SG 4.16 to 4.19. These constants are close to those for almandine but the colour helps to distinguish them. The absorption spectrum is also distinctive. Though faint almandine bands may be seen in many spessartines, the strongest bands, as always with manganese minerals, occur in the violet and ultra-violet. There are two weak bands in the blue at 495 and 485, a stronger one at 462, and a really powerful band at 432 nm in the violet. Further bands at 424 and 412 nm can only be seen in pale specimens. To the expert, inclusions in spessartine are often distinctive such as the lace-like feathers seen in Plate 31, others consist of wavy liquid feathers rather like those seen in rubellite tourmaline. No other garnets show such structures.

Grossular

The type of grossular most commonly used in jewellery is the orange-brown variety known as **hessonite** or cinnamon stone. In earlier days the term 'jacinth' was often used for this garnet in the trade, though properly this belongs to zircon of similar colour. The peculiar granular appearance of hessonite, arising from numerous small transparent inclusions chiefly of apatite and calcite, is a feature that assists the observer to recognize it at sight. The appearance of these tell-tale inclusions under a low-power microscope is shown in Figure 22.3 and in Plate 32. In addition, the internal structure of hessonite has a curious treacly or fused appearance caused by swirls of finely divided calcite. Inclusions formerly accepted as diopside (with which grossular is often associated) have been proved to be apatite by Dr E. Gübelin.

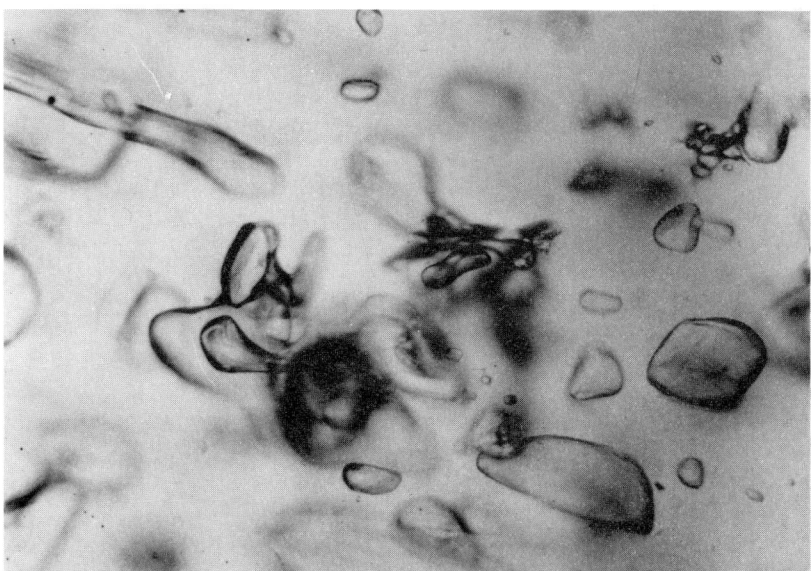

Figure 22.3 Inclusions in hessonite garnet

Hessonite comes almost exclusively from Sri Lanka, and the most typical constants are 1.743 for the refractive index and 3.65 for the SG. The B–G dispersion is 0.027. Sometimes there is a certain amount of the spessartine molecule mixed with the grossular, resulting in higher constants and absorption bands in the blue and violet, due to manganese. Hessonite itself shows no absorption bands when pure.

A massive green variety of grossular found at Buffelsfontein in South Africa has been misleadingly called 'Transvaal jade', and in recent years pink to rose-red varieties of this have been recovered and polished to form attractive beads and cabochons. The material contains up to 5 per cent water, and is more correctly named hydrogrossular. Samples of hydrogrossular from Pakistan can resemble jadeite more closely than the Transvaal type, more particularly in those pieces where a white matrix contains patches of brilliant green.

Both the Transvaal and the Pakistan stones have been found to vary a great deal in SG from sample to sample, and this variation is due to the admixture in varying proportions of massive idocrase (californite) with the hydrogrossular molecule. Chemically, the two minerals are closely similar, both being essentially silicates of calcium and aluminium. Robert Webster found the SG range to be 3.36–3.57 for the Transvaal stones, while the author's results on Pakistan material vary from 3.28 to 3.52. The samples of lower SG consist mainly of idocrase, and show the idocrase absorption band at 461 nm in the blue quite strongly. There is a weaker band in the green at 530 nm. In green hydrogrossular the colorant is mainly chromium, and an absorption band at 630 nm in the orange is noticeable. This has a sharp edge on the green side. Such mixtures are interesting scientifically but are difficult to label with a name which is sufficiently accurate and yet also a possible 'selling' name.

A few transparent green grossular garnets from Pakistan have also been seen. These had refractive index 1.738 and SG 3.63, and showed a chromium absorption spectrum. Their dispersion being only about half that of demantoid, they are perhaps not a serious rival to that superlative gemstone, but if available in quantity they should prove very popular.

Commercially more important finds of transparent green grossular were made in East Africa in the early 1970s. As in the case of the blue zoisite discovery (1967) in Tanzania, Tiffany & Co. were the first to recognize the commercial potential of these attractive stones, and to provide them with a new varietal name 'tsavorite' in reference to the Tsavo national park region in which the stones are found. The trace elements vanadium and chromium are both found in the brightest green stones and are responsible for the fine colour. The refractive index of the stones (1.74) and the density (3.68) are similar to most fairly pure grossulars and should make identification quite simple. Grossular garnets of various colours are shown in Plate 37.

Andradite

The green variety of andradite known as **demantoid** is the most highly prized of the garnets in commercial use, and fine specimens are exceedingly hard to obtain and are increasingly costly. In the trade the name 'olivine' was formerly used for this green garnet – an unfortuante misnomer, since olivine is a name well established by mineralogists for quite another mineral, of which peridot is the gem variety. Demantoid has the high refractive index of 1.89 and a dispersion higher than diamond, giving it liveliness and fire. The SG is 3.85, and these properties vary only slightly, as this species alone among the gem garnets does not suffer extensive replacement by other garnet molecules. The colour of demantoid is yellowish to bright leaf green (see Plate 40), always with more tendency to yellow-green than the bluer green of emerald, from which its appearance differs quite sufficiently to make confusion unlikely by those familiar with both these gemstones. There are, indeed, few other species for which the demantoid is likely to be mistaken by anyone who is at all experienced. The transparent green forms of grossular garnet mentioned above may act as rivals to demantoid, but have so far not been found in so fine a colour, nor have they the high dispersion which gives demantoid so much liveliness and fire. On the refractometer grossular will give readings near 1.74, while demantoid with an index of 1.89 will give a 'negative' reading.

Far more likely to be mistaken for demantoid on the grounds of appearance are the artifical 'garnets' popularly known as 'YAG' which, when coloured green by chromium, resemble demantoid very closely and will also give a negative reading on the refractometer. The distinctive demantoid signs of 'horsetail' fibrous inclusions and the strong absorption band in the blue-violet will, of course, be missing from the man-made material.

Green zircon is, perhaps, the stone most likely to be confused with demantoid, since the colour is rather similar, and there is enough dispersion (0.038 as against 0.057 for demantoid), to give the stone considerable life. When green, zircon is almost invariably of the low or intermediate type (see Chapter 17), and with these double refraction may not be visible with a lens, though sharp extinction between crossed 'Polaroids' will be observable in most cases. The spectroscope will reveal a rather woolly band in the red with these zircons, and possibly other zircon bands in addition. Some demantoids may show an absorption 'doublet' in the deep red and fainter bands in the orange – all due to chromic oxide to which the fine green colour can be attributed. However, the most constant feature of the demantoid spectrum is an intense absorption band

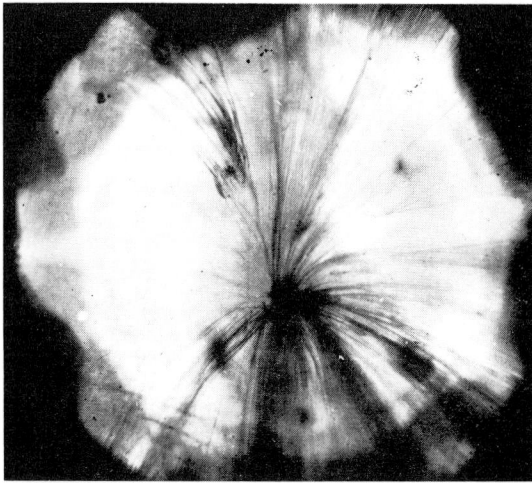

Figure 22.4 Asbestos fibres as inclusions in demantoid garnet

at 440 nm in the violet, which may only be visible as a sharp cut-off at the end of the spectrum.

A peculiarity of demantoid is the nature of its inclusions. Careful scrutiny with lens or microscope will reveal these as wisps of tiny silky asbestos fibres, often radiating from one or more centres.

If a parcel of demantoids is carefully examined, almost all specimens will be found to contain at least some of these fibres, and they provide a very practical aid to identification. The microscope will, of course, show the inclusions more plainly; a photomicrograph of a typical specimen is reproduced in Figure 22.4 and will give the reader a good idea of their nature (see also Plate 33).

Sphene has the brightness and fire of demantoid but is seldom of the same colour: a glance through a lens will at once reveal the strong double refraction of sphene and make distinction easy. The colour of **peridot** is rather close to that of demantoid, but the stone is lacking in lustre and fire and once again the doubling of the back facets can easily be seen with the lens.

The artificial rare-earth 'garnets' should be briefly mentioned here. These contain no silica but have the garnet crystal form and crystal structure. Only two of these – YAG, which is an yttrium aluminate, and GGG (gadolinium gallium garnet) – have so far entered the lists as gem materials. A description of these and other synthetic garnets will be found in Chapter 9.

Mixed varieties of garnet

It has already been explained that minerals of the garnet group are seldom found in the 'pure' condition, but usually consist of mixtures of several of the six recognized species of the group. Samples containing 98 per cent of grossular, 100 per cent of andradite, and 97 per cent of the spessartine molecule have been analysed, but 87 and 77 are the highest percentages of almandine and pyrope, respectively, in any one sample. During the past decade, gem-quality stones showing new variations on the garnet theme have been reported in the journals. As with so many new sub-varieties of gemstone, these garnets, mostly notable for their unusual colour range, are found in East African deposits (sediments and pegmatites) connected with Basement Complex rocks originating in Mozambique and Tanzania. These have been examined in depth by P. Zwaan, E. A. Jobbins, J. M. Saul, and other workers.

These strange garnets have been found to consist essentially of mixtures of pyrope, spessartine, and (more surprisingly) grossular molecules, often with traces of vanadium, which may have influenced the colour. In one case a garnet recovered from the Umba River valley (not far from the mine where pastel shades of gem corundum have been found) could be described as greenish-blue in daylight and magenta by tungsten light. This particular specimen had a refractive index of 1.757 and a density of 3.816. The absorption spectrum showed a broad vague band centred near 575 nm, a strong band in the violet centred near 432 nm due to the spessartine content, and other weaker bands.

The name 'rhodolite', applied originally to an almandine–pyrope type of garnet found in Carolina which had an attractive 'rhododendron' colour, has already been briefly mentioned, and has attained renewed popularity in the trade following the discovery, in various deposits near the Kenya–Tanzania border, of garnets having the required pinkish-violet colour. These stones have a refractive index range from 1.750 to 1.760 and density from 3.83 to 3.90, and all show an almandine absorption spectrum of moderate strength. Samples from the 'type locality' of rhodolite from North Carolina have very similar constants.

Magnetism in gemstones

There are occasions in testing gemstones when the oddest and most unconventional tests can prove highly useful. One such test is for magnetism in gemstones, and it is one which powerful small modern magnets and the coming of the 'aperiodic' type of balance makes it easy to apply on a semi-quantitative basis.

The garnets happen to be a group of gemstones which show magnetism to a considerable extent, and in testing for which a simple test for magnetism can be of real value: hence its inclusion in this chapter.

Iron is by far the most strikingly magnetic of the elements, and, on the whole, it is the gemstones containing considerable amounts of iron which are attracted by a magnet. However, the author found that the presence of manganese also gave rise to a surprisingly strong pull when approached by a magnet.

Apart from the 'hematine' imitation of hematite and the magnetic hematite from Brazil, only one gemstone responds with sufficient strength to the pull of a strong pocket magnet as to be actually picked up when the latter is applied to it, and that is the type of diopside showing a four-rayed star. In this presently rather popular material the asterism is caused by inclusions of magnetite needles, and it is these that account for the strong reaction to a magnet. However, a cabochon of almandine balanced on its back on a polished surface will 'swivel' under the influence of the magnetic pull and a suspended stone will also show a slight but perceptible effect. It is by using a balance, however, that the actual strength of the attraction can be most conveniently assessed.

Any pull due to the metal of the balance pan has to be guarded against, and a convenient arrangement is to place a large cork on the balance, on which the stone (previously weighed) can rest. The stone and cork together are then put into balance with weights, rider, etc. on the other beam, and the small horseshoe magnet, held in the left hand, gradually brought near the stone from above until it almost touches the surface. If the stone is appreciably magnetic the poise of the balance will begin to be disturbed when the magnet is about 1 cm distant from the stone. The minimum weight which can be 'held' by the magnet can be taken as the required reading, and the difference between this and the true weight obviously represents the magnetic 'pull'. This figure is fairly easy to establish when using an 'aperiodic' balance, in which the weight can be read continuously on a scale which is projected from a graticule attached to the pointer of the balance. With a balance which employs a rider for its fractional weights, this would have to be moved experimentally a little at a time before testing whether the magnetic pull would succeed in compensating for the shift. Lacking a rider, fractional weights would have to be subtracted in stages, applying the magnet after each alteration to test whether the balance could be restored by the pull.

In his experiments, the author worked with a small 'Eclipse' horseshoe magnet, made from 'Alnico' alloy, and costing very little at most large ironmongers. Tests showed that, to achieve any sort of consistency with specimens of the same type but of different sizes an 'allowance' has to be made for the size of the stone, or in practice, the *weight* of the specimen.

Finally, the empirical formula:

$$\frac{\text{Magnetic loss in weight} \times 100}{\sqrt{\text{Weight}}}$$

was taken as representing the 'magnetism' of the stone: i.e. its magnetic susceptibility. The units of weight used do not matter; either metric carats or milligrams leading to the same result, as in determining SG. The more important results obtained are as follow:

	Almandine	290–410
	Spessartine	250–360
Strongly magnetic	Rhodochrosite	270
	Rhodonite	280–370
	Hematite	220–310
	Demantoid	120–200
	Epidote	100
	Pleonaste	80–130
	Peridot	50–75
Moderately magnetic	Pyrope	40–100
	Dark green tourmaline	50–70
	Hessonite	40
	Indicolite	40
Weakly magnetic	Brown sinhalite	15
	Green tourmaline	10–20

The really practical value of even a crude test with a magnet has been experienced by the author in separating the following stones, which can be similar in appearance and are not easily distinguished by more normal methods. In each case the non-magnetic (or much less magnetic) substance is mentioned first:

Orange metamict zircon from spessartine
Green metamict zircon from demantoid
Sinhalite from brown peridot
Black diamond from pleonaste or hematite
Red spinel from pyrope garnet

The effect of magnetic 'pull' can, of course, be observed even with mounted stones, though the lack of exact knowledge of the specimen's weight debars the use of the empirical formula suggested. However, it will be noticed that one is dividing the magnetic pull × 100 by the *square root* of the weight: thus the difference between the case of a 4-carat and a 9-carat stone will only be as 2:3, which does not affect the 'magnetic susceptibility' figure very seriously.

At this point it must be mentioned (see also Chapter 9) that inclusions of flux are not uncommon in synthetic diamond grit, and these may give rise to a distinct magnetic effect. Natural diamonds and diamond-grit are non-magnetic.

23

Tourmaline, peridot, and spinel

Tourmaline is one of the most attractive of minerals and is very widely distributed, though only in relatively few localities is it found in transparent pieces suitable for cutting into gemstones. Its striated prismatic crystals, with their rounded triangular cross section, are very easily recognized (see Figure 23.1).

Figure 23.1 Specimens of tourmaline showing crystal form and colour zoning

301

The mineral has an exceedingly complex chemical composition, and it cannot easily be expressed as a formula: but this apparent complexity is due to the many substitutions of one kind of atom for another that can take place in the crystal lattice. Broadly speaking, it can be described as a borosilicate of aluminium containing variable amounts of magnesium, iron, calcium, and the alkali metals.

In the past, mineralogists have been accustomed to sub-divide the tourmaline group of minerals into three main categories, based on their composition, as follows: an iron tourmaline, **schorl**, which is mainly black in colour; magnesium tourmaline, **dravite**, which is commonly brown; and **elbaite**, in which the alkali metals sodium and lithium play a prominent role. The elbaite tourmalines are predominantly pink, red, or pale green (often partly pink and partly green), and virtually all gem tourmalines belong to this group. In the past few years detailed research by P. J. Dunn and others has resulted in the addition of two group names to the above long-established trio. The first of these, **uvite**, can be described as a dravite tourmaline in which calcium is present to a greater degree than sodium. Like dravite, uvite is commonly brown, and cannot be distinguished from the former by ordinary gemmological tests. The second, christened **liddicoatite**, differs from elbaite only in being more calcium- than sodium-rich. Schorl is an old miner's term for common black tourmaline, the names dravite, elbaite, and uvite are geographically based, while liddicoatite was chosen as a means of honouring Richard T. Liddicoat, President of the Gemological Institute of America, for his outstanding services to gemmology. Liddicoatite is, in fact, essentially a gem tourmaline, specimens being often zoned in pink and green, but also being seen in transparent brown and blue forms.

Since these subvarieties of tourmaline cannot be reliably distinguished by density or refractive index measurements, the gemmologist and jeweller can feel free to class them all simply as 'tourmalines' with an appropriate colour description (see Plate 38). The variety names 'rubellite' for red tourmaline and indicolite for blue varieties have never enjoyed universal currency and are really unnecessary. Generally speaking, however, the alkali tourmalines are pink, red, green, or parti-coloured pink and green, and have the lowest SG of the group, varying between 3.01 and 3.06; magnesium tourmalines are brown, with SG 3.04–3.10, while the black iron-rich schorl varieties have an SG varying from 3.08 to 3.20. Yellow tourmalines and deep blue 'indicolite' types, neither of them very common, are among those with rather high SG, about 3.10.

The tendency of the elbaite and liddicoatite types (that is, the alkali tourmalines) to show parti-coloration has been mentioned briefly above, and is a very special feature of these tourmaline crystals, the colours concerned being almost invariably pink and pale green. The colours may appear in more or less concentric zones having as common axis the main axis of the crystal, or the whole crystal may change its colour along its main axis.

Besides the colours mentioned, completely colourless ('achroite') tourmaline is sometimes seen, while tourmalines of an almost emerald green have been found in Tanzania and Burma. Some of these contain vanadium and others owe their colour to chromium. The chrome tourmalines, true to form, show a narrow absorption doublet near 675 nm in the red, a broad absorption band centred near 610 nm in the orange, and have a bright red appearance through the Chelsea colour filter.

The refractive indices of tourmaline are usually near 1.62 and 1.64, with an average separation of 0.018 between the two readings. It must be remembered that it may be necessary to try the stone in different orientations on the refractometer to obtain the maximum double refraction. At right angles to the optic axis there is full double refraction. In the direction of the optic axis there is only single refraction, giving a reading for the higher index, corresponding to the 'ordinary' ray. This edge remains constant if the stone is rotated on the instrument, while the edge corresponding to the 'extraordinary' ray will be seen to move to a minimum position and then back again. The extent of the double refraction in tourmaline is of great practical importance, as it serves to distinguish this mineral from others of similar refractivity but lower birefringence, notably topaz and the rare minerals andalusite (0.01), apatite (0.003), and danburite (0.006), not to mention the singly refracting pastes which quite often have a refractive index in this region. Textbooks often quote 0.025 as the double refraction of tourmaline, but in transparent stones so high a value is seldom encountered, though in black, opaque tourmalines the birefringence may be 0.03 or even higher.

One exceptional case, however, should be mentioned. Small crystals of tourmaline, free of matrix, from Narok in Kenya were brought to light by Dr John Saul some years ago. These were transparent but of very deep red. Their density was normal (about 3.07) but their refractive indices and particularly their birefringence were unusually high, figures for refractive index being 1.626 and 1.658, giving a birefringence of 0.032. A partial analysis gave 3.6 per cent as the total iron content, but a complete analysis (and thereby perhaps an explanation of the high birefringence) is still lacking. At the lower end of the scale two brownish-red tourmalines were found to have the low DR of 0.014, and these showed a distinctly biaxial interference figure.

A curious effect is seen when some green tourmalines are tested on the refractometer: instead of two shadow edges, four edges are clearly observable or, in some cases, as many as eight! Attention was first drawn to this puzzling effect by Dr C. J. Kerez, and was reported in 1967 by R. K. Mitchell, who suggested that the phenomenon be known as the 'Kerez' effect. Since then, work on the subject has been carried out by C. A. Schiffmann, and by Professor H. Bank and others, and it has been shown that the anomalous 'extra' shadow edges are no longer visible after the stones have been repolished; as expected, the false readings are due to alterations that are only skin deep. The effect is thought to be due to local overheating when the stone was being polished originally.

Putting the matter briefly, any transparent gemstone having a mean refractive index of about 1.63 and a birefringence of between 0.015 and 0.020 may safely be taken as a tourmaline. The fact that only the lower index edge moves when the stone is rotated is an additional sign. If the table facet should happen to have been cut at right-angles to the optic axis, both edges will remain stationary during rotation, the edges being in their position of maximum separation the whole time. If, as often happens in practice, the optic axis is parallel to the length of the cut stone, then only a single edge will be visible in this direction, and maximum separation will be found when the stone is rotated to a position 90° from this.

The double refraction of tourmaline is strong enough to cause a doubling effect of the back facet edges when carefully examined with a lens through the front of the stone, unless the specimen is very small.

The effect is masked in a deep brown or green stone, due to the complete absorption of the 'ordinary' ray. The strong absorption of this ray can be seen in the dichroscope with brown stones, when one image of the window will be practically black. In the nineteenth century mineralogists took advantage of this effect in brown tourmaline by mounting two sections of the mineral, sliced parallel to the optic axis and set at right angles to each other, in circular wooden frames in what were known as 'tourmaline tongs'. These acted as a crude form of polariscope working on the same principle as Land's remarkable 'Polaroid'. Polaroids mounted in the manner of the old tourmaline tongs would, as a matter of fact, be quite useful to gemmologists.

In paler stones the dichroism is not so marked, though still distinct. It is weakest in the pale green stones, almost like pale emerald in tint, which have been produced by heat-treatment of suitable material from Namibia.

Tourmaline is very seldom free from flaw-like liquid inclusions. Although these really consist of thin transparent films, in certain directions they give the effect of being black and opaque, due to total reflection of light at their surfaces. Typical inclusions of this kind are shown in the photomicrograph reproduced as Figure 23.2.

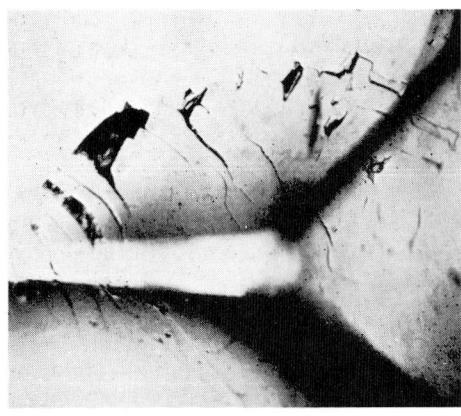

Figure 23.2 Typical inclusions in tourmaline

The absorption spectrum of tourmaline has been described in Chapter 10. It varies according to the colour of the specimen (as in other allochromatic minerals) and many varieties show no distinctive bands. It should be remembered that red tourmalines may show two narrow bands in the blue at 458 and 450 nm, which must not be carelessly confused with the far stronger bands at 476.5, 475 and 468.5 nm in ruby. The diffuse absorption region in the green has within it a distinctive fine line at 537 nm. These bands may be due to manganese. If a refractometer is not handy and a quick distinction from ruby is desired, it may be remembered that ruby is strongly red-fluorescent while tourmaline shows no fluorescence at all.

Before leaving tourmaline, one curious feature of the stone deserves mention. It is strongly 'pyro-electric' – which means that when undergoing a change of temperature opposite ends of the crystal become electrically charged in opposite senses. Pretty experiments can be carried out to display this effect, but the phenomenon makes itself known without any such devices when tourmaline jewellery is displayed in an illuminated window or display case and the stones are near enough to a light bulb to become perceptibly warmed thereby. Under these conditions, tourmaline attracts to itself an amazing amount of dust, becoming completely coated if left for any length of time.

Although the question of their identification has been incidentally discussed in previous chapters, two other gem minerals, peridot and spinel, are sufficiently important to warrant specific mention.

Peridot is the name given in England to the green transparent variety of the important rock-forming mineral olivine, which is a silicate of magnesium and iron, termed 'chrysolite' in older American mineralogy. The name 'olivine' once unfortunately gained wide currency in trade circles for quite another mineral, the green demantoid garnet from the Urals. Since the name olivine has been established for the magnesium–iron silicate for some 150 years there can be no question of its abandonment now. Thus, to prevent confusion, jewellers should avoid this name altogether, calling the green garnet 'demantoid' and using 'peridot' as a safe appellation for the true olivine.

Peridot is one of the easiest of all stones to recognize, both on account of its distinctive yellowish green colour and because its strong double refraction gives a clear 'doubling' effect of the back facet edges when viewed through the table with a lens. As far as colour goes, the only close match is provided by paste imitations, and these, of course, are easily distinguished by their single refraction, not to mention their other physical characters.

Recently, synthetic corundum rather similar in colour to peridot has been produced, but when compared with the actual gem, the colour is distinctly different, while sythetic spinel soudé triplets have also been made in 'peridot' colour.

Peridot is a representative of an isomorphous series of minerals of which the end members are a magnesium silicate (forsterite) and an iron silicate (fayalite). The SG and refractive indices of the mineral naturally vary according to the proportion of iron to magnesium present. The green types most commonly used in jewellery (from the Red Sea, Burma, or Arizona), all contain much the same amount of iron (about 10 per cent), and their properties do not vary greatly from 3.34 for SG, 1.654 and 1.690 for lower and upper refractive indices and 0.036 for birefringence.

The Island of St John in the Red Sea is the classic source for peridot and most of the stones used in jewellery originally came from there. Burma is the richest modern source and provides stones of fine colour and important size. Probably the most prolific source at present is in the desert region of Arizona and New Mexico, where ants often bring to the surface bright green pebbles, indicating the presence of peridot in the decomposed volcanic rocks below the sand. These can be clear and bright, if rather pale in colour, but also include darker green brownish stones which are richer in iron. A distinguishing feature of Arizona peridots consists of 'lotus-leaf' or 'waterlily leaf' inclusions (see Plate 41) consisting of delicate droplets surrounding a central rounded crystal, which Dr E. Gübelin has identified as chromite.

Small pebbles of peridot found on the Hawaii beaches have a beautiful colour, enriched in this case by traces of chromium. These stones are volcanic in origin and contain nearly spherical droplets of volcanic glass bearing a startling resemblance to gas bubbles seen in glass (Figure 23.3). Sometimes these

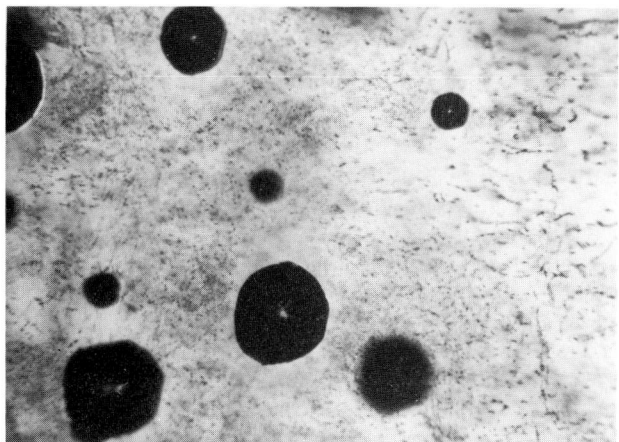

Figure 23.3 'Bubbles' of glass in Hawaii peridot

'bubbles' are oriented parallel to the orthorhombic outline of former crystal faces. In the main, peridots from all these sources, despite the fact that they belong to an isomorphous series, have very similar constants, the average figures for which are as given above. It should be noted that the difference between the intermediate refraction index ('beta') and the upper and lower extreme values is very close, and this fact can be used to distinguish brown samples of peridot from the mineral **sinhalite**, whose existence as a separate species was first established in 1952. The absorption spectrum of peridot also helps here, the spectrum being due to iron and consisting of three evenly spaced bands in the blue region, centred at 493, 473, and 453 nm. In sinhalite, although there are three bands in very similar positions to those in peridot, an extra band at 463 nm can be clearly seen. Further discussion on the distinction between brown peridot and sinhalite, on the rare occasions when their constants are nearly the same, will be found in Chapter 20.

Spinel also belongs to an isomorphous group of minerals but the stones used as gems are mostly almost pure magnesium aluminates in which traces of chromic oxide and ferrous oxide produce red and blue colorations, respectively (see Plate 42). More common than pure colours are less attractive intermediate shades to which it would be hard to give a name. Absolutely colourless spinel is practically unknown in nature, though stones of very pale mauve or pink are sometimes encountered.

Spinel is one of the most neglected and underestimated of all the stones suited for use in jewellery, probably because in the past it was, in its most attractive red variety, thought of as an inferior sort of ruby – a mineral with which, in fact, it has very little in common except a certain similarity in appearance. The term 'balas ruby', formerly often used for red spinels (including the famous 'Black Prince's ruby' in the Imperial State Crown), has long been abandoned, but no suitable 'selling name' exists to help its popularity with the jewellery-loving public.

The means of distinguishing red spinel from ruby and from pyrope garnet have already been described in Chapters 13 and 22. Under the microscope,

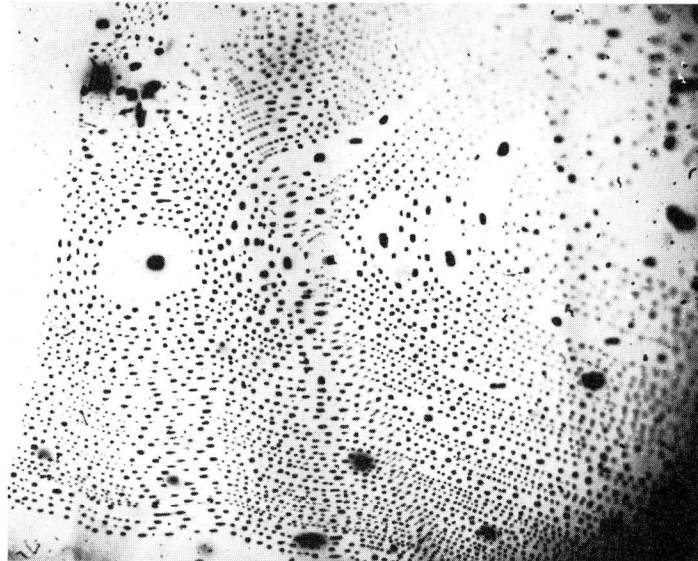

Figure 23.4 A curious ghost-like 'feather' in red spinel from Burma

spinel can usually be recognized by means of its typical octahedral inclusions. An interesting ghost-like 'feather' consisting of thousands of such small crystals is shown in Figure 23.4; this structure was found in a red spinel from Burma. Blue spinel is often very similar in appearance to indicolite tourmaline. Occasionally, blue spinels from Sri Lanka are found in which there is substantial replacement of magnesium by zinc and the SG and refractive index of such specimens are considerably higher than normal. In recent years a few natural cobalt-blue spinels have been shown to contain cobalt, a feature formerly regarded as being confined to synthetic blue spinel. Natural blue spinels show absorption bands at 434, 460, and 480 nm. Pure spinel has SG 3.58 and refractive index 1.715, while the highest zinc-rich '**gahnospinel**' may have corresponding figures up to SG 4.06 and RI 1.7542. There is nothing in the appearance of these zinc-rich spinels to distinguish them from the normal types.

Very occasionally, cut specimens of virtually pure zinc spinel, **gahnite**, have been encountered, but are too rare to be anything but collectors' items. Those seen have been transparent and tourmaline-green in colour. Figures for the density and refractive index of artificially produced pure zinc spinel have been given as 4.58 and 1.805, respectively. In nature there is some variation, owing to isomorphous replacements. A stone from Madagascar measured by the author had a density of 4.64 and refractive index 1.798, while a Brazilian specimen measured by Professor H. Bank had SG 4.55 and RI 1.792.

The absorption spectrum of these stones was typical of the ferrous iron spectrum of normal blue spinels, but with a different emphasis on certain of the bands. The specimen from Madagascar mentioned above had a weak band at 585 nm in the yellow, a moderately strong broad band at 553 nm, another at 509 nm, one centred at 460 nm, which was very strong, and finally a moderately strong band in the violet at 433 nm.

The most valued spinels are those approaching ruby-red in colour. In fact the colour, though attractive, never quite attains the rich crimson of ruby. This is fortunately very noticeable under the microscope when testing parcels of small rubies, and this fact, together with the lack of dichroism and difference in inclusions, are also signs which save the practised gemmologist from making a mistake when a spinel 'intruder' makes its appearance.

An excellent test for any pink or red spinel is to illuminate the stone with an intense beam of light which is concentrated onto the specimen through a flask of strong copper sulphate solution, and examine its fluorescence spectrum through a small prism spectroscope. The 'organ pipe' group of bright red lines seen in the spinel spectrum is completely distinctive. Garnet will, of course, show no fluorescence of any kind, while with ruby the fluorescence is concentrated almost entirely into the fluorescence doublet which, in a small prism spectroscope, appears as a single strong bright line. It should be noted, however, that **synthetic red spinels** show a fluorescence spectrum which to the eye resembles that of ruby, though the main line is at 684 nm instead of 693.5 nm. It is not quite clear how or why these synthetics are being made. Spinel boules, if they are to grow well under the Verneuil blowpipe should have a considerable excess of alumina compared with the 1:1 ratio of Mg:Al which is found in nature, and with boules of the usual 1 MgO:$3\frac{1}{2}$ Al$_2$O$_3$ formula the effect of adding chromium to the raw materials is to produce a green colour, not red. It would scarcely seem worth taking the special precautions necessary to grow red spinel boules of the 1:1 formula, considering the small demand for natural red spinel compared with ruby. However, some stones of this type *have* been produced, and one must keep a watchful eye open for them. Curved lines and bubbles can be seen in the majority of these stones, very similar to those in synthetic ruby. A dark purplish-red synthetic recently seen did not show curved striae, but contained numerous comma- and tube-like gas inclusions, typical of synthetic spinel. Because of their lower alumina content, these red synthetics have a refractive index of only 1.715 compared with the 1.727 one usually associates with synthetic spinel.

Blue synthetic spinel is very extensively manufactured, particularly in pale shades which imitate aquamarine rather effectively. Their orange colour when viewed through the Chelsea colour filter, cobalt absorption spectrum, and typical anomalous birefringence giving a 'tabby extinction' effect between crossed 'Polaroids' are all diagnostic.

Synthetic spinels also have a marked cubic cleavage, not seen in the natural stones, and this may manifest itself as cracks crossing one another at right angles in carelessly cut stones.

Natural blue spinels are seldom really blue in the sense that sapphire is blue; they tend to be greyish, greenish, or inky-blue. Through the Chelsea filter they do show a reddish tinge, especially cobaltian specimens, but this is far removed from the striking orange hue seen with synthetic blue spinels. The colour of blue spinel is probably due to ferrous iron, and this gives rise to a series of absorption bands. The key features are a broad absorption band in the blue, centred at 459 nm with a narrow band on the green side of this, at 480 nm. Weaker bands in the yellow and green are variable in intensity and position, but give the general impression of a rich and complex spectrum which is characteristic enough to be recognized by the practised gemmologist.

This 'blue spinel' spectrum is seen to some extent in the purplish types to which no exact colour description can be given. Such stones are lacking in beauty, and have small commercial importance. Spinel is one of the harder gemstones, being placed equal with topaz, the standard 8 on Mohs' scale. It takes a good polish, and in its pale varieties displays distinctly more fire than the corundum gems, though the numerical difference in this property is quite slight (0.020 for spinel; 0.018 for sapphire). In synthetic colourless spinel the 'fire' is quite perceptible: one of the factors which make it a popular substitute for diamond in cheap jewellery.

24

Jade and jade-like minerals

All the stones so far considered are transparent, and their beauty largely depends upon that factor. Certain minerals, however, of which jade, turquoise, and lapis lazuli are the chief, have a sufficiently beautiful colour and are sufficiently rare in pieces of fine quality to earn a place in jewellery despite the fact that they are translucent or opaque. The true jades and the many 'greenstones' which have passed as jade will be the subject of this chapter, while turquoise and lapis lazuli will be dealt with in Chapter 25.

Jade is a term which even mineralogists consider permissible for two distinct minerals. Of these, jadeite or 'Chinese' jade, is the rarer and more precious, though nephrite or 'New Zealand' jade has also been greatly favoured as a material for carvings by the Chinese, by whom the jade minerals are held in particularly high esteem. Jadeite is a silicate of sodium and aluminium and belongs to the important pyroxene group of rock-forming minerals, while nephrite is a silicate of calcium, magnesium, and iron, and belongs to the equally important amphibole group. The minerals owe their extreme toughness, which is one of their outstanding characteristics, to the fact that they consist of a mass of small interlocking crystal fibres or grains (see Plate 48). A number of other compact green materials have been used in place of jade, though chiefly for carvings, small bowls, bead necklaces, etc., rather than in actual pieces of jewellery. The properties of these materials will also be briefly noted below, to assist the reader to distinguish them from the true jade minerals.

Jadeite, because of its beautiful colouring and texture, is in great demand for use in rings, brooches, pendants, ear-rings, necklaces, and so on, and fine translucent green pieces command high prices. Though bright green is the most favoured colour, jadeite is also found in a wide variety of hues, including black, white, mauve, orange, and brown. The colour of green jadeite varies greatly in saturation, from palest shades down to a deep emerald hue. The colour is often mottled, and veins or small patches of bright green in pleasing contrast to a background of almost pure white jadeite are fairly common, and highly characteristic of the material. The texture (see Plate 46) of the stone is often more granular than fibrous but the grains themselves commonly show a fibrous interior. Slight differences in the hardness of the constituent grains cause it to possess a peculiar dimpled or shagreened surface (see Plate 47) when polished – an effect which seems only to add to the charm of its appearance. However, modern polishing using diamond powder produces an even, mirror-like surface – the dimpled charm is often absent in contemporary carvings. In some pieces a bladed, gently shining, crystalline texture is plainly visible to the naked eye, or under a pocket lens. The better qualities of jadeite are appreciably translucent, and owe their bright green colour to chromic oxide. Such jadeite has an absorption spectrum rather similar to that of emerald, though the narrow bands in the red are not so clearly defined and are in rather different positions. It may be added that green jadeite, unlike emerald, does not show red under the Chelsea filter. In addition, as described in Chapter 10, jadeite often shows distinctive absorption bands in the violet, and these may be very valuable in confirming its identity. By far the stronger of these is at 437 nm, and is an intense and very well-defined narrow band. In the paler varieties of jadeite it can be clearly seen by reflected light – or by transmitted light if the piece is thin and translucent enough to let through sufficient light. In brown or dark green pieces there may be too much general absorption in the region for the band to be visible. A dark green coarse-textured form of jadeite sometimes known as 'Yunnan jade' presents an unusual spectrum when enough light can be passed through a thin portion. The SG of jadeite of all colours is remarkably constant, and is almost exactly the same as that of di-iodomethane (3.33), so that a trial in this liquid will act as a useful check where its characteristic appearance is not sufficient to establish its identity. In the case of the jade minerals, refractometer measurements cannot always be relied upon as a means of distinction, as stones are usually cut with rounded or carved surfaces. Where the back of a cabochon piece is sufficiently flat and polished, a fairly good reading can be obtained, and jadeite often shows the two edges due to double refraction, which is rather surprising in a crystal aggregate. The refractive indices are 1.654 and 1.667, and the hardness nearly 7 on Mohs' scale.

The distant vision method can usefully be employed to obtain an approximate refractive index reading on polished cabochon surfaces of jadeite, and provides a valuable confirmatory test.

Jadeite has a comparatively low melting point (lower then nephrite) and a splinter fuses easily to blebs of colourless glass under the blowpipe.

Although jadeite has been found in Japan, California, the USSR, and Guatemala, and the presence of celts of jadeite which were left in various parts of Europe by Neolithic man argues the presence of other localities for the mineral, there is only one source for the varieties of jadeite used in modern jewellery, and for carvings and *objets d'art*. This is in Upper Burma, in the neighbourhood of Myitkyina, where jadeite occurs in dykes and is recovered in boulders from the local streams. The coarse-textured dark green jadeite often known as 'Yunnan jade' probably also emanates from Upper Burma. The trade is mainly in Chinese hands, hence the popular name 'Chinese jade' for the material.

In 1984 the General Electric Company of America announced the sythesis of jadeite. They produced various colours, including greens and lavenders. The textures obtained resemble natural material in many respects. There are no significant differences in refractive indices, fluorescence or specific gravity, but the 437 nm absorption band was not seen in synthetic material. The hardness was 7½–8 in contrast to the 6½–7 of natural jadeite. At present General Electric has no plans for commercial production, but a similar process has been patented in Japan. In view of the high cost of synthetic production and the ready availability of natural material, it seems unlikely that synthetic jadeite will be encountered on a commercial basis.

In fact the Chinese made no use of jadeite until the eighteenth century, all their ancient jade carvings being executed in **nephrite**, to which the Chinese, who venerated the material, gave the name *Yu*, which is also their name for precious stones in general. The name *fei-ts'ui*, 'kingfisher stone', is now used by the Chinese for jadeite only, though originally given to certain fine green nephrites.

The Chinese sources for nephrite were chiefly in East Turkestan (Khotan, Yarkand), from whence boulders of the amphibole jade have been transported for centuries over an incredibly arduous transcontinental route.

New Zealand 'greenstone' is a rather unsatisfactory name often used for the true nephrite jade originally used and worked by the Maori from boulders found in the rivers Arahura and Teremelau in the South Island. Modern carvings in New Zealand nephrite are mostly carried out in Idar-Oberstein, where skilled craftsmen copy Maori 'Tikis' and other designs, but pleasing Maori designs are now produced in New Zealand. The stone has a fine spinach-green colour and contains rather more iron than do the Chinese varieties of the same mineral.

Mineralogically speaking, nephrite can be equated according to its iron content either with tremolite, which is white in colour ('mutton-fat jade') or with the green mineral actinolite, which contains appreciable iron and supplies some of the asbestos of commerce. Both these minerals belong to the amphibole group, of which hornblende is the most important rock-forming member.

Large quantities of nephrite have come from Wyoming, California, British Columbia, and Alaska. Other well-known sources are in the Lake Baikal region of Siberia, where gigantic boulders are found, and Jordansmühl in Silesia. Siberian nephrite is a translucent spinach-green containing black spots of magnetite or graphite, while the Jordansmühl material has an attractive range of colour from ivory-white to a fine translucent green, most specimens being markedly veined or parti-coloured.

During recent years, Taiwan has become well known as a source of good-quality green nephrite jade but supplies have now dwindled. More surprisingly, green translucent stones showing a distinct chatoyancy have emanated from the same source. A natural tendency to market these stones as 'nephrite' cat's-eyes has been discouraged by leading gemmologists, for although the composition and properties of these stones are akin to nephrite, their structure consists of strictly parallel fibres to which the chatoyancy is due, whereas in nephrite the fibres are in random orientation as a tangled mass in typical jade fashion. Analysis has shown that the most accurate description of these cat's-eyes would be 'tremolite cat's-eyes'. Samples tested were found to have a density of 3.01 and refractive indices 1.615–1.631.

The SG of nephrite is markedly lower than that of jadeite, an average figure being 2.95, and the range being between 2.80 and 3.03. The refractive index is also lower, giving a shadow edge near 1.62 on the refractometer. The hardness can be given as 6½ on Mohs' scale, though the surface hardness of some old specimens seems higher, and other specimens may give anomalously low results, being easily marked by the point of a pocket-knife or needle. Nephrite is more fibrous than jadeite, and polished pieces lack the dimpled surface lustre of the latter.

Though so far only white ('mutton fat') and sombre green have been mentioned as colours for nephrite jade, shades of brown, reddish-brown, or yellow are sometimes seen, particularly in ancient, weathered pieces. The green colour is mainly due to ferrous oxide, which is an essential constituent; but the more richly coloured pieces may also contain a little chromium, and rather vague narrow bands in the red can be detected in consequence, though they are never so clearly developed as in jadeite. A weak line at 498 nm where the green changes over to blue can sometimes be seen, perhaps accompanied by even weaker lines nearby, but there is no absorption band in the violet beyond a general absorption of the whole region.

Before proceeding to describe the many minerals which have been or may be confused with either of the two jades, warning must now be given of a new advance in the gentle art of faking. Methods have been developed for dyeing pale specimens of true jadeite so that they have the appearance of valuable green jadeite (and lavender) of the highest quality. According to information received from Hong Kong, two organic dyestuffs, one blue and the other yellow, are used to obtain the desired effect. In some samples the colour can be seen, on close inspection, to be concentrated in the small cracks or veins in the mineral. But in others the colouring is surprisingly uniform, as though the dyestuffs had been able to penetrate the tiny crystals themselves. Many of these dyed jadeites fade on exposure to light. Some specimens are dipped in wax to produce a better 'polish'.

The other form of colour faking for jadeite is more elaborate, and involves careful workmanship, but as far as appearance goes the result is outstandingly effective. Here a piece of pale jadeite is hollowed out to form a thin-walled cabochon shell. A second piece of similar material is made to fit this exactly, and the two are cemented together with a layer of green dyestuff between them. In the specimens seen, no attempt has been made to conceal the fact that the whole is a composite stone, since a ridge can be seen running round the base of the cabochon. However, in mounted pieces this would not be detected, and high prices might very well be paid in error for stones which quite clearly show the characteristic texture and lustre of true jadeite and have a very fine green colour.

Fortunately, there are tests which can reveal the fakes once suspicion has been aroused. First, the dyed stones tend to show red or pinkish under the Chelsea filter, which natural green jadeite does not. Second, the spectroscope reveals the presence of dyestuff by means of a fairly strong though rather woolly absorption band in the red, while the narrow chromium lines which untreated jadeite of this colour would show, are missing. John Koivula found that many lavender jades of a pinkish-purple shade lose their colour when heated in the range of 220–1000°C. Furthermore, these jades showed a strong bright orange fluorescence under long-wave ultra-violet light. Lavender jades of a bluish-purple colour do not bleach and show only a weak brownish-red fluorescence in these conditions. If, therefore, a lavender jade fluoresces bright orange and shows suspicious traces of dye it should be treated with caution. It is a regrettable fact that all finely coloured pieces of jadeite must now be viewed with suspicion and preferably submitted to a laboratory test before good money is paid for such articles.

The colour of these stained jades can be removed by treatment with nitric acid or by less harsh reagents, but such a test is not a very satisfactory means of learning the truth.

Of the many jade-like minerals, the only stones likely to be seen in rings or brooches which might be mistaken for jadeite are the green chalcedony minerals **chrysoprase** and **chrome chalcedony**, **green-stained chalcedony**, and perhaps translucent **emerald**, which is sometimes cut in cabochon form. A distant-vision refractometer reading should reveal that their refractive index is considerably lower than for jade, and the spectroscope should also prove helpful in distinguishing them from jade and from each other. Chrysoprase owes its colour to nickel and shows only a weak and vague absorption line in the orange-red. The others all show the narrow lines in the deep red due to chromium, but in sufficiently different fashion for the practised eye to recognize the particular aspects of each.

Emerald has by far the clearest and crispest spectrum, and, being monocrystalline, this shows considerable variation with the direction of viewing or when a Polaroid disc is turned over the eyepiece of the spectroscope. The chalcedony minerals, in common with jade itself, are polycrystalline, which entails a less well-defined set of lines and a lack of variability with direction or under polarized light. Jadeite of a good chrome-green can show lines somewhat resembling those of emerald – a strong doublet in the deep red with two weaker lines on the orange side of this. The chrome chalcedony recently found and exploited in Zimbabwe shows a fairly clear-cut doublet but little besides, while the stained green chalcedony shows a fuzzy three-band spectrum consisting of a doublet and two weaker lines with transparency patches alongside them on the long-wave side. This spectrum may, however, show as a broad band in the red-orange, although stronger lighting on a thin edge may help to resolve it.

Though seldom met with in jewellery, green varieties of the zinc carbonate mineral **smithsonite** should be mentioned here as they can have a fine jade-like colour and translucency. At one time smithsonite was marketed under the trade name 'bonamite', after a firm called Goodfriend. It hardness (5) is much lower and its SG much higher (4.35) than for jadeite, and it shows no distinctive absorption bands. Being a carbonate, it lends itself to one simple if crude test: it 'fizzes' when touched with a drop of weak hydrochloric acid. The test is a useful one for many of the ornamental stones, but obviously, as with hardness tests, must be used with care and discretion. A small drop of quite dilute acid will suffice, and a lens is used to verify that effervescence is actually taking place. The drop of acid should be removed as quickly as possible: it will inevitably leave a dull mark on a polished stone.

Probably a safer manner of carrying out the test is to take a minute scraping of powder from the specimen, place this on a microscope slide, and to this add a small drop of dilute acid by means of a thin glass rod. The reaction can then be clearly observed under a low-power microscope.

This test for carbonates is only mentioned here on account of its extreme simplicity and frequent usefulness. There are many other chemical tests which can successfully be employed on the micro scale when dealing with the non-transparent gemstones. Some of these are described in Webster's *Gemmologist's Compendium*.

Many other green or greenish minerals have been used for beads, carvings, and small ornaments, and the name 'jade' with some place name attached has often been misleadingly applied to these. 'Transvaal jade', for instance, was a name applied to massive forms of green, pink, or orange-red translucent garnet found at Buffelsfontein in the Transvaal. This can resemble true jade rather closely and has gained some popularity as an ornamental stone. Gemmologists have been accustomed to referring to this material as massive grossular: but since it may contain as much as 5 per cent of water, replacing some of the silica in the grossular molecule, it is more correctly termed **hydrogrossular**. The same can be said of even more jade-like garnet rock found recently in Parkistan.

Correct designation of these attractive stones is made even more difficult by the fact that the garnet is mingled with varying proportions of **idocrase**, to which the name **californite** is commonly given when it occurs in this massive green form. The two minerals are in fact closely similar both chemically and structurally. Nevertheless the gemmologist, by careful study of the absorption spectra and SG of these mixed pieces, can obtain a fairly clear idea as to which species predominates. The SG of pure grossular can be taken as 3.60, while that of idocrase is considerably lower, perhaps 3.32. For the jade-like pieces from the Transvaal, Webster found that the majority came within the SG range 3.36–3.57; the author found a similar variation (3.28–3.52) for Pakistan material. It was noticeable that when the Pakistan 'greenstones' were studied in order of increasing SG, the stones with the lower SG showed a strong band in the blue at 461 nm and a weaker band at 530 nm – both typical of idocrase – while with the stones of higher SG these bands were hardly visible, and in their place a broad band centred at 630 nm in the orange, characterized by a rather sharp edge on the green side, became visible. This band is typical for hydrogrossular coloured green by chromium.

Massive grossular is known to show an orange glow under X-rays – a most useful aid to identification for laboratory workers. This glow seems to persist even when the sample consists mainly of idocrase. The cause for the fluorescent glow is not yet known.

The pale apple-green form of serpentine known as **bowenite** (see Plate 49) makes an attractive substitute for jade, and was wrongly identified as nephrite by the Dr Bowen whose name it commemorates. Bowenite also occurs in cream, yellowish, and brown varieties. This is curiously harder than other forms of serpentine, and also more translucent. The SG of those specimens examined by the author have been consistently near 2.59, with refractive index 1.55; but higher values have been recorded in the literature, presumably for specimens from different localities. Little groups or chains of greenish flakes of chlorite are often enclosed in bowenite, and are a helpful characteristic in the recognition of this very pleasing material. The Maoris gave to the exceptionally translucent variety of bowenite found in Milford Sound the poetic name 'Tangiwai', meaning 'tears'. Quite large pieces are found in China and carved into imposing figures, plates, cups, and so on. A rather softer but handsome oil-green serpentine is **williamsite**. This contains black octahedral inclusions. It SG and refractive index are slightly higher than for bowenite. A much softer green mineral sold as 'Styrian jade' is more properly called **pseudophite** or clinochlore, and belongs to the chlorite family, which is closely related to the serpentines. While bowenite can just be scratched with a needle, pseudophite is softer than even calcite.

Prehnite is another jade-like mineral, having a rather oily pale yellowish-green colour and a radiating fibrous structure which is rather distinctive. It is highly translucent; in rare cases, indeed, it is quite transparent and colourless. **Amazon stone** is not unlike jadeite, though its bluish green colour and typical feldspar sheen and 'shredded' structure are easily recognizable to the practised eye. The massive green idocrase, **californite**, has been used as jade, and so has the impure secondary mineral **saussurite** (see Plate 52), derived from the decomposition of basic rocks. It is a compound of plagioclase feldspars and zoisite/epidote. Saussurite may vary in colour from whitish to mottled shades of green and brown. The pattern is variable and a carving may resemble jadeite in one area and nephrite in another. The SG may vary from around 2.75 (whitish) to 3.25 in darker specimens. The hardness is usually around 6–7 and the refractive index around 1.55–1.56 for whitish areas and 1.68–1.70 for dark areas. A nearly white saussurite vase recently tested showed clearly the absorption band at 455 nm in the blue-violet, belonging to clinozoisite and the related mineral epidote. The green **aventurine quartz** (see Plate 51) has been called 'Indian jade', though its spangles of fuchsite mica render it easily distinguishable.

A type of aventurine quartz in which the spangles are not visible is sometimes encountered and might well be mistaken for malachite, as it has a rather similar tone of green with bands of darker and paler colour. Its SG refractive index, and hardness, however, are those of quartz (aventurine is really a quartzite containing chrome mica). Moreover, it shows red under the Chelsea filter and a chromium absorption spectrum, with a doublet in the deep red near 680 nm and another line on the orange side of this, is visible.

One more name should be mentioned. The mottled green material known as **verd-antique** is a serpentine marble. Its mottled and veined appearance, its effervescence when touched with acid, and a strong absorption band in the blue at 465 nm serve to identify this rather attractive rock (one can hardly call it a mineral).

Connemara marble, **Iona stone**, and **ophicalcite** are all names which have been used, according to its occurrence or the taste of its user, for this same material. **Verdite**, a variegated light and dark green rock sometimes with brown areas, is composed of a fine-grained chromian muscovite mica with some fine-grained rutile. Ruby is sometimes present (see Plate 50). It should not be confused with verd-antique mentioned above. Its extreme softness (3) as well as its appearance should make any confusion with jade unlikely. Verdite has a SG near 2.9 and refractive index 1.58. It is found in Zimbabwe and South Africa, and attractive carvings have been produced in recent years.

'Maw-sit-sit' is a convenient locality name for a brilliant green rock, often mottled with dark, almost black veins and spots, which was found by Dr Gübelin, who collected it, studied it, and had it analysed, to consist mainly of albite feldspar coloured with ureyite, a chromium analogue of jadeite. Once seen, maw-sit-sit is rather easily recognized. It can be distinguished from the true jade minerals by its lower refractive indices (near 1.53 in the paler parts) and SG (average 2.77). Its hardness, as one might expect, is that of feldspar (6 on Mohs' scale).

Steatite, or soapstone, is often encountered in Chinese carvings. It is often veined or mottled in yellows, browns, greenish and greyish shades. It is composed essentially of talc (H = 1) but admixture with other minerals may increase the hardness and SG. The base of a carving is usually scratched and the material will yield to the fingernail. The SG is about 2.7 and the RI approximately

1.55. Agalmatolite is a sack name given to various materials resembling steatite. Much agalmatolite is steatite, some is essentially fine-grained mica (muscovite) often with admixed quartz or feldspar.

Before closing the list of minerals which can be reasonably confused with jadeite or nephrite, one should mention translucent forms of **emerald**, and also **fluorite**. Both these attractive materials have been used for small carvings in the manner of jadeite, and have a rather similar colour. A SG test, or trial on the refractometer, may not be feasible. The emerald will show dichroism, and appear dim red between crossed filters, while both it and fluorite will probably show pink under the Chelsea filter. Fluorite is much softer than jade, it may have a marked violet fluorescence under ultra-violet light and an easy octahedral cleavage which may manifest itself in any surface chips or as incipient cleavage cracks. **Imitation jade** made of glass may be either very crude or quite effective. Scrutiny with a lens will reveal bubbles at or near the surface.

Long though it is, the above catalogue of minerals resembling jade could be considerably extended. The physical properties of the materials mentioned are summarized in Table 24.1 for convenience, though it must be realized that these minerals, unlike transparent gemstones, are seldom pure, and therefore these properties may vary considerably.

Table 24.1 Properties of jade and jade-like materials

Species	SG	Mean RI	H
Jadeite	3.33	1.66	7
Nephrite	2.8–3.1	1.62	6½
Amazonstone (microcline)	2.56	1.53	6
Aventurine quartz	2.66	1.55	7
Bowenite (serpentine)	2.6	1.55	5½
Californite (idocrase)	3.4	1.72	5½
Chaldedony	2.6	1.54	7
Emerald	2.7	1.57	7½
Fluorite	3.18	1.43	4
Hydrogrossular (garnet)	3.48	1.73	6½
Prehnite	2.87	1.63	6
Pseudophite	2.7	1.57	2½
Saussurite	2.7–3.2	1.55–1.70	6½
Smithsonite	4.35	1.62[a]	5
Steatite	2.7–8	1.55	1–2½
Verdite	2.75–3.25	1.55–1.70	6½
Williamsite (serpentine)	2.62	1.57	4

[a] The second RI is 1.849 but a vague shadow edge at 1.62 is all that can be seen on the refractometer.

The jeweller must not feel too appalled by the number of minerals mentioned. Few of them are actually used in articles of jewellery, and the main question, 'is it true jade or not?' can usually be answered either by inspection or after simple tests. In confirming the identity of jadeite the spectroscope can be very helpful, since it can be used no matter what form the specimen takes, and is effective for mounted stones as well as unmounted. As has already been stated in Chapter 10, a green jadeite shows well-marked chromium bands in the red. These are narrow but unlike those in emerald (which they otherwise somewhat resemble) in being rather diffuse and showing no alteration with direction or through a rotated Polaroid disc. Paler types of jadeite show a strong, narrow band in the violet at 437 nm. This is present in green types also, but may be masked by the general absorption of the violet. If the specimen is sufficiently thin and translucent these bands are ideally seen by transmitted light, but can also be observed by strong reflected light, with the slit of the spectroscope held close to the brightly illuminated specimen. A flask of copper sulphate solution will make observations easier by concentrating the light and removing the glare from the unwanted end of the spectrum.

If it is desired to identify one of the less easily recognizable jade substitutes it will be wiser to have it tested in a well-equipped laboratory or mineralogical museum having facilities for X-ray analysis.

It is of great assistance to have small samples of the various jade-like materials (most of which can be procured from a good mineral dealer) which will give the reader a better idea of their appearance and reactions to various tests than can be gained from written descriptions.

25

Turquoise and lapis lazuli

Turquoise is a hydrous phosphate of aluminium and copper, in which some of the aluminium is usually replaced by iron. The sky-blue colour to which the gem chiefly owes its attraction is due to a copper compound, while iron when present tends to impart a far less desirable greenish tint. For practical purposes, turquoise is amorphous, though small triclinic crystals of the mineral have been found. In thin section it is seen under the microscope to consist of crystalline particles, interspersed with amorphous whitish material. Specimens containing much amorphous matter are more porous, less dense, paler in colour, and softer than purer types.

Of all the stones commonly used in jewellery, turqoise probably presents the greatest difficulty to the gemmologist. Being subtranslucent its internal features cannot be studied; being multicrystalline its refractive indices yield only an average figure of no exact significance; and being porous it lends itself only too readily to impregnations with wax, plastics, or silicate solutions. Moreover, there are several natural minerals which resemble, or can be made to resemble, turquoise in appearance, and all manner of sheer imitations for the gemmologist to contend with.

As with all such problems, the best line of approach is to make a close study of the genuine material from all obtainable sources. The surface structure in particular of true turquoise is rather distinctive, the amorphous-looking pale blue background being interspersed with tiny shreds and speckles of whitish material, while small brown veins of limonite are often to be seen. In 'Egyptian' turquoise, which is of more intense blue and more noticeably translucent than other turquoise, small dark blue discs are noticeable against the paler background. The first step, then, when testing turquoise, is to examine carefully its surface under lens or microscope.

The next significant test is to study the absorption spectrum of light reflected from the stone. Light from a properly housed 250- or 500-watt projection lamp or other really powerful lamp, concentrated onto the stone through a flask of strong copper sulphate solution or any other good blue filter makes the detection of the characteristic bands in the blue a fairly simple matter. The strongest and most significant visible band in turquoise is in the violet at 432 nm, and is fairly narrow and distinct in most cases: a weaker and broader band in the blue near 460 nm helps to make a characteristic pattern. Though photography reveals a further absorption band at 420 nm equal in strength to the 432 band, this is not observable visually and thus has small diagnostic importance. None of the minerals similar in appearance to turquoise and none of its imitations show these turquoise absorption bands. Their presence thus forms a most rapid and useful check that turquoise material is present – and if the surface structure is 'right' and there are no signs of 'bonding' (see below), the test can be considered conclusive.

On the refractometer the 'distant vision' method should give a reading near 1.62 or 1.61 for turquoise, and provides a helpful check, or a warning where some distinctly different reading is obtained. Where stones are free from their setting a SG test can be very revealing: it serves as a guide to the quality and origin of the sample and separates true turquoise from similar minerals and from its imitations. Turquoise from the Sinai Peninsula (mentioned above as 'Egyptian' turquoise) has the highest SG, near 2.81. Iranian turquoise, which still comes from the ancient mines near Nishapur, has a SG only a little lower. Neither of these types is found in large pieces. Asiatic and various North American sources can, at their best, provide fine turquoise in pieces of considerable size: poorer qualities which are soft and porous are also plentiful, and these are impregnated in various ways to make them saleable. Even good-quality American turquoise is treated with wax to improve the polish and thereby the colour – without the addition of any actual colouring agent. The SG range of this type of turquoise is about 2.75–2.65: and 'turquoise' of SG lower than 2.6 must be treated with great suspicion.

The hardness of the best turquoise is about 6 or a little less – and it can thus be scarcely scraped with a penknife or marked with a needle, but care should be taken that it is not surface waxing which is being scraped. Lower grades and 'made-up' imitations are decidedly softer. Another point worth noting in turquoise is the waxy lustre it reveals at any chipped or broken surface. In this it differs from most of its substitutes.

The **synthetic turquoise** made by Pierre Gilson has a characteristic surface structure, and is described in Chaper 9. **Imitation turquoise** has been made from suitably coloured plaster of Paris, and from various strongly compressed powders, some of which may have a composition similar to that of turquoise itself. These lack the typical surface structure, and the absorption bands of turquoise, and are softer, more porous, and of lower SG. **Glass imitations** will show small bubbles just below the surface when diligently examined, and so will **porcelain**; these, and **enamel** and **stained chalcedony**, lack the surface structure of turquoise and have a vitreous lustre. Specimens of the hydrated borosilicate mineral **howlite** have recently been dyed blue to form a reasonably effective substitute for turquoise. Robert Webster found the SG of this material to vary between 2.50 and 2.57, and its hardness to be about 4½, which is softer than turquoise but harder than reported in textbooks. The blue dye used to colour the normally off-white mineral provides a broad absorption band in the green but no bands in the blue or violet. On the refractometer a vague reading at 1.59 was obtained.

Several types of 'bonded' turquoise have been produced in the USA, making use of natural turquoise too pale and soft to have any place as an ornamental stone unless it is treated in some such way. Some of the crushed material has been bonded with polystyrene resins: such pieces are of lower SG than true turquoise, tend to be sectile under a knife, and yield a characteristic odour when a fragment is heated in a small glass tube – though a destructive test of this kind is against normal canons of gemmology. More difficult to detect, because the SG is nearer that for untreated turquoise, is a similar mass bonded with water-glass (sodium silicate). Such bonded stones are dangerous in that they may show the absorption bands of true turquoise (some may even show an enhanced spectrum!), but the lack of the normal surface structure should serve as a warning, and any doubts be resolved by sending such difficult stones for a laboratory test.

Of the natural minerals resembling turquoise, **odontolite** or 'bone turquoise' is a fossil ivory or bone which has become stained blue by the iron phosphate mineral vivianite. Odontolite invariably contains some calcium carbonate and thus effervesces when a drop of dilute hydrochloric acid is applied to it. It can be distinguished both by its higher SG of from 3.00 to 3.25 and by its organic structure when examined under the microscope. The hardness is about 5 on Mohs' scale. **Variscite** is a green hydrous aluminium phosphate, hardness 5, and SG near 2.5. The mean refractive index is also lower than that of turquoise, being about 1.58. **Amazon stone** feldspar, already mentioned under jade, may also be mistaken for turquoise of poor quality, but again, its lower SG and refractive index as well as its sheen and 'shredded' structure will serve to separate the two. The mineral **lazulite** (not to be confused with lapis lazuli) is not often seen in jewellery, but should be mentioned here, as it sometimes closely resembles an inferior turquoise. It is rather more translucent, has a very similar refractive index, but a higher SG (3.1). Under the microscope it lacks the surface structures of turquoise mentioned above.

The hydrated copper silicate, **chrysocolla**, can form sky-blue cabochons resembling turquoise, though much softer (about 3) and of lower SG (2.2). When impure, chrysocolla can vary from blue to brown and black. When impregnating rock crystal its properties become nearly those of pure quartz.

Lapis lazuli

The use of **lapis lazuli** as a stone for adornment goes back to very early times, and the ancient mines in Afghanistan are still providing the finest material after being worked more or less continuously for some 6000 years. Almost alone among gem materials, lapis lazuli is not a single mineral but a *rock* in which blue minerals such as lazurite, haüyne, and sodalite are inextricably mingled with calcite, diopside, pyrites, and other species. In fine specimens the blue minerals sufficiently predominate to give an almost homogeneous appearance to the polished stone, though brassy specks of pyrites are almost always to be detected, and may form a useful sign that the stone is genuine. Poorer-quality material such as that from Chile contains considerable white areas, and even the blue regions are diluted to a paler tint than the true deep ultramarine.

The SG of lapis lazuli has often been given in textbooks as 2.38–2.45, but these figures are based on measurements originally carried out nearly 100 years ago on small fragments of the included blue minerals. The SG of fine pieces of lapis lazuli actually averages about 2.80, and 2.7–2.9 may be taken as the normal range unless much pyrites is present to raise the SG to higher values. True lapis lazuli may thus be distinguished from its commonest imitation, the stained jasper known as 'Swiss lapis', which has a SG of only 2.58. This 'Swiss lapis' material should actually be distinguishable at sight, since its colour is an inferior blue, it has a more vitreous lustre, and it contains little patches or veins of transparent quartz in contradistinction to the brassy specks of pyrites seen in genuine lapis lazuli.

In 1954 a new and effective imitation of lapis lazuli appeared. This was a form of synthetic spinel coloured heavily with cobalt, which was manufactured by heating the ingredients (oxides of magnesium, aluminium and cobalt) at temperatures rather below the 2130°C at which spinel actually melts. The result is a granular crystalline product, formed in the solid state, that is, 'sintered'. **Synthetic sintered spinel** is thus the correct description of this effective imitation lapis, which at first caused some stir but is now very seldom seen. Dr Jaeger was the inventor of this process, and the stones were marketed by the firm of Degussa in Frankfurt.

Specimens of this sintered imitation lapis are rather expensive: they are sold as small flat ovals or rectangular tablets ready for use as seal stones, for which their hardness, lustre, and fine colour are well adapted. If desired, small specks of gold can be inserted to simulate the brassy specks of pyrites which are known to be a characteristic feature of genuine lapis lazuli.

This imitation is not difficult to recognize, since its structure and all its properties are different from those of the mineral it represents. Examination under a lens reveals a granular texture, and under the Chelsea filter it appears a brilliant red (true lapis lazuli is a dull brownish-red under the filter). Readings on the refractometer are not clear-cut, but a shadow-edge near 1.725 can be seen (lapis lazuli gives vague readings near 1.50). The hardness of the sintered spinel is 8 on Moh's scale, and its SG is 3.52. By reflected light it shows a strong cobalt absorption spectrum, with bands at 650 nm (strong) in the red, 580 (weak) in the yellow, 532 (strong) in the green, and 480 and 452 in the blue.

The material is only feebly translucent, showing a peculiar reddish-purple colour when a strong beam of light is transmitted through a thin piece.

Other imitations of lapis lazuli are unlikely to prove deceptive. One blatant fake is a blue aventurine glass, spangled with triangular crystals of copper. A more ambitious imitation uses crushed Chile lapis lazuli bonded with plastics, and containing a fair sprinkling of actual pyrites particles to add verisimilitude. The pyrites particles, however, under a lens have clearly not crystallized *in situ*, and a suspected piece will yield a smell of plastics when touched with a hot needle.

A more ambitious attempt to produce a true synthetic lapis lazuli has been carried out by Pierre Gilson, as described in Chapter 9. The Gilson 'synthetic' lapis lazuli is handsome in appearance and may contain introduced fragments of iron pyrites to enhance the effect; it is, however, notably porous, low in density, and effervesces strongly when a small drop of dilute hydrochloric acid is applied, with a strong smell of sulphuretted hydrogen. Furthermore, it contains appreciable amounts of hydrous zinc phosphates and is, therefore, an imitation.

Two other blue non-transparent minerals which are sometimes cut and polished should be mentioned here. One of these, **sodalite**, belongs to the same group of minerals as the blue material in lapis lazuli, and has rather a similar appearance, though the colour of sodalite is bluer and less saturated than that of fine lapis lazuli. The hardness of sodalite (6) is the same as that of lapis lazuli, but the SG is distinctly lower, 2.30 being an average value. The refractive index (1.48) is also rather lower, but good readings are not easy to obtain.

The other mineral, **lazulite**, has already been mentioned under the minerals which might be confused with turquoise, which it resembles much more closely. It is the similarity in name that makes it seem worth bringing into the present context. The name 'lazurite' has often been used by mineralogists to represent lapis lazuli, or the blue content of the stone, and this is simply asking for trouble, since a real feat of memory is needed to remember that the '-lite' mineral is something quite different from lapis lazuli. Perhaps an easier mnemonic is to remember that the '-rite' mineral name is the 'right' one in this instance.

Finally, two recently discovered ornamental minerals may be described here. One of these, essentially a silicate of calcium and potassium, was found in the USSR near the Charo River, from which the name **Charoite** was derived. Charoite is opaque and unattractive as found but when polished its lively and unusual mottled purple colour is revealed and makes it a highly suitable material for carvings of various kinds (see Plate 27). Charoite has a hardness of 5–6, RI 1.55 and SG 2.68, and can be recognized by its distinctive and often fibrous appearance.

The other new mineral was discovered in Japan in 1976, and named **Sugilite** after a Professor Sugi. This gave no indication of gem potential until a purple manganoan variety appeared in the Wessel Mine, Hotazel, in the Northern Province of South Africa. Here it formed purple masses from which translucent pieces of fine violet-purple could occasionally be extracted and used for cabochons, beads, and carvings which have found great favour when marketed under the names 'Royal Azel' or 'Royal Lavulite' in the USA.

Sugilite is a complex silicate of alkali metals, iron, aluminium, etc., hexagonal in structure. The hardness is 6, density 2.74, and refractive indices 1.610 and 1.607 for the ordinary and extraordinary rays. Absorption bands due to manganese have been observed in the violet at 411, 419 and 437 nm.

In composition, structure and properties, sugilite closely resembles another rare mineral, sogdianite, for which it was at first mistaken.

26

Cat's-eyes, star stones, and others

Gems which owe their attraction to special optical effects can now be considered. Minerals which contain thin crystalline rods or fibres, in parallel formation, or canals where these have formerly been, are found to display a band of light at right-angles to the direction of the inclusions when they are cut *en cabochon*. This is due to reflection from the surface of the thin crystals, and the result is popularly known as a 'cat's-eye', the general name of the phenomenon being 'chatoyancy'. The more numerous and the finer the inclusions, the more perfect is the resulting ray (see Plate 54). The **cat's-eyes** most frequently seen are those of **chrysoberyl** and **quartz**. Of these, chrysoberyl provides stones which have a more sharply defined and slightly iridescent ray. It also takes a higher polish. When this is combined with a translucent body colour of honey-yellow or greenish-brown the resulting stone can be one of the most lovely of all gems, and of considerable value. As with all cat's-eyes, the ray appears at its best under a single light or in direct sunlight. Multiple or diffused illumination spoils the effect. The **alexandrite** variety of chrysoberyl is occasionally chatoyant and true synthetic alexandrite cat's-eyes have now been made by Kyocera of Japan.

Quartz cat's-eyes in their natural state are usually pale fawn or yellowish-brown in colour. The structures producing the ray are not fine canals as in chrysoberyl but fibres of asbestos, often too coarse to give a really silky ray. At their best, quartz cat's-eyes are very difficult to distinguish by sight from their more precious counterparts, while there are many inferior chrysoberyl cat's-eyes which present an untypical appearance. Where the stones are unmounted a quick and certain test is to place the stone in bromoform or di-iodomethane, in either of which quartz will float and chrysoberyl rapidly sink. Where the stone is mounted, a refractive index reading by the 'distant vision' method will settle the issue with equal certainty, quartz giving a reading around 1.55, while with chrysoberyl the shadow-edge will be near 1.75. Where enough light can be transmitted through the specimen, an absorption band in the blue-violet at 455 nm will prove the stone to be a chrysoberyl. Other indications are the higher lustre of the chrysoberyl stones and the tendancy in cutting to leave the base 'lumpy' to conserve the weight, where quartz cat's-eyes are more often neatly finished at the base. **Tourmaline, beryl,** and **chrome diopside** are among the species in which the chatoyancy effect is not very uncommon. Chrome diopside is a deep and sleepy green. Its SG is always near 3.30. Green or brown **enstatites** can also appear as cat's-eyes and have rather similar appearance and properties to diopside; but the absorption line at 506 nm is stronger and sharper. Very perfect cat's-eyes have been found in pale pink and deep blue or violet **scapolite** from Burma (see Plate 56). These scapolites have a SG near 2.61 and refractive indices 1.54–1.55; they are decidedly rare. **Apatite** is yet another gem mineral from which cat's-eyes have been cut. These are usually greenish and translucent, with a SG near 3.20 and a distinct didymium absorption spectrum, with fine lines in the yellow and green.

The occurrence of brownish-green transparent kornerupine in the gem gravels of Sri Lanka was first established by the writer and his colleague C. J. Payne in 1936, and since that date specimens, often of more attractive colour, have been recovered in Burma, Tanzania, and Kenya. Nevertheless, the mineral has always been considered more a collector's item than a commercial gemstone. Recently, however, **kornerupine cat's-eyes** showing remarkably sharp chatoyancy have been on sale in Sri Lanka in considerable quantity, and are said to be derived from gravel in the Matara district. The stones are, for the most part, small (averaging less that a carat) and dark green in colour. In common with other kornerupines they show strong pleochroism and have a density near 3.33, which conveniently matches that of di-iodomethane very closely. According to Professor P. C. Zwaan, the chatoyancy can be ascribed to included rutile needles in parallel formation. Graphite inclusions were also noted.

Another newly marketed form of cat's-eye consists of a fibrous green amphibole similar to nephrite in composition, emanating from Taiwan. These stones are best described as '**tremolite cat's-eyes**', since the term 'nephrite' should be reserved for the jade material in which the fibres are felted in random orientation, whereas an essential feature of the cat's-eyes is the strictly parallel orientation of the individual fibres whereby they produce a handsome cat's-eye effect. These tremolite specimens have an average density of 3.01 and their refractive indices average 1.615 and 1.631.

The handsome golden-brown chatoyant stone known as '**tiger's-eye**' quartz, or more commonly '**crocidolite**', was originally a blue mineral of the amphibole group, in which the long fibrous crystals have been replaced by silica, and oxidation of the residual iron content caused a change to its present attractive colour. Crocidolite is really the name of the original unaltered mineral. The appearance of this stone with its golden and brown banded structure make it easy to recognize at sight: its properties are similar to those of the other quartz minerals. A similar material, which has retained its blue colour, is found near Salzburg in Austria. Grey varieties, and paler types which have been stained in a number of unlikely colours, are sometimes seen in cheap jewellery. All these are apt to be included under the rather vague term '**Hungarian cat's-eyes**'. They are usually cut in long oval cabochons with very steep sides and a flat base. The ray is sharp and clear, but the stones are virtually opaque, and there is none of the subtle translucent background or living movement in the silvery ray that makes a fine chrysoberyl cat's-eye incomparable in its class.

Though not a mineral, and devoid of chatoyancy, what is sometimes called **shell** or **Chinese cat's eye** should be mentioned here. This is one form of **operculum** – a domed piece of shell which serves as a lid to the door or orifice of a snail-like (gastropod) shellfish. Those mostly used are from the species *Turbo petholatus*, which is found in the seas north of Australia as far as Vietnam. They are used in their natural form in brooches or bracelets, etc. The upper surface is domed, with a porcelain-like lustre and shades of colour grading through browns, yellows, and greens. The base is flat and lacking in lustre, and shows spiral lines of growth. The SG of operculum is 2.70 or a little higher, and the hardness about 3½ on Mohs' scale.

Asterism, or the star stone effect, is a phenomenon essentially similar to chatoyancy, the difference being that instead of a single shimmering band of internally reflected light, two, three or even six such bands are seen, which, by crossing, produce four-, six-, or twelve-rayed stars.

The inclusions, oriented parallel to important symmetry directions of the mother crystal, can usually be seen under the microscope if the illumination is suitably arranged.

The most beautiful **star stones** are those found in the corundum gems, **ruby** and **sapphire** (see Plate 2). In these the star normally has six rays, intersecting at angles of 60°. The clearest stars are usually found in translucent sapphires of pale greyish-blue, star rubies or sapphires of fine colour being extremely rare. The star is always displayed to its best advantage in direct sunlight, or under the light of a single bulb – and diffused daylight will show it at its worst. Very occasionally a twelve-pointed star may be seen in sapphire, the rays being alternately sharp and diffuse (Figure 26.1). Formerly a corundum showing a star

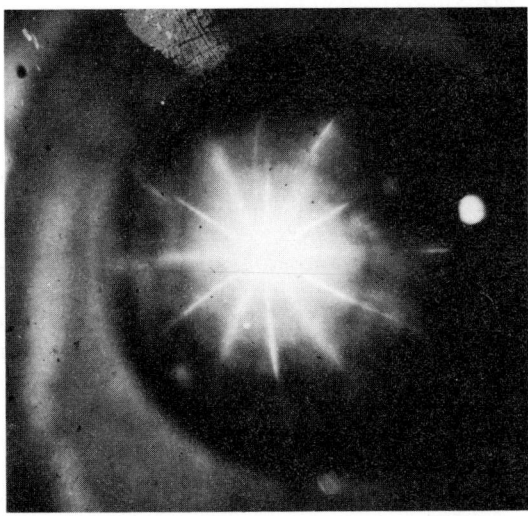

Figure 26.1 A twelve-pointed star sapphire

could with certainty be considered a natural stone. The **synthetic star rubies** and **sapphires** referred to in Chapter 9 have altered this comfortable state of affairs, and it is now necessary to examine each star stone critically to ensure that it shows the zoned structure of the native star corundums, and not the bubbles or curved colour bands of Verneuil synthetics (see, for example, Figures 26.2 and 26.3). Some synthetic star sapphires are obviously not natural, being almost opaque and with the rays bright and wire-like, and as though they were 'painted

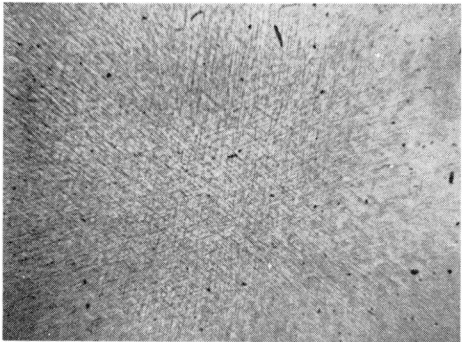

Figure 26.2 'Silk' in synthetic star sapphire

Figure 26.3 Zoned bands of 'silk' in natural star sapphire

on'. Others are much more natural-looking and may need very careful examination in di-iodomethane before revealing their nature. Almost always, the neatly ground-off base is a useful clue. Other stones to display asterism (see Plate 55) are **rose quartz** (six rays), and **almandine garnet**, which may show a series of faint four- and six-rayed stars. An ingenious form of doublet in which the brilliance of the rose quartz star is greatly enhanced has already been described in Chapter 21. A type of star stone which has been widely used in jewellery since its discovery some years ago is **star diopside** (see Plate 2). These stones are virtually black and opaque, but show a very brilliant four-rayed star under a single light. The SG of 3.35 is slightly higher than normal for the mineral, and the refractive indices are 1.674–1700, though usually, of course, only an approximate distant vision reading can be taken, as the stones are cut *en cabochon*. These stones are, however, easily recognizable at sight by the oblique angle of about 75° at which the arms of the cross intersect, and by the characteristic rod-like inclusions, some fine, some rather coarser, which can be seen under the lens. These inclusions are magnetite, and render the stone capable of being picked up by a strong pocket magnet.

One intruder into the star stone market has nothing natural about it. It consists of a cabochon of blue **glass** on the base of which fine lines, intersecting at 60°, have been scored. This is backed with a metallic mirror, as in the rose quartz doublet, and the star produced is remarkably effective. While these may be popular for costume jewellery, they can only deceive the unwary and the unforewarned. Glass is also used, especially in India (Cathay), to produce cat's-eyes in a variety of realistic and improbable colours.

The beautiful and unique sheen displayed by the feldspar known as **moonstone** is also due to the internal structure of the stone, which is, in fact, constructed of extremely thin layers of two sorts of feldspar, orthoclase and albite (see Plates 43–45). Light is reflected from these layers in lovely elusive silvery gleams, which have a decided blue moonshine tint in the finest specimens, which come from Burma. No substitute can really imitate this effect; opalescent glass or milky quartz or white chalcedony could only deceive those who are unacquainted with the real thing. Moonstone also occurs in other colours ranging from greenish-greys to buff and orange. Some of these stones are chatoyant, and recently four-rayed stars (not quite at 90°) have been marketed. Ordinary moonstone has a SG of 2.57, but **'black moonstone'** from Burma showing a similar effect though with more play of colour is a variety of **labradorite feldspar**, and has the higher SG of 2.69. A lovely form of labradorite is now mined in Finland and sold as 'spectrolite' when it is used in jewellery. **Sunstone** or **aventurine feldspar** is a translucent, pale brownish speckled and spangled material, seldom used in jewellery and having SG 2.64. For other aventurines, see Chapter 21.

The most effective substitute for moonstones, perhaps, is a milky heat-treated form of white synthetic spinel which has recently been marketed. Heat-treated amethyst, milky white in colour, is also used in Idar-Oberstein as a cheap substitute for moonstone. A SG test in suitable diluted bromoform will distinguish true moonstone from these inferior stones.

Brief mention may also be made of two opaque stones which are currently being used in cheap jewellery on account of their brilliant metallic lustre. These are hematite and iron pyrites, the latter being sold in the trade as 'marcasite', which is the name properly belonging to another less common mineral of similar composition. **Hematite** is an oxide of iron, of which it is an important ore. It has been mined, among other places, in Cumberland. When cut and polished, hematite displays a brilliant black metallic lustre, but its true colour is red, as can be seen when its powder is made to form a 'streak' by rubbing a specimen across an unglazed porcelain plate. The well-known polishing agent, 'jeweller's rouge', is made from powdered hematite. Spheres of hematite have occasionally been used to represent black pearls, but the resemblance between the brilliant metallic-looking surface of the mineral and the pearly lustre of the latter is very remote. The surface is sometimes cleverly 'dulled' to make the deception more plausible, but the high SG of hematite (4.95–5.16) and its red 'streak' serve as distinguishing factors if its appearance leaves any doubt in the mind of the observer. Some hematite is magnetic.

The name 'Hemetine' is the trademark of a clever substitute for hematite manufactured in the USA. Recent samples have a similar streak and SG to the true material, but although those seen by the author were strongly attracted by a pocket magnet, which may provide a rapid means of distinction from the natural mineral, others are reported as being difficult to identify.

Psilomelane is a manganese oxide (containing barium) which has been made into beads. It resembles hematite but has a more silvery lustre and has a brownish-black streak in contrast to the dark red of hematite. The hardness is around 6 and the SG around 4.35 in worked samples.

Pyrite or **iron pyrites** is the commoner of the two crystallized forms of iron disulphide. The brassy colour of this mineral has frequently caused it to be mistaken by the ignorant for gold. Seams and flakes of pyrite are often to be seen in household coal, where its presence in any quantity may cause minor explosions in the grate. Allusion has already been made earlier in this chapter to the specks of pyrites seen in most specimens of lapis lazuli. True **marcasite** is far less common, and is more liable to decompose than pyrite. Practically all 'marcasite' used in the trade to embellish inexpensive jewellery has been cut from crystals of pyrite. Its SG is nearly the same as that of hematite: 4.9–5.1. The most plausible substitutes for 'marcasite' are brass and stainless steel, but the low cost of the real material makes any such deception hardly worth while.

To conclude this chapter, a short description may be given of some opaque or translucent ornamental stones which are occasionally used in jewellery.

The appearance of **malachite**, with its handsome bandings and circular markings in dark and pale green, is familiar to most people. It is hydrated carbonate of copper, with hardness only 4 and a SG of near 3.8. Plastic imitating malachite is now manufactured; its lower SG should be a ready distinction. Synthetic malachite has been reported from Russia. **Azurite** is a dark blue copper carbonate containing rather less hydroxyl, and is sometimes interbanded with malachite in the same specimen. A decorative rock consisting mainly of azurite has recently been heralded in the USA as possessing an exceptionally attractive colour, hardness, and cutting quality compared with azurite or azurmalachite from other localities. The source is the Copper World Mine in California and the material is marketed as 'Royal Gem Azurite'. In this material, azurite assumes a more important role as a gemstone than heretofore.

Due to the carbonate content, malachite, azurite or mixtures of the two effervesce when a small drop of dilute hydrochloric acid is applied to some inconspicuous spot and the drop turns yellow. The density of azurite is variable, but well above 3.0, and thus distinctly higher than that of lapis lazuli.

Rhodochrosite is a translucent pink carbonate of manganese, sometimes marketed under the trade name 'Rosinca', and used for clock cases and the like. It is difficult to distinguish from the manganese silicate **rhodonite**, which has a very similar colour. The SG of each is variable and very similar (near 3.6) and a hardness test may here be justified.

Rhodochrosite is often banded parallel to the walls of the vein where it was formed, and the crystals are then seen to be perpendicular to the bands. In rhodonite the crystals are in random order, the material is less translucent and traversed by black veins.

Varieties of decorative **marble** are extensively used for clock-cases, ashtrays, bookends, inkstands, etc. Deposits of stalagmitic marble are mined in parts of Algeria and also in Egypt, Mexico, and the Argentine. Often these marbles are banded, and this has given rise to such terms as 'Mexican onyx', 'onyx marble', and the like, which are now frowned on: the term 'onyx' is better reserved for certain types of agate, a far harder material consisting of silica, whereas these marbles consist of calcium carbonate in the form of calcite. **Alabaster** is another name wrongly used for some types of marble. This term properly belongs to a white rock-like form of **gypsum**, a crystalline calcium sulphate containing water in its composition. Alabaster, like marble, is very suitable material for carving: in each the translucency and soft colouring are most attractive. Alabaster is even softer than marble (2 on Mohs' scale as against 4), and the SG is lower (2.3 as against 2.7). Marble gives a fairly good edge on the refractometer at 1.49; alabaster at about 1.52.

Most of the above minerals, it may be noted, are carbonates, and it is worth remembering that as a subsidiary test they will all effervesce when touched with a drop of dilute hydrochloric acid. Inevitably a dull spot will show where such a test has been made, so a safer procedure may be to scrape a trace of powder from an inconspicuous part of the specimen and carry out the acid test under the microscope on a glass slide.

Chrysocolla is a soft hydrated copper mineral which, when embedded in quartz, can provide attractive turquoise-blue cabochon stones. These have virtually the properties of quartz. Chrysocolla is also one of the mixtures of copper minerals which form the variegated blue and green rock popularly known as 'Eilat Stone', which is found near Eilat on the Gulf of Aqaba. This has a density of 2.8–3.2 and, unlike the azurmalachite it resembles, does not effervesce when touched with acid. Because of its copper content, however, it turns the droplet yellow.

It may be noted that, after detailed study of thin sections under the microscope, Professor W. F. Eppler prefers to classify 'Eilat Stone' as an impure form of turquoise.

27

Amber, tortoiseshell, coral, jet and ivory

Amber differs from most gem materials in being of vegetable origin. It is a fossil resin, originally exuded from various species of trees (including conifers) at different times over the past 250 million years and often still containing insects and other creatures trapped within it. Several types of amber are recognized, though their properties are practically identical; of these, Baltic amber, or 'succinite', is by far the most important. Baltic amber is commonly some shade of yellow, ranging from whitish- to brownish-yellow; it may be transparent, cloudy, or almost opaque due to the inclusion of myriads of small air bubbles. British ambers show similar variations (see Plate 39). Sicilian amber (simetite) and Romanian amber (rumanite), are seldom yellow, shades of brown, reddish-brown, and even black being more common. Romanian amber is often extensively cracked, but can nonetheless be turned and polished successfully. Burmese amber or burmite is generally brown. It contains many insects and sometimes veins in which calcite can be seen. Since World War II amber has been produced in considerable quantities from several localities in the Dominican Republic. Much of this amber is pale in colour but darker browns and reddish-browns also occur; darker material may bleach in sunlight. Some Dominican ambers fluoresce bluish or bluish-green in daylight and all fluoresce strongly in these colours under ultra-violet radiation.

Slight differences of composition are known to exist between these varieties of amber, all of which are essentially hydrocarbons. Succinite, as its name suggests, contains more succinic acid than the other ambers, giving rise to characteristic choking fumes when it is strongly heated. Gedanite is an amber-like fossil resin which is found with succinite. It contains no succinic acid and is soft enough to be scratched with the fingernail.

The hardness of the main varieties of amber is from 2½ to 3 on Mohs' scale, Burmese amber being the hardest type. The SG ranges from 1.04 to 1.10; the presence of numerous air bubbles in cloudy amber makes its SG slightly lower than that of the clear specimens. Being amorphous, amber has only a single refractive index, which averages 1.54. It softens at about 180°C, melts between 250° and 300°, and burns with a characteristic aromatic odour. One of the best-known properties of amber is the ease with which it becomes electrified when briskly rubbed – attracting small pieces of paper or other light objects. In this respect several of its imitations behave in a similar manner, so that this cannot be regarded as a distinctive test, except that where *no* frictional electricity is developed (as seems to be the case with the casein plastic described later) the material is certainly *not* amber.

First among the imitations of genuine block amber must be reckoned **pressed amber**, or **ambroid** as it is often called. Since 1881 this material has been extensively made from fragments of Baltic amber which are too small to be used as individual pieces. There are softened by heating between 200° and 250°C and pressed through a fine steel sieve or mesh to become amalgamated into a coherent mass which has very much the same properties and appearance as block amber. The best method of detection is by examination under a lens or microscope, or simply with the naked eye. Pressed amber shows a flow structure, globules of clear amber among the cloudy mass following definite lines. There may also be elongation of included bubbles parallel to one direction, whereas in untreated amber bubbles are usually spherical. One should add that some clear samples of pressed amber recently examined were virtually transparent, though with a treacly or oiled appearance. This material showed no bubbles and no sign of flow structure. The density was slightly lower (1.06) compared with a group of true amber pieces, as easily demonstrated by flotation in a suitable brine solution, and the 'strain birefringence' between crossed polars showed an invariable transmission of light compared with the bars of extinction crossing the stone as it was rotated, which is typical of block amber. Pressed amber may also show disc-like spangles or stress figures (see Plate 61). However, the most striking feature of this modern pressed material was the *brilliant chalky blue fluorescence* shown under long-wave ultra-violet light, under which illumination granular structures were also noted.

The more recent natural resin **copal** or **kauri gum**, which is extensively used in New Zealand, has very similar properties and appearance to amber. It can be distinguished by its readier fusibility when a hot needle is placed on some inconspicuous part of the specimen (comparison with amber will be necessary to make this difference noticeable) and by its greater solubility in ether. When a drop of this liquid is placed on copal resin it becomes quite sticky, and a dull spot is left on the surface when the liquid has evaporated. Amber is unaffected by this treatment, and so is pressed amber, despite statements to the contrary.

Copal resin can nearly always be distinguished by its tendancy to 'craze' at the surface. It is also much softer than amber when gently touched with a knife. Included insects are not always a sign of true amber. They may be found trapped in copal resin also (see Plate 60). Insects imbedded in plastics are usually far too 'perfect' to be mistaken for the real thing. Also they never show any sign of struggle, as they were dead before they were trapped.

Amber is effectively and frequently imitated by various artificial resins which are generally grouped together under the inclusive term 'plastics'. The earliest of these, the cellulose nitrate known as **celluloid**, is now less often used on account of its dangerous inflammability, though safety celluloid (cellulose acetate) bearing names such as 'Cellon' and 'Rhodoid' does not suffer from this disability. The casein plastics 'Galalith', 'Erinoid', etc. are more suitable, and most used of all are the phenolformaldehyde condensation products 'Bakelite' and 'Catalin'. Some actual determinations on the SG and refractive index of imitation amber and tortoiseshell made from these plastics materials are shown in Table 27.1, the properties of block amber, copal, and tortoiseshell being included for comparison. Further reference to tortoiseshell will be made later in this chapter. Due to differences in the filling materials used to give body and colour to these plastics, there may be some variations in their SG and refractive index, but the figures given are typical.

Table 27.1   Properties of amber, tortoiseshell and plastics

Material	SG	RI	Under knife
Amber	1.08	1.54	Splinters readily
Copal resin	1.06	1.53	Splinters readily
Tortoiseshell	1.29	1.55	Sectile
Casein, imitation amber	1.33	1.54	Sectile, rather tough
Casein, imitation tortoiseshell	1.32	1.53	Sectile, rather tough
Casein, imitation tortoiseshell	1.34	1.54	Sectile, rather tough
Bakelite, imitation amber	1.28	1.64	Sectile, tough
Bakelite, imitation amber	1.26	1.66	Sectile, tough
Cellon, imitation tortoiseshell	1.26	1.48	Sectile
Rhodoid, imitation tortoiseshell	1.28	1.49	Sectile
Celluloid, imitation tortoiseshell	1.38	1.49	Readily sectile
Celluloid, imitation tortoiseshell	1.42	1.50	Readily sectile

It will be noted that the SG of all these plastics materials, though low compared with gemstones, is considerably higher than that of amber, and this provides one of the most useful and reliable methods of distinction. It would, of course, be possible to dilute sufficiently one of the usual heavy liquids such as bromoform with toluene to make a solution of SG about 1.12, suitable for separating amber from these imitations, but this would be both wasteful and unnecessary, since a strong solution of common salt will answer the purpose equally well. Ten level teaspoonfuls of salt in an ordinary tumblerful of water (50 g salt in 250 ml of water) provides a liquid of the required SG. In this brine solution all specimens of amber, pressed amber, and copal resin will float, while all the usual plastic imitations will sink.

Two types of plastics not given in Table 27.1, since thay have not yet been used to imitate amber, should perhaps be mentioned here. One of these, the remarkably transparent 'Perspex' or 'Diakon', is a polymerized acrylic ester, and has SG 1.18 and refractive index 1.50. Most of the modern 'ivory' and coloured telephones are moulded in Diakon. The other, a polystyrene product now made in Britain under the name 'Distrene', has the very low SG of 1.05, though its refractive index is comparatively high (about 1.58). Distrene would, of course, float in the brine solution together with amber and copal resin, but the fact that is peels under the knife, its much lower softening point (70–90°C), and its ready solubility in benzene would enable it to be distinguished from amber if it should ever be used as an imitation.

In cases where it is not possible to obtain an unattached specimen, a SG test is still often feasible, especially with strung beads, where the weight of the thread is negligible. The necklace can be made into a 'bundle' secured with thread and weighed in air and water. The far greater SG of plastics will quite easily be apparent. Other tests are available. These, unfortunately, involve marking the specimen to a slight extent, but with care such damage is almost imperceptible when carried out on an inconspicuous part of the specimen such as the edge of the stringing hole of a bead. One such test, already implied in Table 27.1, is to determine the sectile qualities of the material by carefully controlled application of the sharp blade of a pocket-knife. It will be found that amber, pressed amber, and copal resin break away in chips or splinters under the blade, while plastics materials peel away in shavings. Bakelite is extremely tough and resistant to the knife; celluloid, Rhodoid, and Perspex yield much more readily, while Erinoid is intermediate in its toughness. The jeweller is advised to practise carrying out this test on known pieces of poor-quality amber and on any obtainable plastics before attempting to use it on specimens of possible value. Any peelings or chips removed can further be tested in the flame of a spirit lamp on the blade of a knife. Amber, pressed amber, and copal all burn with emission of aromatic fumes. Fine chippings or peelings can also be heated in a small test tube. The various ambers, copal, and plastic imitations all behave in characteristic ways, and this can be observed by careful heating in the tube. Amber, etc., all melt and give whitish fumes with an acrid smell, whereas the commonly used phenolformaldehyde plastics smell of phenol (or carbolic). Celluloid is extremely inflammable, but Bakelite and Erinoid, the latter smelling of burnt milk, only char under these conditions.

Glass imitations of amber are sometimes seen, and from a distance may look effective, but their coldness to the touch, hardness, and high SG compared with amber all serve to avoid any real confusion.

Though it can hardly be described as a gem material, a few words on **tortoiseshell** may be useful here, as it is very extensively imitated by the plastics which have just been described. The properties of the casein plastics and those of the safety celluloid type are very similar to those of true shell, as can be gathered from Table 27.1. Under the microscope, the dark patches of tortoiseshell will be seen to contain swarms of spherical reddish particles, whereas in the plastics the imitative dark patches lack this structure, and the edges of the dark areas are more sharply defined. Chips of tortoiseshell fuse to a black mass smelling of burning hair, while casein plastics char and smell of burnt milk.

Coral, like pearl, is a product of the sea. It is formed from the calcareous skeletons of myriads of tiny polyps which live in vast colonies in warm waters of moderate depth. Many forms of coral exist, but only the so-called noble or precious coral is at all extensively used for adornment. It is composed chiefly of calcium carbonate in the form of calcite, arranged as fibres radiating from the central axis of the curving coral branches. The SG is about 2.68 and the hardness a little less than 4 on Mohs' scale. Red and pink are the most popular shades, and staining is sometimes resorted to for obtaining a desired tint. So-called coral has been made by Gilson from crushed calcite, but it lacks the typical coral structure. Imitations of coral have also been made in galalith of suitable colours, but these can easily be distinguished by several tests. The SG of coral is far higher, and it effervesces briskly when a small drop of hydrochloric acid is applied to the surface. The grained structure of coral is also distinctive (see Plates 63, 64), and this is of assistance in separating it from the **pink conch pearl**, which is of higher value. Pink pearl also has a higher SG (2.84) than coral since its chief constituent is aragonite in place of calcite (see Chapter 28). Golden and black corals are also used nowadays, the former often showing a 'plucked chicken' structure (see Plate 65) and the latter showing annular rings resembling tree trunks.

Jet, like coal, is a form of fossilized wood and woody structures can sometimes be seen under the microscope. Its hardness is 3½ on Mohs' scale, and its SG about 1.30. Its former popularity as a material for mourning jewellery has dwindled practically to nothing in modern times. It can be distinguished from vulcanite imitations by the absence of sharp edges in the latter, which are moulded and not cut, and from black glass and black onyx by its lower hardness, lower SG, and warmth to the touch. A trial with a hot needle in some inconspicuous part of a jet ornament will evoke a smell of burning coal. Vulcanite subjected to this treatment will emit the distinctive odour of burning rubber.

Ivory is a favourite material for small carvings and *objets d'art*, and is also sometimes used as bead necklaces and other forms of inexpensive jewellery. When fresh it is creamy white in colour but tends to become yellow with age, as can be seen in the case of the keys of old pianos. It lends itself to intricate carving and takes a beautiful soft polish. Ivory varies considerably in quality: at its best it displays a fine texture and lustre, seeming almost moist and alive, while other samples may be blanched and dead-looking, resembling bone. The student who once defined ivory as 'animal dentures of all kinds' was not so far from the mark – substitute 'dentine' for 'denture' and his description would have been broadly correct. As everyone knows, the main source of ivory is the tusks of elephants, especially of the African elephant, but the teeth of other large mammals such as the walrus, narwhal, and sperm whale are also forms of ivory – although these are usually placed in a different category and the latter are conveniently referred to as 'marine' ivory.

There is also 'fossil' ivory from the tusks of the woolly mammoth, an animal that has been extinct since Ice Age times but whose remains have been very plentiful in Siberia and elsewhere. Actually no fossilization has taken place in mammoth ivory, and its composition and properties differ very little from those of modern elephant ivory. The teeth of the hippopotamus (which are also classed as ivory) are denser and finer in texture than ivory from the elephant.

In composition ivory consists chiefly of calcium hydroxyphosphate together with gelatinous organic matter. The hardness is 2½ on Mohs' scale and the refractive index about 1.535, although a good reading is difficult to obtain on the refractometer. The density of elephant ivory is usually between 1.71 and 1.78, the values for marine ivory and especially for the hippopotamus variety being rather higher (1.80–1.95). Under long-wave ultra-violet light all forms of ivory (as with human teeth) show a whitish to violet-blue fluorescence; this is brighter in samples from hippopotamus and walrus than with those from the elephant.

Carvings from elephant ivory are almost always identifiable, on careful inspection by eye or with a lens, by the curved crossing lines to be seen in transverse sections of the original tusk. These structures (sometimes referred to as the 'lines of Retzius') have often been likened to the 'engine-turned' pattern sometimes seen on the backs of watches, but this is only a convenient simile: the lines in ivory have quite clearly an organic and not a mechanical origin. They differ slightly in colour and translucency and are exceedingly distinctive. In some samples they are quite obvious to the eye, while in others they must be searched for. An excellent idea of their nature can be gained from Figure 27.1.

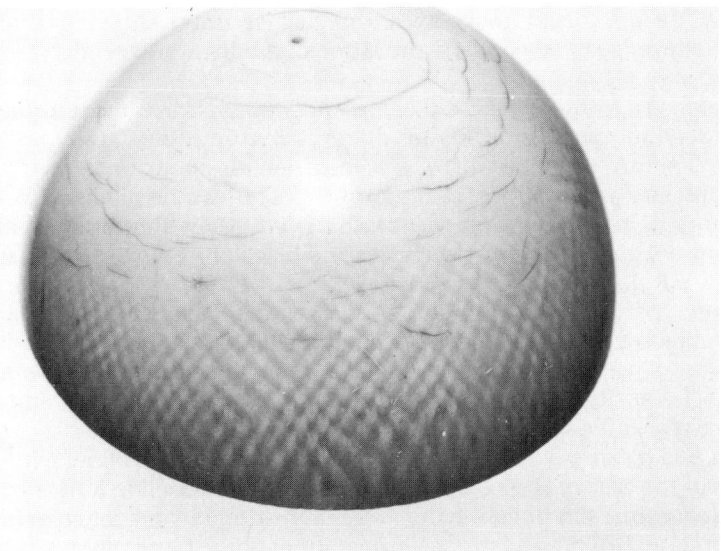

Figure 27.1 Characteristic structure seen only in elephant ivory (from Cavenago-Bignami, *Gemmologia*, published by Hoepli)

Longitudinal sections of ivory show parallel banding, which should be distinguishable by its irregularity and nature from the parallel lines often seen in plastic imitations.

Although ivory from the African elephant has hitherto been commercially the most important there is a move in certain countries (notably the USA) to ban its importation in an enlightened attempt to preserve a magnificent species now threatened with extinction. This may have the effect of increasing the trade in 'marine' and hippopotamus ivories.

Although the latter ivories do not show the 'engine-turned' structures that are so distinctive a sign of elephant and mammoth ivories in cross section, they may be recognized by the nature of the longitudinal striae seen on a polished surface, and, if need be, by the examination of a thin peeling taken from an inconspicuous part of a carving and examined immersed in a drop of oil or benzol under a low-power microscope. Very fine fine tubular structure lines should be visible in all forms of dentine ivory, usually crossed by the coarser lines of growth. According to Grahame Brown (in the *Australian Gemmologist*), in hippopotamus ivory the tubules show a crinkled structure; in walrus ivory the tubules are straight and of larger diameter; in sperm-whale ivory there are fine branching tubules with prominent growth bands, while in narwhal ivory the fine branched tubules are set in a granular matrix.

Of the natural substances resembling ivory, **bone** is the most common and has a similar composition but the structure when examined in detail is quite different. It is permeated throughout its length by a series of canals known as the 'Haversian' canals, through which in the living animal the fluids necessary for growth and health were able to flow. Under a low-power microscope a transverse section of bone shows a series of irregularly circular sections of the canals, each canal being surrounded by numerous dot-like lacunae, while a longitudinal section shows the canals as long white lines in which dirt has often infiltrated. The density of bone is distinctly higher than that of elephant ivory, averaging about 2.0.

So-called **'vegetable ivory'** consists of seeds of the corozo palm found in South America, or of the African doum palm. These nuts, roughly ovoid in shape, are too small to be of use except as material for small objects like thimbles and buttons. Corozo nut 'ivory' consists largely of albumen and is pure white and free from visible texture. The density of the vegetable 'ivories' is much lower than that of true ivory, being near 1.40.

The most common imitation of ivory is the plastic material known as xylonite or **celluloid**. This was first produced in Britain as early as 1865, by the reaction of nitrocellulose with camphor and alcohol. In the USA a similar discovery was made, stimulated by the desire for a cheaper substitute for the ivory from which billiard balls were traditionally made. Celluloid normally has a refractive index near 1.50 and density near 1.40, but this can be raised to as high as 1.90 for use as billiard balls by adding a suitable filler; zinc oxide is a favourite choice. Unlike ivory, celluloid is easily peeled under a knife and is readily, even dangerously, inflammable. When it is rubbed, the smell of camphor can often be detected.

28

Pearls: real, cultured, and imitation

The pearl holds, and always must hold, a unique position among gems. In its long and illustrious history it has ever been highly prized for its silvery beauty, which needs no enhancement at the hands of the lapidary. While closely connected with the galaxy of precious stones, it stands apart from these by virtue of its origin within the living body of a mollusc.

The structure of pearls is a consequence of their manner of growth within the pearl sac, which takes place in a very large number of successive stages. In each minutely thin layer a delicate membranous network of cells is first formed from the horny material known as conchyolin, followed by the deposition of minute aragonite crystallites within each cell, much as honey is deposited in the waxen framework of hexagonal cells in a honeycomb. This organic framework is on so fine a scale in oriental pearls that it is invisible even under high magnification, but it suffices to maintain the shape of the pearl after it has been completely 'demineralized' by treatment with acid, which removes all carbonate of lime from the framework. The above describes the growth of one pearly film only, and there may be many thousands of these superimposed and overlapping in the completed pearl. How thin these layers are may be gauged by examining the surface of a pearl under the microscope. The surface is seen to be crossed with extremely fine irregular but roughly parallel lines, which are actually the edges of the fine layers of deposition (see Figure 28.1). It is to the combined effect of the superposition of these thin translucent plates and their closely spaced overlapping edges that the 'orient' of the pearl is due. Examination of even the finest imitation pearl shows a complete lack of this highly characteristic structure. The closer the spacing of the serrated plate edges, the finer the lustre of the pearl.

Apart from these small-scale discontinuities of growth there are the seasonal cessations from growth which give rise to the well-known concentric layers which are easily visible to the naked eye in any sectioned pearl and can also be seen by looking down the drill hole of a pierced pearl with a lens (see Figure 28.2).

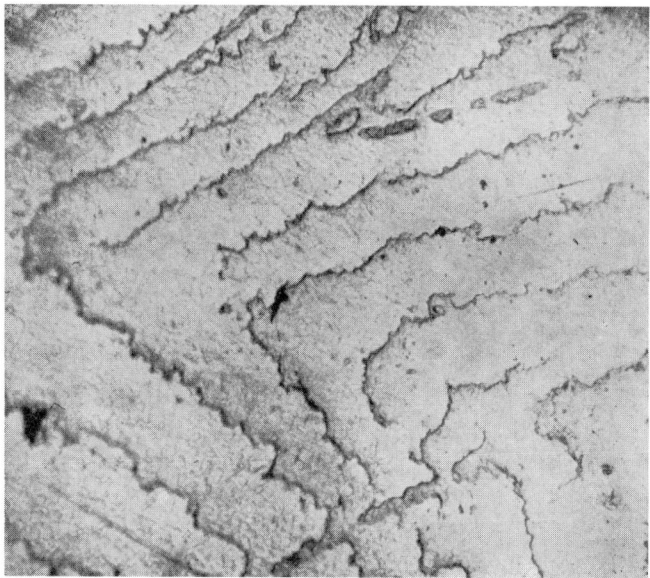

Figure 28.1 Surface structure of mother-of-pearl magnified approximately 100 times. Pearl shows a similar structure (photo: V. G. Hinton)

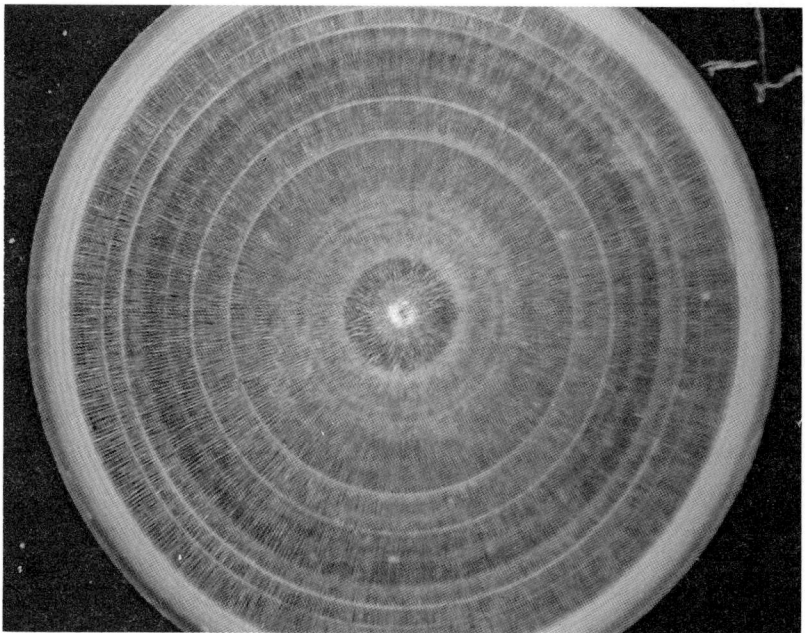

Figure 28.2 Section of natural pearl showing concentric structures composed of minute aragonite crystallites

Figure 28.3 Section of cultured pearl showing parallel-banded mother-of-pearl nucleus with concentric outer layers

In **cultured pearls** the structure is entirely different. In place of the vanishingly small or entirely absent nucleus characteristic of the natural pearl there is a mother-of-pearl bead in which the layered structure is flat or only slightly curved. The great difference between natural and cultured pearl structures can be seen in Figures 28.2 and 28.3. There have been rumours of small natural pearl nuclei having been used, but all supposed specimens of this kind so far examined by the author have proved to have the usual mother-of-pearl bead as nucleus. Occasional use of glass bead nuclei and even nuclei of steatite has been reported.

The thickness of the pearly coating of the bead in cultured pearls varies considerably. In the cheaper varieties often a mere film has been deposited, enabling the observer to detect the tell-tale sheen reflected from the underlying bead, and to see the 'stripes' of the mother-of-pearl structure when the pearls are held against a bright source of light. The most usual thickness is perhaps half a millimetre, and the thickest skins may be rather more than two millimetres thick, showing many layers. Contrary to common belief, it is not necessarily the bead with the thickest skin which has the best appearance, always provided that sufficient thickness has been attained to hide the irregular sheen of the underlying mother-or-pearl.

To one who is accustomed to handling pearls, the difference in appearance and even the 'feel' of cultured pearls is usually quite sufficient to enable them to be distinguished in a moment from natural pearls, especially when a number are seen together, as in a necklace. To put these minute differences into words is not easy, but the following points may be helpful to the beginner.

The lustre of cultured pearls tends to be waxy, as the outer coating is more highly translucent than the substance of an oriental pearl – though natural Venezuelan pearls are also notably translucent. The drill holes of cultured pearls are generally larger than those of new natural pearls, and are frequently crooked. Also, when a broacher is turned in the drill hole, cultured pearls feel perceptibly softer than the natural.

Examination of the drill hole with a good lens under a strong light shining on the side of the pearl will also help to distinguish between the two types. The drill hole should be thoroughly cleaned before making such an inspection. Cultured pearls have a sharp line of demarcation between bead and outer coating, forming a complete unbroken circle (see Plate 67); the junction is not infreqently marked be a black line of conchyolin, deposited perhaps by way of protest by the oyster before getting to work with the nacre. Below this line no further growth lines can be seen, whereas in natural pearls there is a series of these, the layers often getting more yellow or brown in tint as the centre of the pearl is approached.

When faced with a pearl necklace of doubtful origin, it is a good plan to undo the clasp, and to hold the string, fully extended, between the thumb and forefinger of either hand, rotating the beads slowly below a strong light. If the pearls are cultured, some at least will probably be found to reflect the tell-tale gleam of mother-of-pearl, as previously mentioned, when the correct angle is reached. Darkish subcutaneous markings, like varicose veins, are also typical of cultured pearls.

So much for ordinary visual means of distinction. As to scientific tests, various optical and X-ray methods have been evolved which give a sure distinction in all cases, but these require expensive apparatus and expert handling.

The endoscope

Historically, the most successful instrument for the rapid and certain identification of natural and cultured pearls has been the endoscope, which was designed by Chilowsky and Perrin and first produced in Paris in 1926. An improved design was marketed by the pearl dealers Simon and René Bloch, and twenty-five of these instruments were introduced into Britain to be used at the Hatton Garden Laboratory of the London Chamber of Commerce and purchased by some of the large pearl dealers, of which there were many in London at that time.

Endoscopes are still employed in the Paris Laboratory and India only, where several millions of pearls have been successfully tested with their aid. Elsewhere, they are little used, as new machines and the necessary needles have not been available for many years, and the introduction of non-nucleated cultured pearls has reduced their capabilities.

The essential part of this ingenious apparatus is a hollow needle, small enough to enter the drill hole of a pearl, into the end of which is fitted a short metal rod, polished at either end to form two tiny mirrors inclined, in opposite senses, at 45° to the length of the needle. One of these mirrors forms the extreme end of the needle, while opposite the other mirror there is a hole in the needle through which light can pass. Figure 28.4 will make the construction of the needle clear. For ease in handling, each needle is mounted in a flat-sided holder, at the rear of which is a circular orifice which enables it to slide on to a tube on

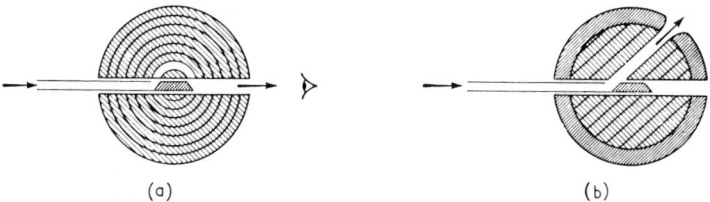

(a)   (b)

Figure 28.4 Principle of endoscope: paths of light in (a) natural pearl and (b) cultured pearl

the body of the endoscope. This tube is on a swivel operated by a lever, and when lowered into the horizontal position transmits a powerful beam of light from a carbon arc, passed into the tube from condensing lenses. Light from the arc passes along the needle until it meets the first of the two mirrors, whence it is reflected through the small hole, in a direction at right-angles to its original path.

When a pearl is threaded onto the needle and passed along it, the light from the needle penetrates into the body of the pearl. The really remarkable discovery made by Chilowsky and Perrin was that *in a natural pearl the tiny beam of light travels mainly round the major concentic layers of growth within the pearl*. It thus happens that, when the two mirrors are equidistant from the centre of the pearl, light travels from one to the other round the curved pearly layers, and can be seen as a little flash or maximum of light by an observer who is watching the end of the needle through a low-power microscope. Figure 28.4 should make this clear.

Full details of the different effects seen when drilled natural and cultured pearls are examined with the endoscope have been given in previous editions of this book, but since only in the Paris and Indian Laboratories is pearl testing by the endoscope used in practice, a detailed description now seems unnecessary. The introduction of the non-nuclear cultured pearl also makes any future use for the endoscope unlikely.

Testing pearls by X-rays

For the majority of part-drilled and undrilled pearls, X-rays provide the most powerful testing methods. X-rays were in fact applied soon after the first advent of cultured pearls in the early 1920s in an attempt to distinguish between these and natural pearls. The techniques were, however, not refined enough, and the method was for many years undeservedly discredited, at least so far as direct radiography was concerned. It is largely to the credit of the American ceramist and gemmologist Dr A. E. Alexander that the method was rehabilitated and the techniques improved sufficiently to make it the most generally favoured process for pearl testing in the different centres where such work is carried out.

In essence, the method is simple enough, but it needs considerable technical skill to obtain good photographs and great care and experience in interpreting results if mistakes are to be avoided. The 'radiability' (transparency to X-rays) of the mother-of-pearl bead used for the nucleus of cultured pearls is very little greater than that of the surrounding nacre. However, the oyster almost invariably deposits a thin layer of conchyolin round the bead before proceeding to add the desired coatings of nacreous substance. Thus a good radiograph of a

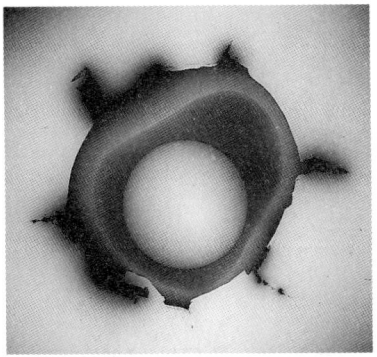

Figure 28.5 Radiograph of cultured pearl

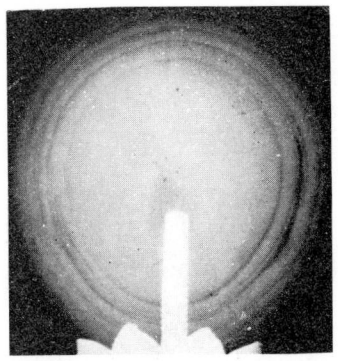

Figure 28.6 Radiograph of natural pearl

cultured pearl necklace will show in almost every instance a clear-cut black line or band round the spherical margin of the bead (see Figures 28.5 and 28.7). When one speaks of a 'black line' the negative picture is, of course, referred to.

In natural pearls, however, the layers of growth which are visible to the eye in a sectioned pearl, or with a lens when looking down the drill hole, are not always visible on an X-ray photograph, however excellently taken, and one is thus sometimes left with only the strong supposition that the pearl is genuine. There are also dangerous cases where the presence of a circular ring in the radiograph of a natural pearl may give the strong suggestion of an included bead. Careful inspection of such apparently perfect rings will usually show a slightly 'broken' margin at one point, and perhaps traces of further structures inside the circle, showing that it cannot be a bead. Further, looked at against a light with the naked eye it can be seen that the pseudo-bead at the centre of the pearl is slightly more transparent to the rays than its surroundings, whereas the core of a cultured pearl when similarly observed is seen to be slightly more opaque than its outer nacreous coating. A radiograph of a natural pearl is shown in Figure 28.6 (see also Figure 28.8).

In the Gem Testing Laboratory of Great Britain it is the practice to use radiography chiefly (1) as a means for condemning complete cultured pearl necklaces, the nature of which has already been virtually determined by visual inspection – this saves the trouble and time of cutting them from their string for an optical test and the expense of restringing – and (2) as a preliminary assay for part-drilled or undrilled pearls. Cultured pearls will usually reveal themselves unequivocally, as already explained, and many natural pearls also, particularly if they be drop- or button-shaped, as is often the case with such pearls. The operator examines the fine-grained negatives with a ×10 lens in front of a good bench lamp, and learns to interpret the fine structure lines which can safely be taken as proof of natural origin. The position and shape of these is often not what one would expect *a priori*. In drop pearls, for instance, one often sees an arc-shaped line crossing not far from the narrower end, which is curved in a similar sense to the domed outline of the top of the pearl.

Those pearls which are not satisfactorily proven by radiography are then subjected to the more stringent test of X-ray diffraction, often loosely though conveniently called the 'Laue' method, though the form of X-ray analysis first initiated by Max von Laue in 1912 dealt exclusively with single crystals – not, as here, with a roughly ordered aggregate of microscopic crystallites.

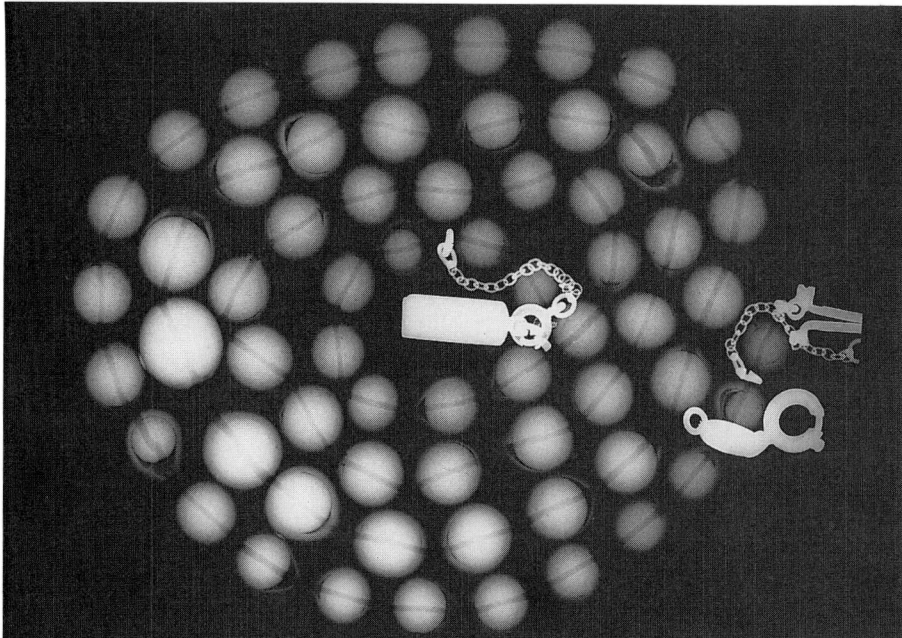

Figure 28.7 Radiograph of cultured pearl necklace

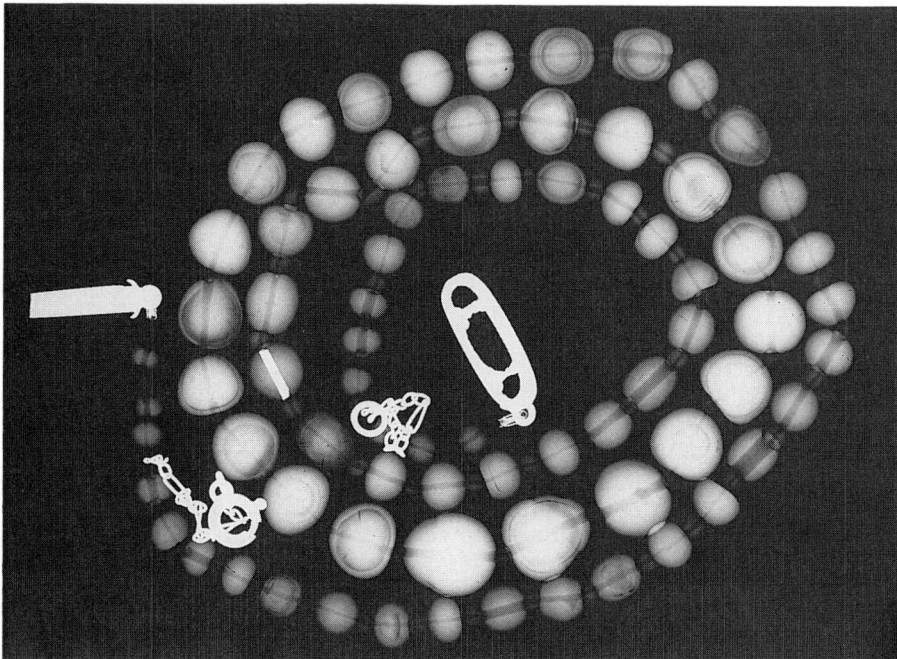

Figure 28.8 Radiograph of natural pearl necklace

When a narrow beam of X-rays is passed through a crystal onto a photographic plate or film a number of faint spots can be seen on the developed negative surrounding the heavy trace due to the undeviated beam. If the direction of the beam coincides with a symmetry axis of a single crystal, the spots are found to form a regular pattern in correspondence with the symmetry of the crystal in this direction. Each spot can in fact be considered as due to the diffraction or reflection of part of the rays by one particular series of parallel planes of atoms within the crystal. Reflection can only take place at certain angles where 'Bragg's law', $n\lambda = 2d \sin \theta$, is satisfied, where n is a small whole number, d the spacing between the atomic planes concerned, θ the angle made with the reflecting plane, and λ the wavelength of the X-rays used. Now if one spot is produced by a given series of parallel atomic planes, and the rays are parallel to a symmetry axis, it follows that other precisely similar spots must be produced in accordance with the symmetry of the crystal. In a hexagonal crystal such as beryl, for instance, a beam of rays passing down the sixfold axis forms a 'Laue photograph' displaying a beautiful sixfold symmetry pattern. A beam passed at right-angles to this would give a picture showing only twofold symmetry.

Now a natural pearl, in addition to its well-known concentric structure, also has a radial structure, since the tiny crystallities of aragonite of which it is chiefly built are disposed at right-angles to the growth layers. Thus a beam of X-rays passing through the centre of the pearl is passing down the pseudo-hexagonal axes of thousands of tiny crystals, and a diffraction photograph of the transmitted beam shows in general a roughly hexagonal series of spots round the main central trace. In some cases the spots are linked together to make a closed hexagon, while in others a broad ring or 'halo' is seen. It is an interesting fact that in button pearls, and to some extent in drop pearls, a halo pattern is seen when the beam has passed down the symmetry axis of the pearl, whereas in pictures taken with the beam at right-angles to this a well-defined spot pattern is the rule.

In cultured pearls the effect due to the outer skin is negligible unless the skin is unusually thick, and the pattern of spots produced will depend upon the orientation of the mother-of-pearl nucleus with respect to the X-ray beam. If the beam is passing at right-angles to the layered structure of the bead this means that it is passing down the pseudo-hexagonal symmetry axis of the crystallities, which are oriented at right-angles to the layers. Thus in this direction, or at small angles to it, a hexagonal pattern of spots is obtained as in a natural pearl. In directions parallel to the layers, or near to this, the beam is traversing the crystallites in a direction of twofold symmetry, and this is reflected in the spot pattern, which shows a maltese cross or rectangular design, with four spots round the main trace as the most important and persistent feature. The photographs reproduced as Figure 28.9 will make clear the nature of the natural pearl and cultured pearl patterns, and the marked differences between them. In testing pearls by the diffraction method, if the four spots of the cultured pearl pattern are clearly shown on the first trial position, there is no need to go any further: the pearl is cultured. If a hexagonal pattern or indeterminate picture is the first result, the pearl must then be turned through 90° and a second picture taken. A second hexagon means the pearl is natural, since a cultured pearl must show the incriminating four spots if turned 90° from a direction in which it does not.

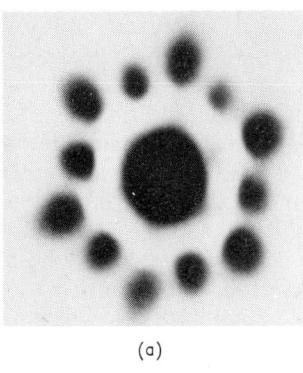

(a)

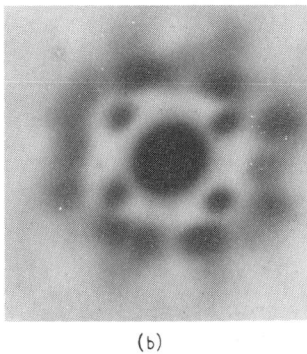

(b)

Figure 28.9 X-ray diffraction photographs. (a) Natural pearl; (b) cultured pearl

There are, of course, difficult cases, where the outer skin of a cultured pearl is abnormally thick, or the nucleus of either a natural or a cultured pearl is displaced from its normal position near the 'centre of gravity' of the pearl. Direct radiography can be used with profit to act as a check in such cases. A great deal depends upon the excellence of the pictures and the experience of the operator in interpreting results. It is worth remembering, for instance, that in a spherical or near-spherical pearl the patterns found when diffraction photographs are taken in two different directions should be practically the same: any irregularities from the hexagonal pattern will be seen in each picture, and this confirms that the pearl is indeed natural. One also gets to recognize the curious 'kite'-shaped pattern associated with a cultured pearl when the beam is passing in an intermediate position.

Many people have condemned this method as unsatisfactory because the diffraction photographs they have seen have been thoroughly bad – usually because the size of diaphragm used is too large and the pearl-to-film distance too small for the outer spot pattern to be clear of the heavy central trace. If the film is too far away, on the other hand, or the diaphragm too small, exposure times will be inconveniently long. In the London Laboratory we found that a pearl-to-film distance of 7 cm and a diaphragm consisting of a lead block pierced with a hole 0.7 mm in diameter gives very good pictures. Fast double-coated X-ray film is used in a metal oral cassette with intensifying screen at front and rear. With a Machlett X-ray diffraction tube with molybdenum target and two beryllium windows from which the rays emerge, and the set running at 50 kV and 20 mA, exposures of from 3 to 6 minutes (according to the size of pearl) are all that is needed, and two pearls can be photographed at the same time. It will be noticed that small pearls give larger patterns than large ones, since in the latter the longer waves emitted by the tube (i.e. the 'softer' radiation) are absorbed by the pearl and the shorter wavelengths produce a smaller pattern.

Reverting for a moment to the conditions for taking good pictures of pearls by direct radiography, one essential, if the delicate structures sought for are to be reproduced, is to use really fine-grained film, on which a ×10 loupe can be used with profit. Kodak Industrex MX film is excellent for the purpose. For cultured pearls, particularly if these have thin skins, it is important not to overexpose or overdevelop the image, as the fine line indicating the junction of the bead with the outer skin may otherwise be lost. On the other hand, with natural pearls a certain degree of overexposure may be advantageous to bring out the fainter internal structures. Makers such as Kodak produce dental films wrapped in pairs, and while these are not fine grained enough to be used for detailed work they can be advantageously used in pearl testing, since one film of the pair can be developed more than the other, thus simulating an exposure range. Intensifying screens should on no account be used in such tests. Dr Alexander and others following him have claimed that better radiographs of pearl are obtained when the pearls are immersed to a depth of about half their diameter in a liquid such as carbon tetrachloride, which has a 'radiability' somewhat similar to that of pearl. This is supposed to limit the fogging due to scattering of soft X-rays and lessen the troubles due to the sphericity of the pearls. There is probably some truth in this, but, after following this inconvenient and messy process (involving a liquid-tight plastic dish in which the pearls can be immersed) for some time it was found that quite comparable results could be obtained by lightly embedding the pearls in a thin layer of modelling clay, or even by placing them on the wrapped film quite 'naked and unashamed'. In the case of a single pearl, clearer results are obtained if the scattering is absorbed by lead sheet cut to surround the pearl.

There is another way in which X-rays can assist in distinguishing cultured from natural pearls, and that is by the fluorescence they excite in the various sorts of pearl. The bead forming the core of a cultured pearl is almost invariably made from freshwater mother-of-pearl. This contains a trace of manganese which causes it to emit a greenish fluorescence under X-rays. This glow can quite easily be seen in the majority of cultured pearls, shining through their thin translucent outer skins, provided that some arrangement is made whereby the effect can be studied in darkness through a heavy lead-glass window, with the whole apparatus properly screened to avoid exposing the operator to the highly dangerous X-radiation. There is also a short phosphorescent glow when the rays are switched off.

The only type of natural pearls to show fluorescence comparable in strength with that shown by cultured pearls with a bead nucleus are those from freshwater mussels; but the glow is rather different here, being yellowish in tint and patently showing on the outside of the pearl rather than from the core within. Non-nucleated freshwater cultured pearls from Lake Biwa and other areas also show a strong overall fluorescence under X-rays.

In cultured pearls with the usual mother-of-pearl bead nucleus the fluorescence test can be made much more specific, provided that the pearl is at least partially drilled, by the following technique. The drill hole of the pearl (cleaned from pearl cement or other foreign substances) is so oriented as to face the observer of the X-ray fluorescence effect. If the pearl is cultured, the glow coming from the drill hole will be brighter than that from the surrounding pearl, since it comes unmasked from the fluorescent bead itself. If this effect is seen (a head loupe may assist one in the observation) the evidence is very strong that the pearl contains a mother-of-pearl bead, and is therefore cultured.

It should perhaps be mentioned, before leaving the subject of radiography in pearl testing, that imitation pearls are immediately revealed by such X-ray pictures. Either they show as opaque discs (see Figure 28.10) – this being the case with solid glass pearls – or they show a thin opaque or nearly opaque circle with a nearly transparent centre – this representing the type which is made from hollow glass beads filled with wax.

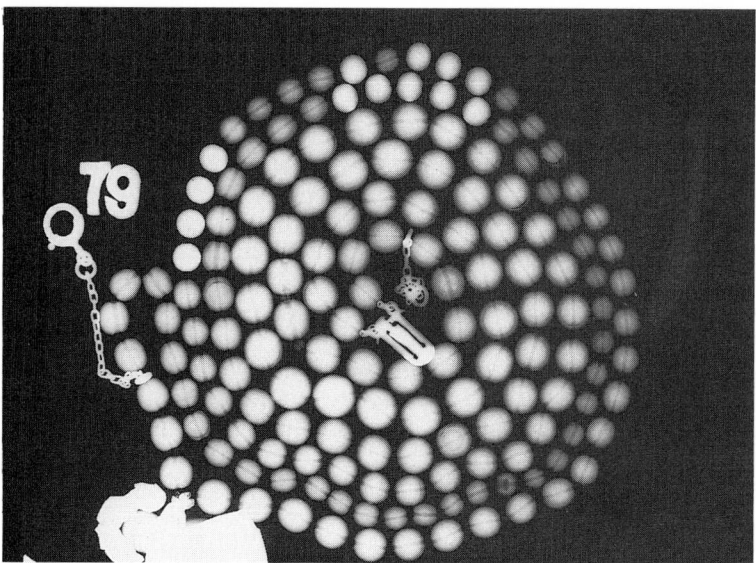

Figure 28.10 Radiograph of mixed pearl necklace showing four imitation pearls immediately right of the 79 number, with nine others present. The necklace also contains natural and cultured pearls

So far, only methods based on visual inspection or those involving expensive instruments and careful training have been suggested to the reader. However, there is one test that is fairly simple to carry out and is quite independent of any judgement based on visual appearances, which can be suggested here. This depends upon the perceptible difference between the *average* density of beaded cultured and natural pearls, and involves the use of the heavy liquid bromoform, already mentioned in Chapter 5. If the test is to be confined to a single liquid it is best to dilute the bromoform with 1-bromonaphthalene until a piece of Iceland spar (pure calcite) used as an indicator, just slowly rises to the surface. Pure calcite has a SG of 2.710, so that we can reckon such a liquid to have a SG of 2.715 or 2.720. In such a liquid the majority of natural pearls will float (on average, about 80 per cent), and any which sink will only do so slowly, whereas the majority of cultured pearls will sink (on average, over 90 per cent), some of them quite decisively. The SG figure 2.74 can be considered the upper limit for Arabian Gulf pearls, whereas with cultured pearls of ordinary quality more than 60 per cent exceed this figure. Venezuelan and freshwater pearls have a less sharply defined range, but on average also have a lower SG than cultured types. Australian pearls alone among natural pearls approach cultured pearls in SG, and even with these the average is definitely not so high, less than 25 per cent exceeding 2.74. In black pearls, on the other hand, the SG tends to be rather low; around 2.65 is a common figure.

Non-nucleated cultured pearls

All the tests described so far have applied to cultured pearls based on a sizeable nucleus of mother-of-pearl. These conventional cultured pearls have altered very little in their nature and properties since they first reached the market in the early 1920s, and they still continue to be the predominant commercial form of cultured pearl.

However, the reader must now be warned of the very considerable production of another type of cultured pearl, in which the bead nucleus is lacking. There have been attempts to persuade the trade to accept these as natural pearls, but since their growth has been initiated by man, and the molluscs producing them are cultivated, it is clear that they are correctly described as 'cultured'.

Most of these non-nucleated cultured pearls are produced on pearl farms in shallow basins round lake shores. Traditionally, the main producing area was Lake Biwa in Japan and non-nucleated cultured pearls were colloquially known as 'Biwa' or sometimes as 'Biwa-ko' pearls, the termination '-ko' being simply the Japanese word for lake. Today, however, large amounts of non-nucleated cultured pearls are produced in China, both in freshwater lakes and seawater. At Lake Biwa the pearls were induced to grow in large freshwater mussels (*Hyriopsis schlegeli*) which are fished from the lake (see Plate 53). These mussels are so full of twisted viscera that attempts to seed them with bead nuclei have not been a commercial success. Instead, small strips of the pearl-secreting mantle taken from one mollusc have been inserted in slits made in the edge of the mantle in another mussel, which is then returned to the water in a cage suspended from a raft. The resultant pearls are usually oval or baroque in shape, and have a distinctive lustre which is bright yet rather oily. In size they are rather small – mostly under five grains. If the mussels are again returned, a 'second generation' of rather better quality can sometimes be gathered.

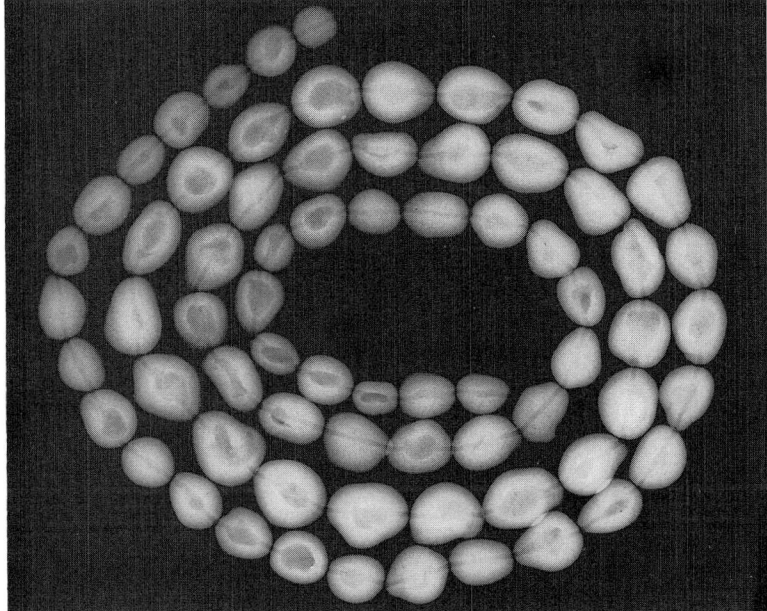

Figure 28.11 Radiograph of non-nucleated cultured pearls

Figure 28.12 The curious central formation seen in halved Biwa cultured pearls

Non-nucleated cultured pearls are not difficult to recognize at sight when one has handled them previously. Proof of their origin is provided by radiography (see Figure 28.11) which reveals small, elongated irregular hollows or conchyolin patches at or near the centre of the pearl (see Figure 28.12). These are unlike any of the structures seen in natural pearls, where any conchyolin 'pips' at the centre are circular in outline and often banded like an agate.

Freshwater pearls contain traces of manganese, which renders them highly luminescent when exposed to X-rays. The Biwa and other freshwater non-nucleated cultured pearls are no exception, and in fact are more fluorescent under the rays than any other, showing in addition a strong and prolonged phosphorescent afterglow. Where X-rays are available this test is the quickest check on their origin, though it does not in itself *prove* the pearls to have been cultured.

There has also been cultivation of non-nucleated pearls in Australian waters using the large pearl oyster, *Pinctada maxima*. These tend to form large misshapen drops and buttons, and radiography reveals tell-tale structures near the centre, similar to, though larger than, those in Biwa pearls. On occasions, non-nucleated cultured pearls are produced unintentionally by the Australian growers. In the effort to grow a beaded cultured pearl a mother-of-pearl bead plus a square of mantle tissue is placed in the gonad of the pearl oyster and it is returned to the sea. The oyster may at this point reject the bead but retain the tissue, and in so doing form a cultured pearl around its unintended irritant.

Byproducts of the culturing process of which this is such an example have been termed 'Keshi' (a term which describes a grain of rice in both appearance and size) by the Japanese. Other examples of how and when the 'Keshi' are said to be produced are (1) within the cultured oyster but before the mollusc is opened to insert the bead and (2) later as a result of the bead insertion but forming individually and separate from the beaded product.

Blister and cultured blister (Mabe) pearls

Natural pearly growths are sometimes found adhering to the nacreous lining of a pearl oyster, and if well shaped can be cut from the shell and used in jewellery.

Good examples can be very effective when suitably mounted, and can command quite good prices. They should, however, always be described as *blister pearls* to distinguish them from true cyst pearls, which alone are entitled to the term 'pearl' (French *perle fine*) without qualification.

Cultured blister pearls have been known for half a century or more, and were marketed before the complete cultured pearl could be successfully produced. The early cultured blister pearls, known as 'Japanese' or 'Jap' pearls, were induced to form by cementing small pellets of mother-of-pearl to the nacreous lining of the shell of a suitable mollusc, which, after being returned to the sea for some years, covered the intruding pieces with a symmetrical dome of mother-of-pearl (see Figure 28.13). These artificially produced blisters were

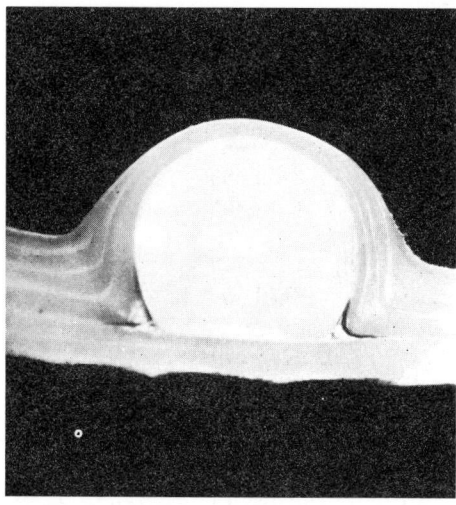

Figure 28.13 Section showing formation of early type of cultured blister pearl before removal from shell

sawn from the shell and the rounded shape completed by cementing to their surface a backing of polished mother-of-pearl. The cultivation of blisters still continues on a fairly large scale even today, but the final processing is now different. The thin domed pellicle of mother-of-pearl formed over the intruding pellet is lined with wax and then backed with a much smaller bead and finished with a domed base of mother-of-pearl. The modern name for these cultured blisters is 'Mabe' pearls. Even when mounted so that the base cannot be seen, the appearance of these cultured blisters is rather distinctive, so perfectly symmetrical is their domed surface. However, radiography can always be called upon to give a definite ruling where there is any doubt. The transparency of the wax filling to the rays makes a radiograph of a 'Mabe' blister pearl quite a striking sight, as can be seen from Figure 28.14.

In addition to the pearls produced by different varieties of the so-called pearl 'oyster' and the freshwater pearl mussels, pearls are also yielded by the shellfish known as abalones, by the Pinna, the giant clam, and the great conch. Pearls from these last two molluscs cannot be classed as true pearls, as they are not nacreous – that is, they have not the fine overlapping plates of nacre which cause the true pearly lustre. Products of the conch, *Strombus gigas*, commonly

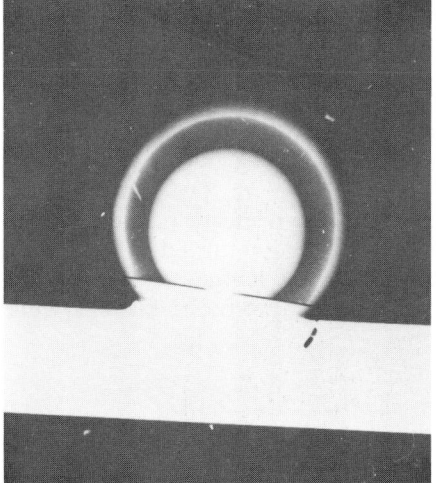

Figure 28.14 Radiograph of 'Mabe' cultured blister pearl (positive print)

Figure 28.15 Surface structure of pink (conch) pearl

called **pink pearls** (sometimes orange coloured (see Plate 59)), have a distinct beauty of their own, and fine specimens are quite valuable. At first sight they look like polished pink porcelain, or polished coral, but they show a silvery sheen at certain angles by reflected light, and under a lens or microscope reveal an attractive flame-like surface figuring by which they can be recognized (see Figure 28.15). Pink pearls can also be distinguished from coral by their high SG (2.83–2.86), the SG of coral being about 2.69. Pearls of a kind, and very similar in nature to the conch pearls just described, come from the giant clam *Tridacna gigas*. These have a SG and structure not unlike conch pearls, but are mainly white. Their patterning is not so attractive as in pink pearl, and they are seldom seen in jewellery.

Shiny, non-nacreous black or dark brown pearls like boot buttons are sometimes seen, which are called clam pearls in the trade, though their precise origin is obscure.

Imitation pearls, however cleverly made, are not difficult to detect on close inspection. Their pearly lustre is due to a mixture containing fish scales (*essence*

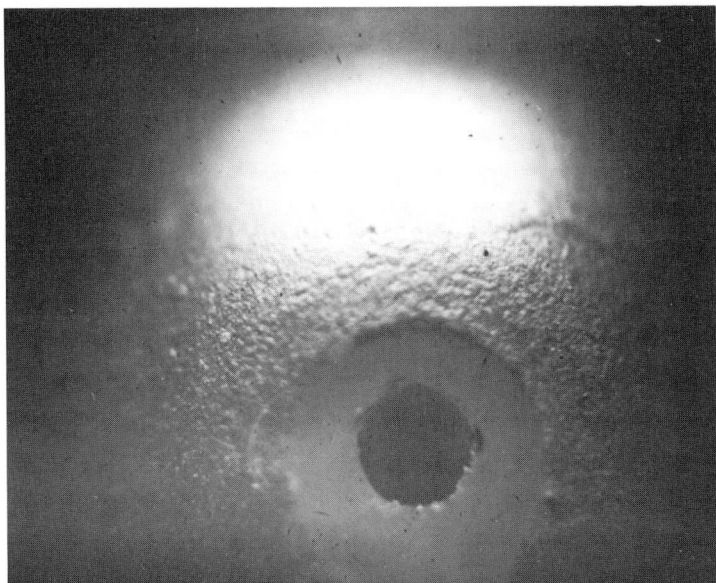

Figure 28.16 Surface of imitation pearl showing underlying glass bead near drill hole

d'orient) which is applied either to the interior of a glass bead or to the exterior of a bead of glass, mother-of-pearl, or other substance. The best-quality imitations are based on beads cleverly shaped to resemble pearls, and receive a number of coatings of the pearly essence. In no case does the surface show the finely spaced parallel serrated markings which are typical of natural and cultured pearls and are a sure sign of growth within the oyster.

Imitation pearls feel smooth when rubbed against the front teeth, oyster pearls feel gritty by comparison. Perhaps the best and simplest test is to examine the edges of the drill holes with a lens, since here the glassy nature of the bead will be revealed, or, if it be some other material, there will be at least some signs of peeling of the applied coating. In Figure 28.16 the surface of an imitation pearl is somewhat unkindly revealed under a magnification of some 15 diameters. It will be seen that the coating has become worn away near the drill hole, revealing the milky glass bead beneath.

Determination by Robert Webster on a large number of imitation pearls of the two main types have shown that the SG of solid imitation pearls is appreciably higher than for oyster pearls, the normal range being 2.85 to 3.18. In more recent tests some pearl beads of inferior quality were found to have the very low SG of 2.33, and others of better appearance have been measured, having a SG of 2.53. The latter have featured as evidence in a notorious murder case.

The wax-filled type have a far lower SG than the solid beads – for the most part below 1.55. Coated plastic beads and increasingly coated mother-of-pearl beads have both been used as imitation pearls.

Polished beads of **hematite** are occasionally seen masquerading as black pearls, but can usually be detected at sight by the lack of pearly lustre. Hematite gives a red 'streak' when rubbed on an unglazed porcelain plate, and, if it can be obtained free from its setting, its high SG of approximately 5.1 is a very distinctive feature.

Stained pearls

Pearls can be stained or tinted by treatment with various dyestuffs. With genuine pearls this is seldom done, except in the case of black pearls. With these, it is not infrequent to find that the colour has been 'improved' – or at least intensified – by treatment with silver nitrate. Any pearl which shows a very homogeneous and intense black colour can be viewed with suspicion: the natural 'black' pearls are seldom really black. Bronze or gunmetal, shades of grey, and a blue-black with subtle iridescence, as on the plumage of a magpie in good condition, are the approved natural colours.

If the pearl is partly drilled, the silver nitrate will have seeped between the pearly layers to some extent, while the conchyolin reduces it to black metallic silver, and an X-ray photograph will then reveal its presence by thin white lines on the negative, as first pointed out by Mr Robert Webster. Another test for staining by silver nitrate is the masking effect this has on the fluorescence of the treated pearls under long-wave ultra-violet light or between crossed filters. Bathed in a strong beam of light filtered through copper sulphate solution, natural black pearls show a dim red glow through a red filter. Stained black pearls remain inert.

It used to be that if black pearls were found to be cultured, this would mean *ipso facto* that they were stained. Today, however, naturally coloured black cultured pearls are grown in the Pacific Islands. It may be useful to warn those who employ radiography in testing pearls that the presence of silver nitrate in the layer of discontinuity between the bead and the outer skin of the stained cultured pearl may largely neutralize the radiability of the conchyolin layer, which normally shows a dark line on the negative, and makes the radiograph less easy to interpret.

Appendix 1

Glossary

absorption spectrum When white light passes through a coloured stone, light of certain wavelengths is absorbed more strongly than others, the colours least absorbed combining to produce the colour of the stone. When viewed through a spectroscope, the colours most strongly absorbed may show as dark bands crossing the spectrum in characteristic positions. Such a spectrum is known as an absorption spectrum, and provides a useful means of identification (Chapter 10).

allochromatic minerals Minerals which would be colourless if chemically pure, but which commonly exhibit a range of colours due to the presence of small quantities of one or more 'colouring' elements. Chief among these elements are those having atomic numbers 22–29: i.e. titanium, vanadium, chromium, manganese, iron, cobalt, nickel, and copper. Corundum, beryl, spinel, and quartz are examples of allochromatic gemstones (see **idiochromatic minerals**).

amorphous Literally, without shape'. An amorphous substance is one in which the internal arrangement of the atoms or molecules is irregular and which in consequence has no characteristic form (see **crystal).**

Ångström unit Unit of length for measurement of wavelengths of visible light and most X-rays but now largely superseded by the nanometre (nm): $10 \text{Å} = 1 \text{nm} = 0.000001$ mm. (See **millimicron, nanometre.**)

anisotropic A term for crystals in which the optical properties vary with direction. All crystals save those in the cubic system are anisotropic and exhibit double refraction.

asterism The star effect seen by reflected light in stones containing suitably oriented rod-like inclusions or channels when these are cut flat or *en cabochon* in the correct direction.

atom A unit which is the smallest part of a chemical element which remains unchanged during all chemical reactions. Although partly composed of electrically charged units such as electrons, protons, etc., atoms as a whole are electrically neutral.

atomic weight The weight of an atom compared with an atom of oxygen = 16.00.

avoirdupois A system of weights in English-speaking countries for general use. There are 16 ounces to the pound avoirdupois, and the ounce is equivalent to 28.35 g or 141.75 metric carats.

biaxial The optical character of crystals belonging to the rhombic, monoclinic, and triclinic systems, which exhibit double refraction but have two directions of single refraction, i.e. two optic axes.

birefringence Synonymous with **double refraction**.

boule The form, shaped somewhat like an inverted pear, assumed by synthetic corundum or spinel when produced under the inverted blowpipe of a Verneuil furnace.

Brewster's angle The angle of incidence at which light reflected from the polished plane surface of a transparent substance attains the maximum degree of polarization parallel to that surface. This was shown by Sir David Brewster to be related to the refractive index of the reflecting medium by the formula $n = \tan i$. The following angles of incidence conforming with Brewster's formula may be quoted for interest: for diamond, 67°30'; for cubic zirconia 65°; for corundum, 60°30'; and for quartz, 57°. An apparatus for the convenient measurement of Brewster's angle has been constructed and might be of value in determining the high index values of diamond and its substitutes, but the production of modern reflectivity meters diminishes the value of this need.

brilliance The degree of brightness in a cut gemstone due to light reflected from the surface and from the back facets.

brilliant A most effective form of cutting for diamond, and so usual for this mineral that the term 'brilliant' is equivalent to 'brilliant-cut diamond' in trade parlance. Brilliant-cut is also used for zircons and other stones. The standard brilliant has 58 facets; 33 in the crown and 25 in the base (pavilion).

cabochon A style of cutting in which the top of the stone forms a curved convex surface. The base may be convex, concave, or flat.

carat The metric carat, equivalent to one-fifth of a gram, is internationally accepted as the standard weight for precious stones. There are 141.75 carats in one ounce avoirdupois and 155.5 carats in a Troy ounce.

chatoyancy The 'cat's-eye effect', due to the reflection of light from fibres or channels arranged in parallel formation within a stone. A silky band of light at right angles to the inclusions is seen when such a stone is cut *en cabochon* in the correct direction.

Chelsea colour filter An effective dichromatic colour fiter transmitting light of only two wavelength regions; one in the deep red, the other in the yellow-green. Useful for discriminating between emerald and its imitations and for detecting synthetic spinels and pastes coloured blue with cobalt (Chapter 4).

chemical element Matter composed of atoms of only one chemical type and which thus cannot be decomposed into simpler substances by chemical means. Diamond is composed largely of the single element carbon, while corundum consists mostly of the elements aluminium and oxygen.

cleavage The tendency shown by certain stones to split along one or more definite directions. Cleavage planes are always parallel to a possible crystal face of the mineral in question.

collimator A device for providing a narrow beam of parallel light for use in spectroscopy. It consists of a tube, at one end of which is fixed a narrow adjustable slit and at the other a lens which has a principal focal length equal to the length of the tube.

colour An effect produced in the normal human eye by light of certain wavelengths (see **spectrum**).

composite stones A comprehensive term which includes doublets, triplets, etc. in which a stone consists of two or more parts either of the same or of different materials cemented or otherwise joined together.

critical angle The angle at which a ray of light in passing from a dense medium, such as a gemstone, into a rarer medium, such as air, is refracted at 90° to the normal. Any rays reaching the intersurface at angles greater than the critical angle are unable to pass into the rarer medium, and are *totally reflected* (Chapter 2).

crossed filters A term coined by the author (by analogy with 'crossed polarizers') to describe a simple but powerful technique for the observation of fluorescence towards the red end of the spectrum. This general idea was first used by Sir G. G. Stokes in 1852. It involves illuminating the specimen by light of one colour (e.g. blue) and viewing it through a filter transmitting a complementary colour (e.g. red). The two filters must be so chosen that when the specimen observed is non-fluorescent no light can be observed through the second (crossed) filter.

cryptocrystalline The structure of a substance such as chalcedony, which consists of masses of exceedingly small crystals but shows no external sign of crystalline structure (see Figure 21.4).

crystal A substance in which the constituent atoms, ions, or molecules are arranged in accordance with a definite regular structure throughout. Under favourable conditions this regularity of internal structure gives rise to a symmetrical external form bounded by plane surfaces known as *crystal faces* (see Appendix 2).

crystal axes Imaginary lines passing through a crystal in important symmetry directions, intersecting in the 'origin' at the centre of the crystal. The axes are usually three in number, and they are chosen to act as a frame or reference by means of which the relative positions of the crystal faces can be described (see Appendix 2).

crystal systems The seven main symmetry groups into which all crystals, whether natural or artificial, can be classified (see Appendix 2).

density The weight of a substance per unit volume, measured in grams per cubic centimetre. Since $1\,cm^3$ of water weighs $1\,g$, the density of a substance is numerically equal to its specific gravity (Chapter 5).

diapheneity See **transparency**.

dichroism The property by which the colour of a stone varies according to the direction in which it is viewed. The two polarized rays passing along a given direction in a coloured doubly refracting mineral are often differently absorbed, and thus differ in colour when they emerge. The two colours can be compared side by side by use of a *dichroscope* (Chapter 4).

dispersion The difference in the refractive index of a stone for light of different wavelengths, giving rise to the flashes of spectrum colours known as 'fire'. The wavelengths usually chosen for measuring the dispersion of gemstones for comparative purposes are 687 nm in the red and 430.8 nm in the violet, corresponding to the B and G Fraunhofer lines of the solar spectrum. The B–G dispersions of some of the gemstones, multiplied by 1000, are as follows: blende, 156; cassiterite, 71; demantoid, 57; sphene, 51; diamond, 44; zircon, 38; peridot and spinel, 20; corundum, 18; tourmaline and spodumene, 17; chrysoberyl, 15; topaz, quartz, and beryl, 14; fluorite, 7 (Chapter 3).

double refraction A ray of light entering any crystal other than cubic in directions other than that of an optic axis is at once split into two polarized rays, each vibrating at right-angles to the other. These rays travel at different speeds through the crystal, and are in consequence refracted by different amounts, the effect being known as double refraction or birefringence. The determination of double refraction in a gemstone is an important test. Where double refraction is strong, the edges of back facets as viewed through the front of a cut stone with a lens appear as double lines (Chapter 3).

doublets Composite stones consisting partly, or sometimes not at all, of genuine material (Chapter 9). The term 'doublets' is also used for closely spaced lines in emission or absorption spectra.

electron volt The energy gained by an electron in passing through a potential difference of 1 volt. A unit much used by physicists to represent the energy of radiation from different parts of the electromagnetic spectrum. It can be linked up with the frequency of light and hence with wavelength by the equation $1\,eV = 8066\,cm^{-1}$ (see **frequency**).

extinction When a birefringent mineral is examined between crossed Nicol prisms or 'Polaroid' discs and no light is transmitted the mineral is said to be in a position of extinction. There are four such positions during a complete rotation of $360°$.

extraordinary ray That ray for which, in minerals belonging to the hexagonal, trigonal, and tetragonal systems, the refractive index varies according to its direction through the crystal.

fire Flashes of spectrum colours from the facets of a cut stone, due to **dispersion**.

fluorescence The emission of visible light by certain minerals when exposed to radiations of shorter wavelength such as ultra-violet light or X-rays (see Chapter 11).

fracture The nature of the broken surface of a solid substance when this does not follow a cleavage plane. The commonest type of fracture is conchoidal (shell-like), typical of glass, quartz, and, to a lesser extent, of several other gemstones.

Fraunhofer lines Dark absorption lines which can be seen crossing the bright continuous spectrum of light from the sun, due to the absorption of light by the vapour of elements in the chromosphere. First observed by the German physicist, Fraunhofer, who designated the principal lines by letters of the alphabet. The principal lines are as follows (wavelengths in Ångström units): A, 7606; B, 6870; C, 6563; D, 5893; E, 5270; F, 4861; G, 4308; H, 3969.

frequency The reciprocal of the wavelength of electromagnetic radiation such as light, and a more direct measure of its energy. The most common unit is the so-called 'wave-number', which is the number of waves per centimetre. Thus $n = 1/$wavelength in centimetres, or $10^7/$wavelength in nanometres or $10^8/$wavelength in Ångströms. Usually written as cm^{-1}. Examples:

$200 \, nm = 2000 \, Å = 50\,000 \, cm^{-1} = 6.19 \, eV$
$400 \, nm = 4000 \, A = 25\,000 \, cm^{-1} = 3.10 \, eV$
$500 \, nm = 5000 \, Å = 20\,000 \, cm^{-1} = 2.48 \, eV$
$600 \, nm = 6000 \, Å = 16\,666 \, cm^{-1} = 2.066 \, eV$

See also **electron volt**.

gauge There are many gauges used in the trade to help in estimating the weight of pearls, diamonds, and other gems. The simplest and the least accurate consist of circular holes suitably numbered, cut out of celluloid or metal. The most elaborate, such as the Leveridge Diamond Gauge, are spring loaded millimetre gauges of special design, together with tables for interpreting the dimensions (diameter, depth, etc.) in terms of weight. A simple sliding vernier millimetre gauge is useful for the jeweller in many capacities. Where measurements can be made, the following formulae may be found useful as a rough guide to the weight of stones:

For a round brilliant, the weight in carats $= 1.7td^2 \times$ density of stone or, for diamond, $6td^2$, where t is the thickness and d the diameter, expressed in centimetres. For a step-cut stone the weight will be $2.5lwt \times$ density, where $l =$ length, $w =$ width and $t =$ thickness.

For spherical undrilled pearls, Mr F. S. H. Tisdall suggests that the formula $28.5td^3$ will give the weight in grains.

grain In Troy weight there are 480 grains to the Troy ounce. Not to be confused with the **pearl grain**, which is one quarter of a metric carat.

gram A metric unit of weight, equivalent to five metric carats. There are 28.35 grams to one ounce avoirdupois.

habit The characteristic crystal form of a mineral. This varies with locality.

hardness The resistance of a mineral to abrasion (see **Mohs' scale**).

heft An old English term, now current in the USA, which is useful in expressing the 'weightiness' in the hand of a stone or other solid, by which one can roughly estimate its density.

homocreate A newly coined term recently proposed by an International Committee on Technical Terminology to denote the restricted use of the word 'synthetic' (as already agreed by gemmologists) to describe man-made gemstones having essentially the same composition, structure, and properties as the equivalent stone found as a natural mineral. Homocreate (pronounce ho*mock*reate) is derived from the Greek *homos*, same, and the Latin *creare*, to make.

idiochromatic minerals Minerals in which the colour is due to some essential constituent of the stone, e.g. malachite, peridot, almandine. In contrast to allochromatic minerals, those which are idiochromatic have a limited range of colour (see **allochromatic minerals**).

imitation stones Materials such as glass or the plastics which may resemble genuine stones in colour and appearance but differ from them in composition and physical properties.

inclusions Particles of foreign matter, solid, liquid, or gaseous, enclosed within a gemstone. The nature of such inclusions provides a powerful clue to the origin of a stone, and enables natural stones to be distinguished from their synthetic counterparts.

inorganic Not produced by vital processes.

interference of light Rays of light pursuing the same path, in which the waves are 'out of phase', suffer interference, and tend to destroy one another, while waves which are 'in phase' (crest corresponding to crest) reinforce each other. The colours seen reflected from opal and labradorite are due to intereference.

ion An electrically charged atom, radical, or molecule.

iridescence The play of rainbow colours due to interference of light as seen in the precious opal, in light reflected from cleavage cracks, on the surface of carborundum crystals, etc.

isomorphous replacement Where one element in a mineral is replaced by another of similar chemical nature and ionic radius without sensibly altering the crystal structure of the mineral. Isomorphous replacement may cause wide variations in such properties as density and refractive index, as may be seen, for instance, in the garnets.

isotropic The term used for those materials in which the optical character is the same in all directions. Such materials are single refractive and show no dichroism. Cubic minerals and amorphous substances are invariably isotropic.

keshi A term commonly applied (if incorrectly) to (small) non-nucleated cultured pearls from Lake Biwa (and China). Keshi is a Japanese word which describes a grain of rice, both in appearance and size. Most pearl dealers apply the term to those small pearls which are the byproduct of the cutting process.

laser A convenient contracted version of the phrase Light Amplification by Stimulated Emission of Radiation. A device using a technique for obtaining exceedingly intense and coherent beams of visible radiation by utilizing the fluorescent properties of ruby, emerald, or other chromium phosphors. A large number of other compounds have also been utilized.

light A form of radiant energy which gives rise to the sensation of sight. Light travels through space with a velocity of 300 000 km (186 285 miles) per second, in common with radio and other waves of a similar physical nature but different wavelengths.

luminescence The effect noticed in some substances of giving out visible light when they are rubbed or scratched (triboluminescence) or when they are irradiated with invisible electromagnetic radiations (**fluorescence, phosphorescence** and thermoluminescence).

lustre The effect produced by light reflected from the surface of a stone. Apart from substances having a *metallic* lustre such as gold or pyrites, the lustre is chiefly determined by the refractive index and perfection of polish possessed by the stone. Since a higher degree of polish can be obtained with hard stones, hardness is also a factor of some importance. The characteristic lustre of diamond is known as *adamantine*, and is displayed in lesser degree by other stones of high refractive index, such as zircon and demantoid. The majority of stones have a *vitreous* or glassy lustre. The terms *resinous, waxy, pearly,* and *silky* lustre are also used.

maser Contracted version of Microwave Amplification by Stimulated Emission of Radiation. A class of amplifier from which the optical **laser** was developed.

mercury spectrum Mercury-vapour lamps of various kinds emit the following powerful lines in the visible and ultra-violet regions (wavelengths in nanometres): 690.7, 623.4, **579, 576.9, 546, 435.8, 404.6, 365, 253.6**. Being rather widely spaced, individual lines in the visible spectrum and in the ultra-violet can be more or less isolated by the use of appropriate filters,

metamict A mineral which has become virtually amorphous due to the breakdown of the original crystal structure by internal bombardment with alpha particles (helium nuclei) emitted by radioactive atoms within the mineral. Many green zircons, especially those from Sri Lanka, which are Precambrian in age, and have thus had over 600 million years of this internal bombardment, owe their low refractive index and density to this cause, and may be termed metamict zircons.

metric carat See **carat**.

micron A thousandth part of a millimetre, 1 micron is symbolized by $1\,\mu$. Now renamed the **micrometre** (μm). This unit is much used in gauging the particle size of diamond powders.

millimicron A thousandth part of a micron, usually symbolized as $1\,m\mu$. Now renamed the **nanometre** (nm). Formerly much used as a measure for the wavelength of visible light. Easily translated into Ångström units merely by the shift of a decimal point, since $1\,m\mu = 10\,\text{Å} = 1\,nm$.

mineral species A homogeneous substance usually produced by the processes of inorganic nature, having a chemical composition, crystal structure, and physical properties which are constant within narrow limits.

Mohs' scale A scale of hardness suggested by Friedrich Mohs over 100 years ago, and still used by mineralogists. The figures in this scale merely denote an *order* of hardness (resistance to scratching) and have no quantitative significance whatever. The minerals chosen by Mohs as standards for his scale are as follows: 1 talc, 2 gypsum, 3 calcite, 4 fluorite, 5 apatite, 6 feldspar, 7 quartz, 8 topaz, 9 corundum, 10 diamond.

molecule A group of two or more atoms in close combination. The smallest quantity of an element or compound which is capable of independent existence.

monochromatic light Light of one wavelength only. The standard monochromatic light used for optical measurements by gemmologists is the yellow light transmitted by glowing sodium vapour, which consists actually of two nearly identical wavelengths having a mean value of 589.3 nm.

nanometre (nm) A unit of length previously known as a **millimicron**. It is equivalent to 10^{-9}m or to 10 Å.

Nicol prism A special prism for producing polarized light, made from two pieces of Iceland spar (calcite) cemented together with Canada balsam. Light entering the prism is split into two polarized rays: of these, the 'ordinary' ray is totally reflected at the balsam layer while the 'extraordinary' ray is able to pass through the prism. In a petrological microscope two Nicol prisms are incorporated.

opalescence A term used both for the milkiness of common opal and (more properly) for the milky iridescence seen in precious opal.

optic axis A direction of single refraction in a doubly refracting mineral. Hexagonal, trigonal, and tetragonal minerals have one such axis, and are termed *uniaxial*; rhombic, monoclinic, and triclinic minerals have two optic axes and are thus *biaxial*.

ordinary ray In uniaxial stones, that ray which travels with constant velocity in any direction within the crystal.

organic Produced by vital processes. In chemistry, the compounds of carbon are termed 'organic' compounds.

orient The characteristic sheen and iridescence displayed by pearl.

ounce See **avoirdupois** and **troy**.

parallel growth A regular grouping of crystals, in which all the edges and faces of the one crystal are parallel to the corresponding parts of its neighbour. A group of crystals in parallel growth is easily recognized by the simultaneous reflection of light from all the parallel faces of the group. This should not be confused with twinning, in which adjoining crystals are symmetrically united about a common plane.

parting In contrast to cleavage, parting is caused by structural *irregularity* in a crystal. The irregularity may be caused by impurities, by twin boundaries, or by stacking faults which can result in a plane of weakness. Regrettably, effects due to a parting or cleavage often look similar in the hand specimen, especially when the parting is good or the cleavage is poor, and the gemmologist must rely on the mineralogical literature to assign the correct term for a particular crystal. For example, calcite has a perfect rhombohedral cleavage whereas ruby and sapphire have a rhombohedral parting.

paste An imitation gemstone made from a certain type of lead glass. Loosely applied to all glass imitation gemstones.

pearl grain See **grain**.

phosphorescence An effect only differing from fluorescence in that the luminous glow persists perceptibly after the removal of the exciting radiation.

plastics Loose term embracing all synthetic resin-like materials which can be moulded under the influence of heat or pressure (Chapter 27).

pleochroism A general term embracing dichroism and the similar effect shown by some biaxial stones in which three distinct colours or shades can be seen (two at a time) by means of the dichroscope.

polarized light Light vibrating in one plane only. Polarized light can be produced by means of reflection from glass plates at certain angles, by absorption of one of the two polarized rays passing through dark-coloured tourmaline plates or Polaroid sheets, or by means of a Nicol prism.

'Polaroid' Sheets of plastic material containing ultra-microscopic crystals of quinine iodosulphate or other material which have the property of transmitting only one polarized ray, the other being almost entirely absorbed by the crystals. 'Polaroid' sheets provide a light and inexpensive substitute for Nicol prisms in the production of polarized light.

radioactivity The spontaneous emission of particles or rays accompanying the gradual breakdown of certain unstable atoms, mostly of high atomic weight. The best-known examples of radioactive elements are radium, thorium, and uranium. The emissions may be of three kinds: alpha particles, which are helium nuclei; beta particles, which are electrons; and gamma rays, which are akin to X-rays of very short wavelength. The first two have little penetrative power, but gamma rays are more penetrating than the hardest X-rays (see Chapters 12 and 17).

reconstructed stones Stones made during the nineteenth century from chips of natural ruby by fusing them together under a blowpipe flame. These so-called reconstructed rubies have been superseded for over fifty years by synthetic stones made by the Verneuil process, for which the term should not be used. Pressed amber is the only truly 'reconstructed' gem material at present being made.

reflection A ray of light incident on a polished plane surface is reflected by it in such a manner that the angle of reflection is equal to the angle of incidence. The proportion of reflected to refracted light in transparent substances increases with the refractive index of the substance and with the angle of incidence. For example, 17 per cent of the light falling at perpendicular incidence on the surface of diamond is reflected, whereas with quartz less than 5 per cent is reflected under these conditions.

refraction When a ray of light passes from air into a denser medium such as a gemstone its velocity is lessened and as a consequence the ray no longer follows its original path but is bent or *refracted* to follow a direction more nearly perpendicular to the surface between the two media.

refractive index A quantity representing the refracting power of a medium, in which air = 1.00 is in practice taken as a standard. The refractive index of a medium may be defined as the ratio of the velocity of light in air to the velocity of light in the medium, or as the ratio of the sine of the angle of incidence to the sine of the angle of refraction when light passes from air into the medium.

refractometer An instrument designed for measuring the refractive indices of various substances (Chapter 2).

sheen The appearance caused by the reflection of light from structures inside a stone, e.g. moonstone.

silk The fine intersecting rod-like crystals or cavities typically seen in Burma ruby, which give a silky sheen by reflected light.

specific gravity The weight of a substance compared with the weight of an equal volume of pure water at 4°C (Chapter 5).

spectroscope An instrument which resolves light into its component wavelengths by refraction through prisms or diffraction by a grating.

spectrum A band of light showing in orderly succession the rainbow colours or isolated bands or colours corresponding to different wavelengths, as seen through a spectroscope, or photographed in a spectrograph. The visible spectrum is only a small region in the vast spectrum of electromagnetic waves, which extend from the longest radio waves to the minutely short waves (gamma rays) emitted by radio-active elements. An *emission* spectrum is produced by the glowing vapour of elements (particularly metals) and consists of bright narrow lines of definite wavelengths. A *continuous spectrum* is the band of all the rainbow colours, red, orange, yellow, green, blue, and violet, merging one into the other, produced by all incandescent solids. An *absorption spectrum* is the series of dark bands crossing a continuous spectrum, seen when white light has been transmitted through a coloured vapour, liquid, or solid.

step- or **trap-cut** A style of cutting used for coloured stones and sometimes for diamond, in which a series of facets both above and below the girdle have edges parallel to those of the rectangular table facet.

symmetry There are three 'elements of symmetry' recognized in crystallography: planes of symmetry, axes of symmetry, and a centre of symmetry. On this basis crystals can be divided into 32 *classes*. These in turn are grouped into seven *systems* (see Appendix 2).

synthetic stones Manufactured stones which have essentially the same composition, crystal structure, and properties as the natural mineral they represent.

transition elements A name applied to a series of metals occupying contiguous positions in the periodic table of the elements, which, due to their atomic structure, absorb light and impart characteristic colours to minerals in which they occur to a major or even an extremely minor extent. In order of increasing atomic number, these are titanium, vanadium, chromium, manganese, iron, cobalt, nickel, and copper.

transparency or **diaphaneity** The degree to which light is transmitted through a substance. A stone is termed *transparent* when objects can be clearly seen through it, as through glass; e.g. quartz, diamond, etc. Where some light is transmitted but no clear outlines can be discerned the stone is termed *translucent*; e.g. jade. Where no light can pass through, the substance is *opaque*; e.g. pyrites.

trap-cut See **step-cut**.

troy weight These were the weights used for precious metals. The equivalents are: 24 grains = 1 pennyweight; 20 pennyweights = 1 oz; 12 oz = 1 lb. Note that the troy grain is the same as the avoirdupois grain, but that the ounce is larger on the troy scale. 1 oz troy = 31.103 g: 1 oz avoirdupois = 28.35 g.

twin crystals Two or more crystals of a mineral which have grown together in symmetrical fashion in such a manner that the parts of the twin have some crystallographic direction or plane in common, but others in reversed position.

ultra-violet light Invisible rays of wavelength somewhat shorter than those of visible violet light. Conveniently classified as long-wave ultra-violet light, e.g. from the 365 nm line of mercury, and short-wave ultra-violet, e.g. the 253.7 nm mercury line.

uniaxial Minerals having a unique direction of single refraction – i.e. one optic axis. Tetragonal, hexagonal, and trigonal crystals are uniaxial.

X-rays Electromagnetic radiations having the same nature as visible light but of much shorter wavelength (usually less than 0.2 nm). Sometimes termed Röntgen rays, after their discoverer.

Appendix 2

The crystal systems

During the past few years there has been a great increase in interest on the part of the public and in the precious-stone trade in well-crystallized minerals in their natural state. In response to demand, a number of little books with coloured illustrations, and some handsome and expensive books also, have been published; crystal specimens have been offered for sale in shop windows in the Hatton Garden area instead of being hidden away in some obscure warehouse, and designers of modern jewellery have even attempted to incorporate small groups of crystals into their creations, untouched by the lapidary.

All this is admirable, and some knowledge of the nature and appearance of crystals should be part of any gemmologist's training. Unfortunately, to study crystallography at any depth is, except for the gifted few, a decidedly difficult matter. It should be unwise for the beginner to go at first to a specialized text in crystallography. Better by far to glance through the sections on the subject to be found in elementary texts on mineralogy when visiting a library, until one is found which by dint of good illustrations and a friendly text seems to make some useful knowledge accessible.

Making a collection of crystals is, of course, an enormous help to understanding, but many minerals crystallize in forms so complex or unevenly developed that they do not appear to conform with the neat geometrical drawings seen in textbooks. Cubic crystals on the whole conform best, and the shapes of the cube itself (pyrite, fluorite), the octahedron (spinel and diamond), dodecahedron and icositetrahedron (garnet) can be readily recognized in the minerals suggested. Hexagonal crystals, represented by beryl and apatite, and trigonal crystals, by quartz, calcite, and tourmaline, also commonly show 'explicable' forms, and good tetragonal idocrase crystals are easily obtained. Good crystals of white topaz are not uncommon which show the symmetry of an orthorhombic crystal well, but they can be very complex in the number of forms displayed. Crystals of selenite (gypsum) are the simplest for displaying typically monoclinic symmetry, while home-grown crystals of copper sulphate will adequately show the forms of that least symmetrical of systems, the triclinic. A series of coloured slides showing drawings of crystals alongside actual specimens has been prepared by R. K. Mitchell, and is exceedingly helpful for the beginner. These are obtainable through the Gemmological Association of Great Britain.

The brief summary of crystal classification which follows may be found useful as a guide and for reference. In order to conform with the courses and examination syllabus of the Gemmological Association of Great Britain, seven systems are here recognized in place of six, as in early editions. This merely means that the trigonal or rhombohedral

363

system is regarded as distinct from the hexagonal system instead of being treated as a subdivision of the latter.

Just as the position of a point on a plane surface can be defined by its perpendicular distance from *two* intersecting lines of reference, or 'axes', so can the position of a plane be defined by the intercepts it makes with *three* intersecting axes. Crystals which have grown under favourable conditions are bounded by flat planes known as *crystal faces*, which are symmetrically disposed. The *crystal axes* chosen as a frame of reference for these faces are imaginary lines passing through the crystal along the directions of greater symmetry and intersecting in the *'origin'* at the centre of the crystal. All crystals are found to fall into one or other of seven main symmetry groups called the *crystal systems*, which are characterized by the nature of their crystal axes. In any mineral, the intercepts made by similar faces with the crystal axes are the same, no matter what the relative dimensions of the faces may be. Further, the angle between two given faces is always the same for all specimens of any one mineral, even if the crystal appears to lack symmetry due to uneven growth.

The typical crystal form assumed by a mineral is known as its *'crystal habit'*, or simply *'habit'*, and is of great assistance in recognizing the mineral in its rough state. The habit of a mineral may vary somewhat with its locality, but the forms it displays will always belong to the same crystal system. Almost all minerals are crystalline. Non-crystalline materials such as the glasses and resins, which have no regular internal structure and thus no characteristic external form, are known as *amorphous*, meaning 'without shape'.

Minerals are often found to be composed of two or more crystal individuals growing together in a symmetrical manner. Such composite crystals are known as *'twins'*. The word *'macle'* is also used, especially by diamond cutters for twinned crystals.

The following is a summary of the seven crystal systems, with a list of the most important gem minerals which crystallize in each.

1. **Cubic system.** All forms in this system can be referred to three equal axes at right-angles to each other. Cubic minerals are isotropic. Common forms are the cube, comprising six square faces; octahedron (eight triangular faces); rhombic dodecahedron (12 rhomb-shaped faces), and icositetrahedron or trapezohedron (24 trapezoid faces). Diamond, spinel, the garnets, fluorite, blende, and pyrite crystallize in the cubic system.

2. **Tetragonal system.** Forms can be referred to three axes at right-angles to each other, two of which are equal, the third and principal axis being longer or shorter than these. Tetragonal minerals are uniaxial, the principal crystal axis being also the optic axis or direction of single refraction. Common forms include prisms (four rectangular faces), basal pinacoids (two square faces), and bipyramids (eight triangular faces). Zircon, scapolite, idocrase and cassiterite belong to this system.

3. **Hexagonal system.** Forms in this system are referred to four axes, three of which are equal and intersect at an angle of 60°, the fourth and principal axis being perpendicular to these and of different length. Hexagonal minerals are uniaxial, the optic axis being identical with the principal crystal axis, which is an axis of sixfold symmetry. Common forms are hexagonal prisms (six faces), basal pinacoids (two faces), and hexagonal bipyramids (12 faces). Beryl and apatite belong to this system.

4. **Trigonal or rhombohedral system.** This is often regarded as a subdivision of the hexagonal system, since trigonal crystals often display hexagonal forms. The main axis is, however, one of threefold (not sixfold) symmetry, which is at right-angles to and of different length from the other three equal axes, which intersect at 60°. The primitive and distinctive form is the rhombohedron – a six-sided figure which can be regarded as a cube which has been compressed along one of its diagonal threefold axes. This simple form can be seen in a cleavage block of Iceland spar (calcite). Many important gemstones have trigonal symmetry, including corundum, tourmaline, quartz, calcite, hematite, and phenakite. Trigonal crystals are uniaxial, the optic axis being the main crystal axis.

5. **Orthorhombic system.** There are three axes at right-angles to each other, all unequal. Orthorhombic minerals are biaxial, that is, they have two optic axes, neither of which coincide with a crystal axis. Common forms are prisms (four faces), pinacoids (two faces), and bipyramids (eight faces). Peridot, chrysoberyl, topaz, andalusite, danburite, and iolite crystallize in this system.
6. **Monoclinic system.** There are three unequal axes, two of which intersect at an oblique angle, the third being perpendicular to these two. Monoclinic crystals are biaxial. Common forms are prisms (four faces), domes (four faces), and pinacoids (two faces). Jadeite, nephrite, sphene, and spodumene are monoclinic minerals.
7. **Triclinic system.** There are three unequal axes, all obliquely inclined. Triclinic crystals are biaxial. Common forms are pinacoids, hemiprisms, and hemidomes (two faces in each case). Gemstones in this system include kyanite, axinite, turquoise, and labradorite feldspar.

In addition to their classification according to their crystal axes, crystals can be grouped into 'classes' within each system, according to their degree of symmetry. There are three so-called 'elements' of symmetry, consisting of a centre of symmetry, a plane of symmetry, and an axis of symmetry. On this basis, crystals are divided into 32 classes, five of which are cubic, seven tetragonal, seven hexagonal, five trigonal, three orthorhombic, three monoclinic, and two triclinic.

The names given to crystal faces and forms are often confusing to the student. The following brief descriptions may help to make things clear.

A **pyramid** is a form each of whose faces cuts all three crystal axes. In the hexagonal system any pyramid face cuts three axes, but not necessarily four.

A **prism** is a form each of whose faces cuts two lateral axes and is parallel to the vertical axis. In the hexagonal system any prism face cuts two lateral axes, but not necessarily all three.

A **dome** is the name given to a form of any face of which cuts the vertical axis and one horizontal axis and is parallel to the other horizontal axis. It can be regarded as a horizontal prism.

A **pinacoid** is one of a pair of faces which cut one crystal axis and are parallel to the other two.

In the cubic system the symmetry is such that the above terms are not relevant. The axes are all equal, and each form consists of equivalent faces which result in a complete solid: thus we have the tetrahedron (four faces); hexahedron or cube (six); octahedron (eight); dodecahedron (12); icositetrahedron (24) and tetrakis hexahedron or four-faced cube (24), the names of which indicate the number of faces present in each case. Two or more of these cubic forms may combine to produce a multifacial but essentially symmetrical solid.

The grained structure of a crystal means that even cubic minerals have directional properties (cleavage, hardness) in keeping with their symmetry, which an amorphous substance cannot have.

It is useful to remember that each face of any one crystal 'form' is physically the same, and the striations, growth, or etch marks on like faces help to establish the symmetry when the crystal is examined. Also, it enables the observer to distinguish between an equidimensional tetragonal crystal (such as may be found, for instance, in an apophyllite crystal) and a true cube. In apophyllite the four prism faces are lustrous and strongly striated in the direction of the c-axis, and thus obviously different in nature from the two basal pinacoid faces which have a matt surface. In a true cube all six faces are exactly equivalent and should have the same lustre and markings.

Appendix 3

Alphabetical summary of gem species

The following comprehensive table resembles in many respects the admirable summary given in Kraus and Slawson's *Gems and Gem Materials*. The alphabetical arrangement enables readers to refer quickly to the more important properties of the chief gem minerals, and a page reference in the last column suggests where they may turn for fuller details of any particular gem. Chemical formulae have been written in the 'dualistic' form (e.g. $MgO.Al_2O_3$ in place of $MgAl_2O_4$) as this may be more easily comprehended. Omissions and even inconsistencies are, for the most part, deliberate.

Species, composition and cryst. system	Varieties and colours	H	SG	Mean RI and DR	Notes
Amber Hydrocarbon Amorphous	*Baltic* – Yellow *Sicilian* – Reddish-yellow *Romanian* – Brown *Burmese* – Yellow to brown	2½	1.08	1.54 Isotropic	A fossil resin. Often includes insects, etc. Splinters under knife. Melts at 280°C. Burns with characteristic fumes.
Andalusite $Al_2O_3.SiO_2$ Rhombic	Green, greenish-brown, with reddish tints *Chiastolite* is impure variety showing greyish cross on black or pale ground	7½	3.15	1.64 0.010	Striking pleochroism. Rare. Often confused with tourmaline, but has lower DR and higher SG.
Apatite $Ca_4(CaF)(PO_4)_3$ Hexagonal	Yellow, blue, green, violet	5	3.21	1.638 0.003	Blue type from Burma, strongly dichoric. Yellow commonly shows rare-earth absorption bands (584 nm, etc.). Very low DR and higher SG distinguish it from danburite.

Species, composition and cryst. system	Varieties and colours	H	SG	Mean RI and DR	Notes
Axinite Complex borosilicate of Ca, Al, Mg. Triclinic	Clove brown, violet	7	3.28	1.685 0.011	Occurs in beautiful bladed crystals: hence name.
Benitoite $BaO\ TiO_2\ 3SiO_2$ Hexagonal	Sapphire blue to colourless	6½	3.67	1.78 0.047	Rare and beautiful stone from San Benito County, California, only. Strong dichroism, strong DR and high dispersion.
Beryl $3BeO.Al_2O_36SiO_2$ Hexagonal	*Emerald* -- Green	7½	2.71	1.575 0.006	Emerald usually flawed. South African emerald rather higher SG and RI; Brazilian emerald rather lower.
	Aquamarine – Pale blue to bluish-green	7½	2.69	1.75 0.006	Aquamarine usually flawless. Madagascar type shows strong dichroism.
	Golden beryl (Heliodor) – Yellow	7½	2.68	1.57 0.005	
	Pink beryl (Morganite) – Rose-pink, red. Also colourless	7½	2.80	1.59 0.008	Constants of pink beryl usually high because of rare alkalis present.
Blende (Sphalerite) ZnS Cubic	Brown, yellow, orange, black Transparent to opaque green	3½	4.09	2.37 none	Magnificent colour, lustre, and 'fire'. But too soft to take or retain high polish. Perfect dodecahedral cleavage.
Brazilianite $NaAl_3(OH)_4$ $(PO_4)_2$ Monoclinic	Greenish-yellow Transparent or translucent	5½	2.99	1.612 0.021	Discovered in 1944 in pegmatite in Minas Gerais, and (1947) in New Hampshire, USA.
Cassiterite SnO_2 Tetragonal	Colourless to very dark brown	6½	6.95 to 7.0	2.045 0.096	High lustre and dispersion. Can resemble diamond.

Species, composition and cryst. system	Varieties and colours	H	SG	Mean RI and DR	Notes
Chrysoberyl BeO.Al$_2$O$_3$ Rhombic	Alexandrite – Green in daylight, red in artificial light	8½	3.71	1.75 0.009	Siberian and Brazilian alexandrites show best colour change. So-called synthetic alexandrites are usually synthetic corundum or spinel.
	Cat's-eye (Cymophane) – Greenish or brownish-yellow, translucent and chatoyant Also yellow, greenish-yellow, colourless, and brown	8½	3.71	1.75 0.009	Several other species show chatoyancy, but 'Cat's-eye' without qualification signifies chatoyant chrysoberyl.
Corundum Al$_2$O$_3$ Trigonal	Ruby – Red	9	3.99	1.765 0.008	Burma ruby, bright red, strong dichroism. Contains 'silk'. Thai ruby, garnet red, less dichroism. No 'silk'.
	Sapphire – Blue Also colourless, yellow, pink, green, and violet	9	3.99	1.765 0.008	Sapphire shows strong dichroism. Green sapphire has slightly higher SG and RI than others. Synthetic corundum made in many colours.
	Star ruby, star sapphire – Translucent, showing asterism				Beware of synthetics.
Danburite CaO.B$_2$O$_3$.2SiO$_2$ Rhombic	Pale yellow, colourless	7	3.00	1.633 0.006	Lower SG and DR distinguish it from topaz. Didymium absorption spectrum frequently present. Burma chief gem locality.
Datolite Calcium borosilicate Monoclinic	Colourless or very pale green, also as cloudy nodules in various colours	5	2.99	1.648 0.044	An attractive collector's stone.
Diamond C Cubic	Colourless Shades of yellow and brown Rarely blue, red, green. (Boart and carbonado for industrial use only)	10	3.52	2.418 Isotropic	Hardest known substance. Perfect octahedral cleavage. Very constant properties. Green, brown, yellow induced by atomic bombardment.

Species, composition and cryst. system	Varieties and colours	H	SG	Mean RI and DR	Notes
Diopside Monoclinic $CaO.MgO.2SiO_2$	Pale green to dark green Sometimes chatoyant Also star stones (four-rayed)	5	3.29	1.69 0.030	Distinguished from peridot by different shade of green and somewhat higher RI and lower DR. From enstatite by higher DR.
Enstatite $MgO.SiO_2$ Rhombic	Green, brown, colourless Sometimes chatoyant	5½	3.27	1.67 0.009	Associated with diamond and pyrope in Kimberley district. Also found in larger pieces but of less attractive green in Burma. Absorption line at 506 nm.
Epidote Silicate of Ca and Al Monoclinic	Dark brownish-green	6½	3.45	1.75 0.035	Colour of crystals usually too dark to make attractive gems. Strongly dichroic. Distinctive 'pistachio' green.
Euclase $Be(AlOH)SiO_4$ Monoclinic	Pale green or blue Sometimes yellow, colourless	7½	3.10	1.665 0.019	Very ready cleavage (hence name). Appearance and RIs almost identical with fibrolite. Distinguished by lower SG.
Feldspar group *Orthoclase* $K_2O.Al_2O_36SiO_2$ Monoclinic *Microcline* Same composition *Albite* Triclinic *Plagioclases* Na and Ca aluminosilicates Triclinic	*Yellow orthoclase* *Moonstone* – Colourless with white or bluish sheen *Amazonite* – Bluish green, translucent to opaque *Albite* – Usually colourless; cat's-eyes *Oligoclase* – Pale yellow *Labradorite* – Grey with play of colour, pale yellowish *Sunstone* – Spangled reddish	6 6 6 6 6 6 6	2.56 2.57 2.56 2.58 2.64 2.70 2.64	1.525 0.005 1.53 0.005 1.53 0.008 1.53 0.005 1.545 0.007 1.565 0.010 1.54 0.009	Important group of rock-forming minerals. Yellow orthoclase, found in Madagascar, contains iron. Moonstone with star or cat's-eye rays. Amazonite, somewhat similar to poor-quality turquoise. Only moonstone and labradorite are used at all widely in jewellery.

Species, composition and cryst. system	Varieties and colours	H	SG	Mean RI and DR	Notes
Fibrolite $Al_2O_3.SiO_2$ Rhombic	Pale blue, greenish	7½	3.25	1.665 0.019	Same RIs as euclase, and also has very ready cleavage.
Fluorite CaF_2 Cubic	Purple, blue, green, yellow, pink, colourless *Blue John* or *Derbyshire spar* – Massive banded	4	3.18	1.434 Isotropic	Perfect octahedral cleavage. Often bright fluorescence under ultra-violet light. Very constant SG and RI.
Garnet group Cubic					Almandine and pyrope form continuous isomorphous series of red garnets. Almandine coloured by iron. Finest pyropes coloured by chromium. Almandine characteristic absorption spectrum.
Almandine $3FeO.Al_2O_3.$ $3SiO_2$	Purplish-red	7½	3.9– 4.2	1.76– 1.81	
Pyrope $3MgO.Al_2O_3.$ $3SiO_2$	Deep blood-red	7¼	3.7– 3.9	1.73– 1.76	
Grossular $3CaO.Al_2O_3.$ $3SiO_2$	*Hessonite* – orange-brown. Also green, pink, etc. massive grossular	7¼ 7	3.65 3.49	1.74 1.73	Hessonite also contains almandine. Resembles jade.
Andradite $3CaO.Fe_2O_3.$ $3SiO_2$	*Demantoid* – Green *Topazolite* – Yellow	6½	3.85	1.89	Very high dispersion ('fire').
Spessartine $3MnO.Al_2O_3$ $3SiO_2$	Orange or yellow Flame red	7	4.16	1.80	Rather rare. Often resembles hessonite. Constants near those for almandine.
Hematite Fe_2O_3 Trigonal	Gunmetal black in crystals or when polished	6	5.1	Very high	Brilliant metallic lustre: used in seal stones, and to simulate black pearl. Leaves red streak on unglazed porcelain.
Hambergite Beryllium borate Orthorhombic	Colourless transparent	7½	2.35	1.587 0.072	Striking double refraction.
Idocrase (Vesuvianite) Ca and Fe silicate Tetragonal	Brown, yellow, green; transparent *Californite* – green massive; translucent	6½ 5½	3.38 3.3	1.70 0.005 1.70	Distinct dichroism. Californite variety resembles jade. Absorption band in blue at 461 nm.

Species, composition and cryst. system	Varieties and colours	H	SG	Mean RI and DR	Notes
Iolite (Cordierite, Dichroite) Mg, Fe, and Al silicate Rhombic	Blue	7	2.59	1.535 0.009	Very strong pleochroism: dark blue, pale blue, and pale brown being three colours seen.
Jadeite $Na_2O.Al_2O_3.$ $4SiO_2$ Monoclinic	Green, white, brown, mauve: translucent	7	3.33	1.66 0.012	'Chinese jade'. Fibrous or granular structure. Often shows slightly dimpled surface. Colour often variegated. More highly esteemed than nephrite.
Kornerupine Mg, Al silicate Rhombic	Green, brownish-green, yellow; transparent Also as cat's-eyes	6½	3.32	1.675 0.013	Strongly dichroic. Rare; but attractive stones found in Sri Lanka, Burma, and Kenya.
Kyanite $Al_2O_3.SiO_2$ Triclinic	Blue; sometimes green or colourless	4 to 7	3.69	1.72 0.019	Flaky structure. Sometimes fine sapphire blue. Strong pleochroism. Hardness varies greatly with direction.
Lapis lazuli	Blue; opaque	5½	2.8	1.5	Brassy specks of pyrites frequent. Inferior pieces have white patches. Rock containing lazurite and other minerals.
Lazulite Fe, Mg, Al phosphate Monoclinic	Blue; translucent to opaque	5½	3.1	1.62 0.031	Rare ornamental stone; sometimes resembles turquoise.
Malachite $Cu_2(OH)_2CO_3$ Monoclinic	Green, banded; opaque	4	3.8	1.78 0.025	Ornamental stone with typical concentric bands of dark and paler green. Effervesces with acid.
Nephrite Ca, Mg, Fe silicate Monoclinic	Green, white; translucent to opaque	6½	2.96	1.62	Classed with jadeite as true jade. Tough. Splintery (hackly) fracture. Colour less bright than jadeite.

Species, composition and cryst. system	Varieties and colours	H	SG	Mean RI and DR	Notes
Opal $SiO_2 + nH_2O$ Amorphous	*Fire opal* – Orange. Seldom shows play of colour	6	2.00	1.45 Isotropic	Contains a varying amount of water. Often porous; thus not advisable to test in heavy liquids. Fire opal usually faceted with domed table. Others cabochon cut.
	White opal – Play of colour on pale, translucent background	6	2.1	1.45	
	Black opal – Play of colour on dark background	6	2.1	1.45	
	Water opal – Play of colour within almost colourless stone	6	2.00	1.45	
Pearl $CaCO_3$ with conchyolin and water	'Natural' – White or creamy	3½	2.71	–	Many localities omitted from summary. 'Natural' pearls from Arabian Gulf and coasts of Sri Lanka. Pearls also classified by shape – button, drop, baroque (irregular), etc. Pearls consist of about 90 per cent $CaCO_3$ in form of aragonite. Pink conch pearl and black clam pearl have no pearly lustre and belong to different category. For distinction between real, cultured and imitation types, see Chapter 28.
	Australian – Silvery white	3½	2.74	–	
	Venezuelan – Translucent white	3½	2.7	–	
	Black – Bronze or gunmetal colours	3½	2.65	–	
	Blue – Lead grey, due to dark nucleus	3½	2.6	–	
	Freshwater – Dull, iridescent	3½	2.7	–	
	Conch – Pink; no pearly lustre	3½	2.85	–	
	Clam – Black; no pearly lustre	3½	2.65	–	
Peridot (Olivine) $2(Mg, Fe)O.SiO_2$ Rhombic	Green Brownish or yellowish-green	6½	3.34	1.67 0.036	The mineralogist's olivine. Green type from Red Sea, Arizona, Burma, etc., contains about 8 per cent Fe. 'Doubling' of back facets easily seen.
Phenakite $2BeO.SiO_2$ Trigonal	Commonly colourless	7½	2.96	1.662 0.016	Peculiarly bright and silvery appearance makes it an attractive gem when well cut.
Prehnite Ca, Al Silicate Rhombic	Pale green or greenish-yellow; translucent	6	2.87	1.63 0.030	Ornamental stone. Sometimes resembles jade.

Species, composition and cryst. system	Varieties and colours	H	SG	Mean RI and DR	Notes
Pyrite FeS$_2$ Cubic	Brassy yellow metallic lustre Opaque	6	4.9	–	The 'marcasite' of the jewellery trade. In nature, marcasite is a separate mineral species. Much used in cheap jewellery. 'Fool's gold'.
Quartz SiO$_2$ Trigonal	Rock-crystal – Colourless Amethyst – Purple Citrine – Yellow (Cairngorm) – Brown Morion – Smoky quartz Rose quartz – Pink, cloudy Aventurine – Green, spangled, also brown	7	2.65	1.548 0.009	Transparent varieties of quartz have very constant properties. Name 'topaz' should not be used for the yellow citrine or brown cairngorm. Jasper is a very impure massive quartz usually brown; stained with Berlin blue to make 'Swiss Lapis'.
	Quartz cat's-eye – Pale brown chatoyant Tiger's-eye or 'crocidolite' – Golden brown, chatoyant Rutilated quartz – Colourless with included rutile	7	2.65	1.548 0.009	
Chalcedony group (Cryptocrystalline Quartz)	Chalcedony – Unbanded, grey Cornelian – Red Chrysoprase – Apple green Agate – Concentric bands of various colours Onyx – Straight bands Sardonyx – Red and white bands	7	2.6	1.53 Small DR	The chalcedony minerals are translucent. They may be stained various colours, and are usually so treated. Many variety names omitted from this table for reasons of space.
Rhodochrosite MnCO$_3$ Trigonal	Rose red to pink; translucent	4	3.6	1.71 0.22	Attractive ornamental stone. Soft; effervesces with acid.
Rhodonite MnO.SiO$_2$ Triclinic	Rose red; translucent	6	3.6	1.72 0.011	Resembles above. Harder; black markings common.
Scapolite Group of complex silicates, etc. Tetragonal	Pale yellow; transparent Pink chatoyant Violet chatoyant	6 6	2.70 2.63	1.57 0.021 1.545 0.009	Pale yellow, from Brazil. Chatoyant types, from Burma.

Species, composition and cryst. system	Varieties and colours	H	SG	Mean RI and DR	Notes
Scheelite $CaWO_4$ Tetragonal	Colourless, yellow, orange	5	5.9 to 6.1	1.93 0.017	When well cut can resemble diamond.
Sinhalite $Mg(AlFe)BO_4$ Rhombic	Pale yellow to dark brown or greenish-brown Transparent	6½	3.48	1.685 0.038	Cut stones formerly thought to be 'brown peridot' were found (1952) to be a new species. Named after Ceylon (now Sri Lanka) where first found. Distinguished from peridot by SG, RI and spectrum.
Smithsonite $ZnCO_3$ Trigonal	Yellow or apple green; translucent	5	4.35	1.73 0.23	May resemble chrysoprase or jadeite. Effervesces with acid. 'Bonamite' a trade name.
Sphene (Titanite) $CaO.TiO_2.SiO_2$ Monoclinic	Yellow, green, brown	5½	3.53	1.96 0.134	Rare. Highly valued in spite of softness for its magnificent appearance. Dispersion higher than diamond. Strong pleochroism and large DR.
Spinel $MgO.Al_2O_3$ Cubic	Red, pink, pale greyish-blue, dark greenish-blue, shades of reddish-purple	8	3.60	1.717 Isotropic	Pure spinel has SG 3.58, RI 1.715. Variety names such as 'Balas ruby' are misleading and should be avoided. Ceylonite types contain much iron, raising SG and RI. Synthetic spinels made in colours to represent other species.
	Ceylonite or *Pleonaste* – Black; opaque	8	3.8 to 4.06	1.78 to 1.754	
	Gahnospinel – Shades of blue, containing varying proportions of zinc	8			
Spodumene $Li_2O.Al_2O_3. 4SiO_2$ Monoclinic	*Kunzite* – Lilac *Hiddenite* – Green. Also yellow	7	3.18	1.67 0.015	Easy cleavage in two directions. Pleochroism distinct or strong. True green hiddenite coloured by chromium, very rare.

Species, composition and cryst. system	Varieties and colours	H	SG	Mean RI and DR	Notes
Topaz Fluo- or hydroxysilicate of aluminium Rhombic	Colourless, blue Yellow ⎫ Pink ⎭ Orange Brown	8	3.56 3.53	1.62 0.010 1.63 0.008	True topaz not to be confused with yellow quartz. Perfect basal cleavage. Pink stones produced by heat treatment of brownish-yellow Brazilian material. Much blue topaz irradiated. Distinct pleochroism.
Tourmaline Complex borosilicate of Al, Mg, Fe, alkalis, etc. Trigonal	*Achroite* – Colourless *Rubellite* – Red, deep pink *Indicolite* – Deep inky blue Also green, pink, yellow brown, black and particoloured	7	3.05	1.63 0.018	Strongly dichroic. Pyro-electric. Colour often in segments or concentric zones. Red types have lowest SG. Yellow, blue, higher; black highest. Distinguished from topaz and andalusite by larger DR.
Turquoise Hydrous phosphate of aluminium and copper Triclinic	Dark to pale sky blue, pale greenish-blue Translucent to opaque	6	2.6 to 2.8	1.61	Egyptian (strong blue, translucent) and Iranian (fine blue) are least porous and have SG near 2.8. American types softer, more porous and have lower SG. Veins of dark limonite matrix frequent.
Zircon $ZrO_2.SiO_2$ Tetragonal	Colourless, blue, orange red, yellow, golden brown Green and shades of green	7½ 6½	4.69 4.0 to 4.5	1.95 0.059 1.79 upwards	Heat-treated types from Indo-China and Thailand have properties of normal zircon as given. Strong DR a distinguishing feature. Green types from Sri Lanka are metamict and have lower and variable properties. Blue zircon the only type to show dichroism.
Zoisite (Tanzanite) Silicate of Ca and Al Rhombic	Blue, purple, brown, Massive green *Thulite* – Massive pink	6½	3.35 3.28 3.10	1.695 0.009	Transparent blue zoisite, an important gemstone. Massive types seldom used in jewellery.

Appendix 4

Specific gravity table

Copal	1.06	Lapis lazuli	2.8	Rhodonite	3.6
Amber	1.08	Beryllonite	2.82	Rhodochrosite	3.6
Bakelite	1.26	Conch pearl	2.85	Spinel	3.60
Tortoiseshell	1.30	Prehnite	2.87	Taaffeite	3.61
Erinoid	1.33	Verdite	2.9	Spinel (synth.)	3.63
Celluloid	1.38	Pollucite	2.92	Hessonite	3.65
Vegetable ivory	1.40	Aragonite	2.94	Benitoite	3.67
Ivory	1.8	Datolite	2.95	Kyanite	3.68
Bone	2.0	Phenakite	2.96	Staurolite	3.70
Fire opal	2.00	Nephrite	2.96	Pyrope	3.7
Opal	2.1	Brazilianite	2.99	Chrysoberyl	3.71
Chrysocolla	2.20	Danburite	3.00	Malachite	3.8
Silica glass	2.21	Amblygonite	3.03	Pleonaste	3.8
Sodalite	2.3	Tourmaline	3.06	Demantoid	3.85
Hambergite	2.35	Lazulite	3.09	Anatase	3.88
Petalite	2.39	Euclase	3.10	Gahnospinel	to 3.97
Moldavite	2.35	Zoisite	3.1	Corundum	3.99
Obsidian	2.35	Andalusite	3.15	Painite	4.01
Leucite	2.47	Carborundum	3.17	Willemite	4.03
Variscite	2.55	Fluorite	3.18	Sphalerite (blende)	4.09
Amazonite	2.56	Spodumene	3.18	Zircon (green)	4 to 4.5
Orthoclase	2.56	Apatite	3.21	Almandine	4.2
Moonstone	2.57	Fibrolite	3.25	Spessartine	4.16
Iolite	2.59	Enstatite	3.27	Rutile (synth.)	4.25
Bowenite	2.6	Axinite	3.28	Smithsonite	4.35
Chalcedony	2.6	Ekanite	3.28	Barytes	4.5
Scapolite (pink)	2.63	Diopside	3.29	Zircon (blue, white)	4.69
Sunstone	2.64	Dioptase	3.30	Pyrites	4.9
Quartz	2.65	Kornerupine	3.32	Hematite	5.05
Synth. emerald	2.65	Jadeite	3.33	Strontium titanate	5.13
Coral	2.68	Peridot	3.34	Cubic zirconia	5.6–5.9
Aquamarine	2.69	Zoisite (blue)	3.35	Scheelite	6.0
Beryl (yellow)	2.69	Idocrase	3.38	Cassiterite	6.9
Pseudophite	2.7	Rhodizite	3.40	GGG	7.05
Emerald	2.71	Epidote	3.45	Silver	10.5
Labradorite	2.70	Sinhalite	3.48	Palladium	11.3
Scapolite (yellow)	2.70	Grossular (massive)	3.5	Gold, 9 carat	11.4
Calcite	2.71	Diamond	3.52	Gold, 14 carat	13.93
Pearl (natural)	2.71	Sphene	3.53	Gold, 18 carat	15.4
Pearl (cultured)	2.75	Topaz (yellow)	3.53	Gold, 22 carat	17.7
Talc	2.75	Topaz (white)	3.56	Gold, pure	19.3
Turquoise	2.8	Periclase (synth.)	3.59	Platinum	21.5

Note: Where there is a wider variation in the above values than one or two units in the second decimal place, only one place of decimals is quoted.

Appendix 5

Refractive index table

Stone	RI		DR	Stone	RI		DR
Fluorite	1.434		none	Axinite	1.675	1.685	0.010
Opal	1.45		none	Kornerupine	1.668	1.680	0.012
Silica glass	1.46		none	Spodumene	1.663	1.678	0.015
Sodalite	1.48		none	Diopside	1.672	1.702	0.030
Calcite	1.486	1.658	0.172	Sinhalite	1.670	1.708	0.038
Obsidian	1.49		none	Zoisite (blue)	1.691	1.700	0.009
Moldavite	1.49		none	Willemite	1.691	1.719	0.018
Leucite	1.508	1.509	0.001	Idocrase (yellow)	1.705	1.710	0.005
Petalite	1.504	1.516	0.012	Kyanite	1.715	1.732	0.016
Orthoclase (yellow)	1.522	1.527	0.005	Spinel	1.715		none
Iolite	1.537	1.547	0.010	Taaffeite	1.718	1.722	0.004
Scapolite (pink)	1.540	1.549	0.009	Spinel (synth.)	1.727		none
Oligoclase	1.542	1.549	0.007	Gahnospinel	to 1.75		none
Quartz	1.544	1.553	0.009	Rhodonite	1.733	1.747	0.013
Scapolite (yellow)	1.548	1.569	0.020	Periclase (synth.)	1.738		none
Labradorite	1.560	1.570	0.010	Rhodochrosite	1.597	1.817	0.220
Bytownite	1.564	1.574	0.010	Epidote	1.736	1.770	0.034
Hambergite	1.555	1.629	0.074	Chrysoberyl (yellow)	1.745	1.754	0.009
Emerald (synth.)	1.560	1.563	0.003	Alexandrite	1.746	1.755	0.009
Aquamarine	1.570	1.575	0.005	Pyrope	1.73 to 1.76		none
Emerald	1.579	1.585	0.006	Sapphire (white)	1.760	1.768	0.008
Morganite	1.58	1.59	0.008	Ruby	1.764	1.772	0.008
Ekanite	1.597		none	Benitoite	1.755	1.802	0.047
Anorthite	1.574	1.589	0.015	Pleonaste (black)	1.78		none
Brazilianite	1.604	1.624	0.020	Spessartine	1.78	1.80	none
Tremolite (green)	1.601	1.642	0.040	Painite	1.787	1.816	0.029
Amblygonite	1.612	1.638	0.026	Almandine	1.76 to 1.81		none
Topaz (white)	1.61	1.62	0.010	Zircon (metamict)	1.79		none
Topaz (yellow)	1.630	1.638	0.008	YAG	1.834		none
Tourmaline	1.62	1.64	0.018	Demantoid	1.888		none
Danburite	1.630	1.636	0.006	Scheelite	1.920	1.937	0.017
Apatite	1.634	1.637	0.003	Zircon (normal)	1.925	1.984	0.059
Andalusite	1.634	1.644	0.010	Sphene	1.900	2.020	0.120
Barytes	1.636	1.648	0.012	Cassiterite	2.002	2.100	0.098
Pargasite	1.628	1.651	0.023	GGG	2.03		none
Datolite	1.625	1.670	0.045	Cubic zirconia	2.17		none
Euclase	1.652	1.672	0.020	Lithium niobate	2.21	2.30	0.09
Peridot	1.654	1.690	0.036	Sphalerite (blende)	2.37		none
Jadeite	1.654	1.667	0.014	Strontium Titanate	2.41		none
Phenakite	1.656	1.671	0.015	Diamond	2.418		none
Enstatite	1.663	1.673	0.010	Anatase	2.493	2.554	0.061
Bronzite	1.670	1.684	0.014	Rutile (synth.)	2.610	2.897	0.287
Fibrolite	1.658	1.678	0.019	Carborundum	2.65	2.69	0.043

Appendix 6

Alphabetical table of the principal constants of gemstones

Name	SG	Mean RI	DR	H
Agate	2.6	1.54	0.004	7
Albite	2.58	1.53	0.005	6
Alexandrite	3.71	1.75	0.009	8½
Almandine	3.9–4.2	1.76–1.81	–	7½
Amazonite	2.56	1.53	0.008	6
Amber	1.08	1.54	–	2½
Amethyst	2.65	1.55	0.009	7
Andalusite	3.15	1.64	0.010	7½
Apatite	3.20	1.64	0.003	5
Aquamarine	2.69	1.575	0.006	7½
Aragonite	2.94	1.60	0.155	3½
Aventurine quartz	2.66	1.55	–	7
Azurite	3.8	1.76	0.110	3½
Bakelite	1.26	1.65	–	3
Benitoite	3.67	1.78	0.047	6½
Beryl	2.7	1.58	0.006	7½
Blende	4.09	2.37	–	3½
Bonamite	4.35	1.62	0.23	5
Bowenite	2.59	1.55	–	5½
Brazilianite	2.99	1.612	0.021	5½
Calcite	2.71	1.57	0.172	3
Californite	3.3	1.70	–	5½
Carborundum	3.17	2.67	0.043	9½
Cassiterite	6.9	2.05	0.10	6½
Celluloid	1.38	1.49	–	2
Chalcedony	2.61	1.535	0.004	7
Chrome diopside	3.30	1.68	0.030	5½
Chrysoberyl	3.71	1.75	0.009	8
Chrysocolla	2.1	1.5	–	2
Chrysoprase	2.6	1.53	–	7
Conch pearl	2.84	–	–	3½
Copal resin	1.06	1.54	–	2
Coral	2.68	–	–	4
Cornelian	2.64	1.53	0.004	7
Corundum	4.00	1.765	0.008	9
Crocidolite	2.66	1.54	0.004	7

Name	SG	Mean RI	DR	H
Danburite	3.00	1.633	0.006	7
Datolite	2.95	1.65	0.044	5
Demantoid	3.85	1.89	–	6½
Diamond	3.515	2.418	–	10
Diopside	3.29	1.685	0.029	5
Ekanite	3.28	1.597	–	6
Emerald	2.71	1.58	0.006	7½
Enstatite	3.27	1.67	0.010	5½
Epidote	3.40	1.75	0.035	6½
Euclase	3.10	1.665	0.019	7½
Fibrolite	3.25	1.665	0.019	7½
Fire Opal	2.00	1.45	–	6
Fluorite	3.18	1.434	–	4
GGG	7.05	2.03	–	6½
Grossular (pure)	3.594	1.734	–	7
Hematite	5.1	3.0	–	6½
Hambergite	2.35	1.58	0.072	7½
Hessonite	3.63	1.745	–	7
Hiddenite	3.18	1.665	0.015	6½
Idocrase	3.40	1.708	0.005	6½
Iolite	2.63	1.548	0.010	7
Jadeite	3.33	1.65	–	7
Jasper	2.55	1.54	–	7
Jet	1.33	1.66	–	2½
Kornerupine	3.32	1.67	0.013	6½
Kunzite	3.18	1.67	0.015	6½
Kyanite	3.68	1.725	0.017	5–7
Lapis lazuli	2.8	1.50	–	5½
Lazulite	3.10	1.62	0.031	5½
Lithium niobate	4.64	2.25	0.090	5½
Malachite	3.8	1.78	0.025	4
Marcasite	4.9	–	–	6
Moldavite	2.35	1.49	–	5
Moonstone	2.57	1.52	0.005	6
Nephrite	2.96	1.62	–	6
Obsidian	2.35	1.49	–	5
Odontolite	3.1	–	–	5
Onyx	2.6	1.535	0.004	7
Opal	2.1	1.45	–	6
Painite	4.01	1.80	0.029	8
Pearl	2.71	–	–	3
Periclase	3.59	–	–	6
Peridot	3.34	1.67	0.036	6½
Phenakite	2.96	1.66	0.015	7½
Pleonaste	3.8	1.78	–	8
Porcelain	2.3	–	–	5½
Prehnite	2.87	1.63	0.016	6
Pseudophite	2.7	1.57	–	2½
Pyrites	4.9	–	–	6
Pyrope	3.7–3.9	1.73–1.76	–	7½
Quartz	2.65	1.55	0.009	7
Rhodochrosite	3.6	1.71	0.220	4
Rhodonite	3.6	1.73	0.014	6
Ruby	3.99	1.765	0.008	9
Rutile	4.25	2.75	0.287	6
Sapphire	3.99	1.765	0.008	9
Sard	2.61	1.535	0.004	7

Name	SG	Mean RI	DR	H
Scapolite	2.70	1.555	0.020	6
Scheelite	6.1	1.926	0.017	5
Serpentine	2.6	1.56	–	5
Silica glass	2.21	1.46	–	6
Sinhalite	3.48	1.686	0.038	6½
Smithsonite	4.35	1.73	0.230	5½
Spessartine	4.16	1.80	–	7
Sphalerite	4.09	2.37	–	3½
Sphene	3.53	1.95	0.120	5
Spinel	3.60	1.715	–	8
Spodumene	3.18	1.67	0.015	7
Strontium titanate	5.13	2.41	–	5½
Taaffeite	3.61	1.72	0.004	8
Tektite	2.4	1.49	–	5
Topaz	3.53	1.63	0.008	8
Tortoiseshell	1.29	1.55	–	3
Tourmaline	3.06	1.63	0.018	7½
Turquoise	2.8	1.62	–	6
Uvarovite	3.77	1.87	–	7½
Variscite	2.55	1.56	–	5
Vivianite	2.6	1.60	0.027	2
Yttrium aluminate	4.57	1.834	–	8
Water opal	2.0	1.45	–	6
Zircon (normal)	4.69	1.95	0.059	7½
Zircon (metamict)	4.0	1.79	–	6½
Zirconia (cubic)	5.6–5.9	2.17	–	8½
Zoisite	3.35	1.695	0.010	6½

Appendix 7

Recommended reading

For those who may wish to know more about various aspects of gemmology, the following books may be helpful. Most of them can be found in good reference libraries, or can be purchased from the Gemmological Association of Great Britain, Saint Dunstan's House, Carey Lane, London EC2V 8AB.

The only book covering a similar range to the present volume is *Handbook of Gem Identification* by Richard T. Liddicoat Jr, published by the GIA, 12th edition 1987 (revised 1988). This is an excellent and useful book, and the new edition is greatly improved. It is well illustrated, and in particular contains a fine series of drawings of absorption spectra by G. R. Crowningshield.

For complete and up-to-date coverage of the whole field of gemmology one must turn to Robert Webster's great book *Gems. Their sources, descriptions and identification*, published by Butterworths in two volumes in 1962 and revised as a single volume in 1970, 1975 and 1983.

Two smaller books by the same author have just been revised and can also be recommended: they are *Practical Gemmology*, which is a short course in the subject suitable for the student with an examination in mind, and the *Gemmologist's Compendium*, which contains an informative glossary, all manner of tables, and illustrations of absorption spectra in colour. These are both published by the NAG Press.

Another valuable book which covers the whole subject is *Gemstones* by G. F. Herbert Smith (Chapman and Hall). This famous work was first published in 1912, and was thoroughly revised after the author's death by Dr Coles Phillips for its fourteenth edition in 1972. It still has the stamp of a classic work, the descriptions of gem species being particularly good. Very regrettably, it is now out of print. The Smith version of *Gemstones* has now been replaced by a new book (1988) bearing the same title by M. O'Donoghue. This is a comprehensive work bringing together gemmological, geological, and mineralogical developments that have taken place during the last thirty years.

For those who prefer a less technical approach the author's *Gemstones for Everyman* (Faber & Faber, 1976) will prove easy to read and is amply illustrated. L. J. Spencer's lively work *A Key to Precious Stones* (Blackie, 1946) is now unfortunately out of print.

Readers interested primarily in diamond are recommended to consult *Diamonds* by Eric Bruton, published by the NAG Press in 1978. This is probably the only book available which covers the subject in all its aspects. *The Diamond Dictionary*, published by the Gemological Institute of America, 1660 Stewart Street, Santa Monica, California, 90404, USA, is a valuable work of reference; a new edition has been issued recently.

Every keen gemmologist should have some good general book on mineralogy on his or her shelves, if only for the full treatment of crystallography which it affords. Dana's *Textbook of Mineralogy* revised by Hurlbut is a very good general book on the subject. In addition to a full account of crystals and the optical and physical properties of minerals, the use of the polarizing microscope, etc., the book contains an adequate description of a wide range of minerals. As far as the descriptions of the species are concerned, *An Introduction to the Rock Forming Minerals* by Deer, Howie, and Zussman, published by Longman, London, is available as a paperback, and is a book well worth having.

A textbook of mineralogy is not the best companion to help one to identify minerals in the field, and *A Field Guide to Rocks and Minerals*, by F. H. Pough, has been an outstanding success in supplying that need. It is fully and attractively illustrated in colour and black and white, and is readily portable. The important and fascinating study of inclusions in gemstones is explained in the superbly illustrated account by the acknowledged master of this subject, Dr Edward Gübelin, in *The Internal World of Gemstones* (Butterworths, 1974). This work is now followed by the superlative *Photoatlas of Inclusions in Gemstones*, in which E. Gübelin has collaborated with J. Koivula of the GIA. *Beginner's Guide to Gemmology* by Peter G. Read (Newnes Technical Books, 1980) provides an excellent and well-illustrated introduction to the whole subject, while a detailed account of the methods used in the growth of synthetic gemstones can be found in two books: *Man-made Gemstones*, by D. Elwell (John Wiley and Sons, 1979) and *Gemstones Made by Man*, by Dr Kurt Nassau (Chilton Book Co., Radnor, Pennsylvania, 1980). Each of these authors is actively engaged in the manufacture of gem materials. A superbly illustrated reference book is the *Colour Encyclopedia of Gemstones* by Joel E. Arem. *Gemstone Enhancement* by K. Nassau and *Emerald and other Beryls* by J. Sinkankas are excellent works with self-explanatory titles, and a series of monographs published by Butterworths includes titles on beryl, garnet, pearls, quartz, jet, amber and corundum.

Though new editions of the leading textbooks appear at fairly frequent intervals, the subject of gemmology grows so fast that those really interested should subscribe to at least one of the specialist journals. They also need support if they are to continue publication.

The Journal of Gemmology, published quarterly by the Gemmological Association of Great Britain, contains original articles and also useful abstracts of papers from other journals. *Gems and Gemology* is published quarterly by the Gemological Institute of America, 1660 Stewart Street, Santa Monica, California 90404. In addition to original articles there are valuable reports from the two GIA Laboratories, giving accounts of unusual materials encountered in routine testing. *The Lapidary Journal*, published monthly from 3564 Kettner Boulevard, San Diego, California, is a well-produced journal plentifully illustrated, including colour, which has many articles on general gemmology as well as on gem cutting. First-hand reports on current gemmological problems and how best to deal with them are to be found in *Retail Jeweller* published fortnightly by International Thomson Publishing Ltd, 100 Avenue Road, London NW3 3TP.

Appendix 8

Recommended suppliers

The Gemmological Association of Great Britain, Saint Dunstan's House, 2 Carey Lane, London EC2V 8AB, are now agents for Rayner instruments, and have a good stock of gemmological instruments and apparatus from other sources. The Association can also supply books, and runs correspondence courses in gemmology in preparation for the Association's examinations.

The Gemological Institute of America, 1660 Stewart Street, Santa Monica, California 90404, USA, also supplies books, apparatus, correspondence courses, etc.

Hanneman Gemological Instruments, PO Box 2453, Castro Valley, California 94546, USA, manufacture and supply a range of gemmological instruments of original design.

Rubin & Son, 18 Greville Street, London EC1N 8SU, and in Antwerp and New York, supply instruments and other materials for the diamond and jewellery trade.

Gregory, Bottley & Lloyd, 9/12 Ricketts Street, London SW6 1RU. Traditional suppliers of minerals, rocks, fossils, and geological equipment.

Coventry University

Index